Leon M. Galitsky

Numerical Analysis

Leon M. Galitsky

Numerical Analysis

Interpolation, Differentiation, and Integration

Mathematics Subject Classification 2020
Primary: 65-01; Secondary: 65D05, 65D25, 65D30, 65L06

Author
Leon M. Galitsky
Moravian Academy
Mathematics Department
4313 Green Pond Road
Bethlehem, PA 18020
USA
calcboy2000@hotmail.com

ISBN 978-3-11-914305-9
ISBN 978-3-11-222105-1 (PDF)
ISBN 978-3-11-222134-1 (E-PUB)
DOI https://doi.org/10.1515/9783112221051

Library of Congress Control Number: 2026934622

Bibliographic information published by the Deutsche Nationalbibliothek
The Deutsche Nationalbibliothek lists this publication in the Deutsche Nationalbibliografie;
detailed bibliographic data are available on the Internet at https://dnb.dnb.de.

© 2026 Walter de Gruyter GmbH, Genthiner Straße 13, 10785 Berlin, Germany

De Gruyter and Walter de Gruyter GmbH are part of De Gruyter Brill.
www.degruyterbrill.com

Questions about General Product Safety Regulation: productsafety@degruyterbrill.com

Front cover illustration: kertlis / iStock / Getty Images Plus
Back cover illustration: Leon M. Galitsky

For Mom, Dad, Nana, and Pop,
my first and best teachers

For Rich,
math was never so much fun

Contents

Purpose and prerequisites

Students successfully completing two semesters of college calculus prior to their final year in secondary education has produced a need for follow-up courses in such schools. A typical next step is an analytical option like Multivariable Calculus, Linear Algebra, or Differential Equations, but those do not necessarily spark a novel interest in the wider field of mathematics. Introducing students to a new branch of mathematics and connecting it to spreadsheet or computer programming applications can open doors of interest in various STEM fields and research that have previously remained closed until collegiate studies.

The goal of this work is threefold for the secondary educator. First, nearly all mathematics leading to calculus involve algebraic manipulation towards finding exact solutions. This text introduces numerical routines which accurately estimate problems that cannot be solved analytically. Second, students are guided through the entire mathematician's process of identifying a problem, the motivation behind a solution to the problem, and the rigorous proof of the solution. Most students' exposure to proofs, if at all, occurs in geometry, a very small subset of proving techniques. This text includes several types of other useful proof strategies. All algebraic steps leading to a final method are shown explicitly. Only when the theory behind the methods is understood can students begin to explore ideas for potential improvement. Third, the connectivity of the entirety of mathematics is noted when concepts from previously studied fields are used. Principles in algebra, geometry, calculus, and linear algebra are specifically referenced in an effort to demonstrate that the solution to a problem in one field can use accepted ideas from another.

For the undergraduate college student enrolled in an introductory Numerical Analysis course, this book serves as either a primary source or a valuable companion text, containing the kind of algebraic and graphical details sometimes assumed and omitted in other college textbooks. Additionally, this text contains an extensive and well-developed (not AI-generated) set of practice problems (see below). Students who have taken a numerical analysis course already and are preparing for higher-level graduate work can use this text as a quick refresher on the major topics.

Great care was taken in developing a comprehensive set of practice problems which identify the various outcomes of a particular numerical method. The number of iterations an algorithm requires varies so that a particular number needed to stop it is not falsely assumed. Examples in which algorithms fail are presented to caution the blind application of a particular method without an underlying understanding of how it works, and to demonstrate not every algorithm can solve every problem. Many exercises involve comparing two or more algorithms, highlighting the advantages and disadvantages of each. Functions used are not limited to simple polynomials; rational functions, trigonometric, inverse trigonometric, exponential, and logarithmic functions are all given equal exposure.

https://doi.org/10.1515/9783112221051-204

Round-off errors can be a frustrating barrier to students checking their work. The most meticulous aspect of the solutions to the exercises is, whether the intermediate steps are rounded to the thousandths place, or whether the maximum number of places is retained in a spreadsheet, the final answers should be identical. Hopefully the effort in designing problems this way will alleviate students' uncertainty about the accuracy of their solutions. Note that in the examples in the text, the maximum number of decimal places is retained at each step, while the numbers displayed at each step are rounded to the thousandths place.

https://doi.org/10.1515/9783112221051-205

1 Roots and zeros

Analytic methods for finding exact roots, or zeros, of an nth-degree polynomial $a_n x^n + a_{n-1} x^{n-1} + \cdots + a_1 x + a_0$ for integer $n > 1$ are limited. Second-degree polynomials can be solved with the quadratic equation. The formula for third-degree polynomials is more complicated, roughly a textbook page, while the fourth-degree formula takes several landscaped pages in tiny font. For polynomials of degree five or greater, one algebraic recourse is the rational root theorem, where a zero may exist at $x = k$ for

$$k = \pm \frac{\text{a factor of } a_n}{\text{a factor of } a_0}.$$

But these are merely possible, not definite, zeros for a factorable polynomial, and it has been proved no general formula exists using radicals or nested radicals for all irreducible polynomials of degree five or greater.

The zero-finding problem can be extended to *any* equation simply by moving all terms to one side. This chapter examines several iterative methods for such equations, in addition to the polynomial roots problem. Iterative methods calculate an estimate and refine the solution through repeated calculations using updated information until a criterion is satisfied; a looping structure. The first two methods use a shrinking interval and are reliable albeit somewhat inefficient. The next two algorithms need only a single point on which to iterate, but do not guarantee success. The final two algorithms approximate the single-point algorithms using two closely spaced points.

1.1 Bisection algorithm

The intermediate value theorem (IVT) from calculus states a continuous function $f(x)$ on a closed interval $[a, b]$ equals every value between $f(a)$ and $f(b)$ for some c in (a, b). Specifically, it follows if $f(a)$ and $f(b)$ differ in sign with those given, there must exist a c in (a, b) where $f(c) = 0$, i. e., a polynomial root or a zero for any equation with all terms on one side.

The bisection algorithm starts with an initial zero estimate at x_1 of the midpoint of the given interval $[a, b]$.

$$x_1 = \frac{a + b}{2}$$

The sign of $f(x_1)$ determines which half of the bisected interval contains the unknown zero and removes the other half, shrinking the solution space. If the sign of $f(x_1)$ is the same as $f(a)$, x_1 replaces a, or in general the lower bound x_L in the next iteration. Otherwise, it replaces b, or in general the upper bound x_U. The next iteration uses the new, bisected interval to find the next midpoint and proceeds until a stopping criterion is met,

https://doi.org/10.1515/9783112221051-001

as discussed later. In general, the bisection algorithm zero estimate at x_i for the current interval $[x_L, x_U]$ is

$$x_i = \frac{x_L + x_U}{2}.$$

(1.1)

Figures 1.1 and 1.2 illustrate the first three iterations of the bisection algorithm. The first iteration in Figure 1.1 locates the x-coordinate of the midpoint between the initial lower and upper bounds, at $x_1 = \frac{x_L + x_U}{2}$. Since $f(x_1) > 0$, x_1 replaces x_U because $f(x_U) > 0$ as well.

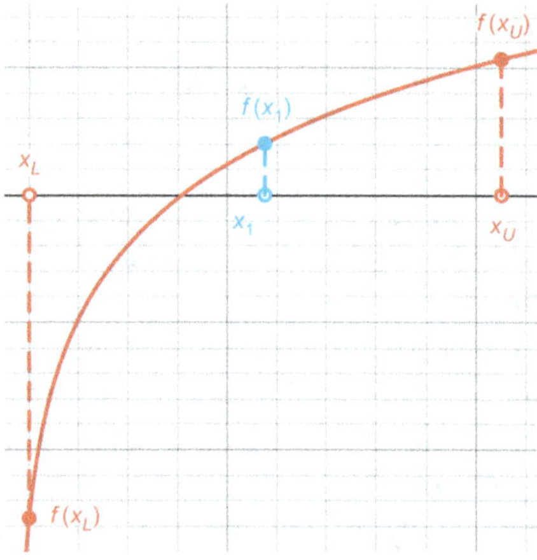

Figure 1.1: The first midpoint of the bisection algorithm at x_1 lies between the initial lower and upper bounds. $f(x_1)$ and $f(x_U)$ are both positive, so x_1 replaces x_U and becomes the new upper bound in the next iteration of Figure 1.2.

Figure 1.2 redefines x_1 as the new x_U, x_L remains the same, and the next midpoint x_2 is generated. This time $f(x_2) < 0$, so in the third iteration x_2 replaces x_L because $f(x_L) < 0$, x_U remains the same, and the third midpoint at x_3 is found.

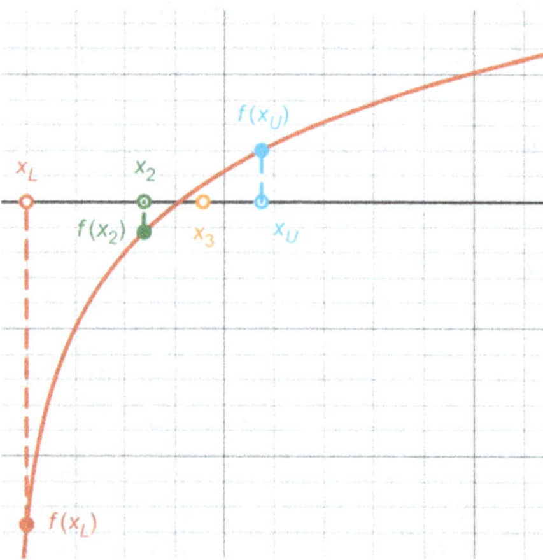

Figure 1.2: The new upper bound $x_U = x_1$ from the previous iteration of Figure 1.1. The second midpoint, at x_2, is located between the current upper and lower bounds. Since $f(x_2)$ and $f(x_L)$ are both negative, x_2 replaces x_L and becomes the new lower bound in the next iteration. The next midpoint of the updated interval is at x_3.

Figures 1.1 and 1.2 demonstrate after each iteration of the bisection algorithm, exactly half of the current interval $[x_L, x_U]$ is always eliminated, while the other half is guaranteed to contain a solution.

Example 1.1. Use the bisection algorithm with three iterations to estimate a zero for $f(x) = x^5 - 3x^2 + 1$ on $[-2, 4]$.

Solution. Using Equation (1.1), the first midpoint is at

$$x_1 = \frac{x_L + x_U}{2} = \frac{-2 + 4}{2} = 1.$$

Since $f(-2) = -43$, $f(4) = 977$, and $f(1) = -1$, $x_1 = 1$ replaces $x_L = -2$ and the new interval is $[1, 4]$. The second midpoint is at

$$x_2 = \frac{x_L + x_U}{2} = \frac{1 + 4}{2} = 2.5.$$

Since $f(2.5) = 79.906$, $f(1)$ is negative, and $f(4)$ is positive, $x_2 = 2.5$ replaces $x_U = 4$ giving a new interval of $[1, 2.5]$. The third midpoint of the last iteration locates the estimated zero at

$$x_3 = \frac{x_L + x_U}{2} = \frac{1 + 2.5}{2} = 1.75.$$

While the bisection algorithm always eventually converges, it is often slow and inefficient. The solution of Example 1.1 is a terrible zero estimate since $f(1.75) = 8.226$; clearly, more iterations of the algorithm are necessary. Also note the inefficiency of the algorithm. The first midpoint at $x_1 = 1$ is actually a better zero estimate than $x_3 = 1.75$ since $f(1) = -1$ is closer to zero than $f(x_3) = 8.226$, although $x_1 = 1$ is still quite inaccurate with $f(1) = -1 \neq 0$. The bisection algorithm sometimes generates estimates which move farther away from the true zero before eventually converging to the solution.

Error analysis

Because the number of iterations for an acceptable estimate is unknown and can be large, simply choosing n iterations as a stopping criterion is not a sound strategy. A better approach is stopping when successive estimates are within a certain small absolute value change δ, or $|x_i - x_{i-1}| < \delta$. However, relative errors are often preferred in analysis because they scale according to the magnitude of the values estimated.

For example, suppose consecutive zero estimates for $f(x)$ are at $x_{i-1} = 0.4$ and $x_i = 0.5$; the absolute value change is $\delta_i = 0.1$. For another $g(x)$, consecutive zero estimates are at $x_{j-1} = 99.9$ and $x_j = 100.0$; the absolute value change is also $\delta_j = 0.1$. Without context, $\delta_i = \delta_j$ suggests terminating the bisection algorithm for $f(x)$ and $g(x)$ in roughly the same number of additional iterations. However, the zero estimate for $f(x)$ has changed dramatically more than it has for $g(x)$, relative to the current estimate. For $f(x)$, the zero estimate has changed

$$\frac{\delta_i}{x_i} = \frac{0.1}{0.5} = 0.2 = 20\,\%.$$

But for $g(x)$ it has changed just

$$\frac{\delta_j}{x_j} = \frac{0.1}{100} = 0.001 = 0.1\,\%.$$

These relative changes indicate the zero estimate for $g(x)$ at $x_j = 100$ has been found, since it has not changed much, relatively 0.1 %, from the previous estimate $x_{j-1} = 99.9$; no more iterations on $g(x)$ are needed. But the zero estimate for $f(x)$ has changed dramatically more between estimates, relatively 20 %, implying $x_i = 0.5$ is not a close enough estimate and the iterations on $f(x)$ should continue.

Correspondingly, another algorithm termination option is stopping after a certain relative change in estimations is met. The estimation relative change error E_c is defined here as the absolute value of the ratio of the difference between the current and previous estimates divided by the current estimate, expressed as a percentage.

$$E_c = \left| \frac{x_i - x_{i-1}}{x_i} \right| \times 100\,\%. \tag{1.2}$$

Example 1.2. Use the bisection algorithm to estimate a solution for $4 + \ln x = 1.3^x$ on $[6, 8]$ until the estimation relative change error $E_c < 1\,\%$.

Solution. First note recasting this equation as a zero-finding problem by putting all terms on one side.

$$1.3^x - 4 - \ln x = 0.$$

The left side can be defined simply as the function in the zero-finding process.

$$f(x) = 1.3^x - 4 - \ln x = 0.$$

In Table 1.1, x_L and x_U are, respectively, the lower and upper bounds of the current interval of the ith iteration, $x_L = a$ and $x_U = b$ in the first iteration, and x_i is the midpoint of the current interval. The five iterations satisfying the criterion $E_c < 1\,\%$ are summarized. The estimated zero is at $x_5 = 6.813$; $f(6.813) = 0.056 \approx 0$.

Table 1.1: After each iteration of the bisection algorithm of Example 1.2, either the lower bound x_L or the upper bound x_U is replaced by the previous x_i value, indicated by the green cell. Whichever of the previous $f(x_L)$ or $f(x_U)$ values has the same sign as the previous $f(x_i)$ value determines which bound is replaced.

i	x_L	$f(x_L)$	x_U	$f(x_U)$	x_i	$f(x_i)$	E_c, %
1	6.000	−0.965	8.000	2.078	7.000	0.329	
2	6.000	−0.965	7.000	0.329	6.500	−0.368	7.69
3	6.500	−0.368	7.000	0.329	6.750	−0.033	3.70
4	6.750	−0.033	7.000	0.329	6.875	0.145	1.82
5	6.750	−0.033	6.875	0.145	6.813	0.056	0.92

Using E_c in conjunction with the bisection algorithm is advantageous because E_c also defines an upper bound. Although the true zero at x_t is unknown, the true relative change error ε_c between the current estimate and the true zero, relative to the current estimate, can be defined

$$\varepsilon_c = \left| \frac{x_i - x_t}{x_i} \right| \times 100\,\%. \tag{1.3}$$

The significance of E_c as an upper bound means ε_c could be smaller than E_c but never larger; E_c represents the worst-case scenario. So if E_c lies below a certain tolerance, ε_c is guaranteed to lie below that same tolerance, possibly by a greater amount.

Theorem 1.1. *The bisection algorithm guarantees $\varepsilon_c < E_c$ at the ith iteration for any $i > 1$.*

Proof. The positive change between the current estimate at x_i and the previous estimate at x_{i-1} is

$$|x_i - x_{i-1}|.$$

Using the bisection algorithm, the estimated zero at x_i is halfway between x_L and x_U with $x_L < x_i < x_U$. It follows

$$x_i - x_L = x_U - x_i$$
$$|x_i - x_L| = |x_U - x_i|$$
$$|x_i - x_L| = |x_i - x_U|$$

The previous estimate at x_{i-1} is at one of the bounds x_L or x_U. Substituting each one separately in the above equation means either

$$|x_i - x_{i-1}| = |x_i - x_U| \text{ or}$$
$$|x_i - x_L| = |x_i - x_{i-1}|.$$

Case 1. The true zero at x_t lies in $x_L < x_t < x_i$.

Since $x_L < x_t$, negating both sides reverses the inequality. Adding x_i to both sides gives

$$x_i - x_L > x_i - x_t.$$

Each side is positive because x_i is greater than both x_L and x_t, so the inequality is also true in absolute value.

$$|x_i - x_L| > |x_i - x_t|.$$

Dividing both sides by $|x_i|$ gives

$$\left| \frac{x_i - x_L}{x_i} \right| > \left| \frac{x_i - x_t}{x_i} \right|.$$

Established before Case 1, $|x_i - x_{i-1}| = |x_i - x_L|$. Substituting on the left side,

$$\left| \frac{x_i - x_{i-1}}{x_i} \right| > \left| \frac{x_i - x_t}{x_i} \right|$$

Multiplying both sides by 100 %, the left side is the definition of E_c of Equation (1.2), and the right side is the definition of ε_c of Equation (1.3), so that

$$E_c > \varepsilon_c.$$

Case 2. The true zero at x_t lies in $x_i < x_t < x_U$.

Since $x_t < x_U$

$$x_t - x_i < x_U - x_i.$$

Each side is positive because both x_t and x_U are greater than x_i, so the inequality also holds in absolute values.

$$|x_t - x_i| < |x_U - x_i|.$$
$$|x_i - x_t| < |x_i - x_U|.$$

Noted before Case 1, $|x_i - x_{i-1}| = |x_i - x_U|$, so substituting on the right side

$$|x_i - x_t| < |x_i - x_{i-1}|.$$

Dividing both sides by $|x_i|$

$$\left| \frac{x_i - x_t}{x_i} \right| < \left| \frac{x_i - x_{x-i}}{x_i} \right|.$$

Exactly as Case 1, multiplying both sides by 100 % concludes with $E_c < \varepsilon_c$. $\qquad\square$

Noteworthy is satisfying the E_c criterion at the ith iteration does not always produce the best zero estimate at x_i for all x_j where $1 \leq j < i$. Because the bisection algorithm estimate sometimes moves farther from the true zero on consecutive iterations, a previous estimate at x_j could be closer to the true zero than at x_i. This seeming contradiction resolves in hindsight after E_c is satisfied, illustrated by Table 1.1 of Example 1.2.

At $x_3 = 6.75, f(6.75) = -0.033 \approx 0$. Because $E_c = 3.7\%$ at x_3, there is no guarantee this estimate already satisfies the criterion of $E_c < 1\%$. At $x_5 = 6.813$ the algorithm terminates because $E_c = 0.92\%$, and $f(6.813) = 0.056 \approx 0$ as well. Only now is the zero estimate at x_3 retroactively verified as the best estimate which satisfies the criterion. The absolute value difference between the estimate at x_3 of 0.033 and zero is smaller than the absolute value difference at any other x value estimate, including at x_5 with a difference of 0.056. Five iterations were necessary to guarantee the criterion was met, even though the best estimate was already located during the third iteration, albeit unverified.

Therefore, once satisfying the stopping criterion for E_c at the ith iteration, the estimate at x_j for $j = 1 \ldots i$ that produces the smallest absolute value difference with zero is the best estimate. Typically, this is at x_i, but not always.

1.2 False position algorithm

Similar to the bisection algorithm, the false position algorithm begins with a bounded interval where the lower and upper bound function values have opposite signs. Instead of using the midpoint to reduce the interval, the x-intercept of the line connecting the points at the bounds of the interval is used as the next zero estimate. Successive estimates proceed as in the bisection algorithm. The lower or upper bound is replaced with the current estimate, whichever has the same function value sign as the function value sign of the current estimate.

Figure 1.3 depicts the first iteration of the false position algorithm. The secant line connecting the bound points $(x_L, f(x_L))$ and $(x_U, f(x_U))$ intersects the x-axis at x_1, the first zero estimate. Like the bisection algorithm, subsequent estimates are based on the sign of the current function value estimate. In this case, since both $f(x_1) < 0$ and $f(x_L) < 0$, x_1 becomes the new lower bound x_L in the next iteration.

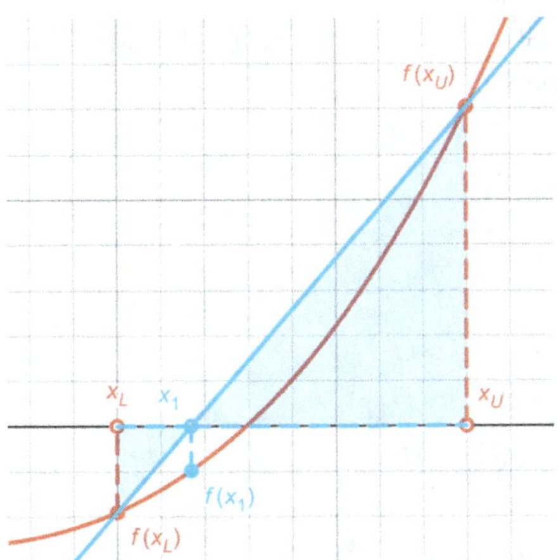

Figure 1.3: Using the false position algorithm, the x-intercept of the secant line between the two bound points is the zero estimate, here at x_1. Since $f(x_1)$ and $f(x_L)$ are both negative, x_1 replaces x_L as the lower bound in the next iteration; see Figure 1.4. The secant line and the x-axis form two shaded similar triangles from which x_1 can be solved.

Figure 1.4 shows the updated interval, in which x_1 is now x_L, x_U remains the same, and the next secant line x-intercept is located at x_2. For a third iteration, x_L would again be replaced, this time by x_2, since both $f(x_L)$ and $f(x_2)$ are negative.

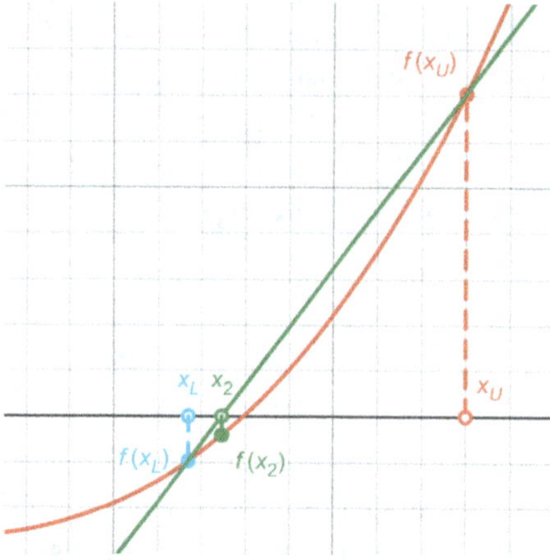

Figure 1.4: In the second iteration of the false position algorithm, the new secant line x-intercept zero estimate at x_2 has $f(x_2) < 0$. So like x_1 in Figure 1.3, x_2 replaces the lower bound since $f(x_L) < 0$ as well.

Solving for the secant line x-intercept employs a well-known theorem from Geometry. The line connecting the points of the bounds of the interval in conjunction with the x-axis create alternate interior angles at the x-intercept and form two similar right triangles. Thus, the ratio of their bases and heights are equal. In Figure 1.3, two similar right triangles are shaded and the blue line connecting the bound points slopes positively so that the left/bottom triangle has a negative base $x_L - x_1$ and negative height $f(x_L)$. This ratio coincides with the ratio of the positive base $x_U - x_1$ and positive height $f(x_U)$ of the right/top triangle. If the line sloped negatively, each triangle would have one negative dimension, and the ratio of their sides again would correspond in sign.

Setting the ratio of the bases equal to the ratio of the heights, using the sides of the left triangle divided by the sides of the right triangle

$$\frac{x_L - x_1}{x_U - x_1} = \frac{f(x_L)}{f(x_U)}.$$

Solving for the estimate at x_1

$$(x_L - x_1)f(x_U) = (x_U - x_1)f(x_L)$$
$$x_L f(x_U) - x_1 f(x_U) = x_U f(x_L) - x_1 f(x_L)$$
$$x_1 f(x_L) - x_1 f(x_U) = x_U f(x_L) - x_L f(x_U)$$
$$x_1(f(x_L) - f(x_U)) = x_U f(x_L) - x_L f(x_U)$$
$$x_1 = \frac{x_U f(x_L) - x_L f(x_U)}{f(x_L) - f(x_U)}.$$

Naturally, this can be used as the general formula so that the false position algorithm zero estimate at x_i for the current interval $[x_L, x_U]$ is

$$x_i = \frac{x_U f(x_L) - x_L f(x_U)}{f(x_L) - f(x_U)}. \tag{1.4}$$

Note the only difference between the bisection and false position algorithms is the method of calculating the ith estimate, effectively the difference between using Equation (1.1) or (1.4).

Example 1.3. Use the false position algorithm to estimate a solution for $4 + \ln x = 1.3^x$ on [6, 8] until $E_c < 1\%$.

Solution. Putting all terms on one side as before, $1.3^x - 4 - \ln x = 0$ so that $f(x) = 1.3^x - 4 - \ln x = 0$. Using Equation (1.4) for the first zero estimate

$$x_1 = \frac{8f(6) - 6f(8)}{f(6) - f(8)} = \frac{8(-0.965) - 6(2.078)}{-0.965 - 2.078} = 6.634.$$

All iterations are summarized in Table 1.2. The lower bound was replaced both times because the new function value estimate was negative both times.

Table 1.2: After each iteration of the false position algorithm of Example 1.3, either the lower bound x_L or the upper bound x_U is replaced by the previous x_i value, indicated by the green cell. Since $f(x_L)$ has the same sign as $f(x_i)$ in the first two iterations, the lower bound x_L was replaced both times.

i	x_L	$f(x_L)$	x_U	$f(x_U)$	x_i	$f(x_i)$	$E_c, \%$
1	6.000	−0.965	8.000	2.078	6.634	−0.192	
2	6.634	−0.192	8.000	2.078	6.750	−0.034	1.71
3	6.750	−0.034	8.000	2.078	6.769	−0.006	0.30

The particular function and starting interval always play a part in the number of iterations needed to satisfy the stopping criterion. Comparing the results generated in Examples 1.2 and 1.3 from the same function, starting interval, and E_c value, the false position algorithm required less iterations and was more accurate than the bisection algorithm, but that is not always the case. In Figure 1.5, the true zero is actually close to

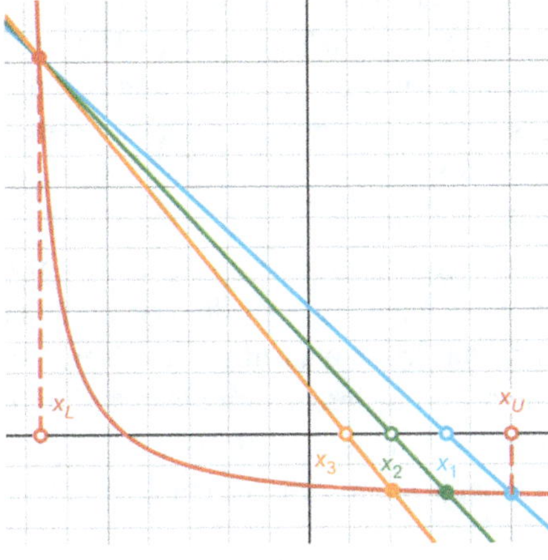

Figure 1.5: The closed circles represent the function values at the endpoints and the correspondingly colored open circles represent the location of both the next zero estimate based on the secant line intersecting the x-axis, and the next endpoint if the algorithm continues.

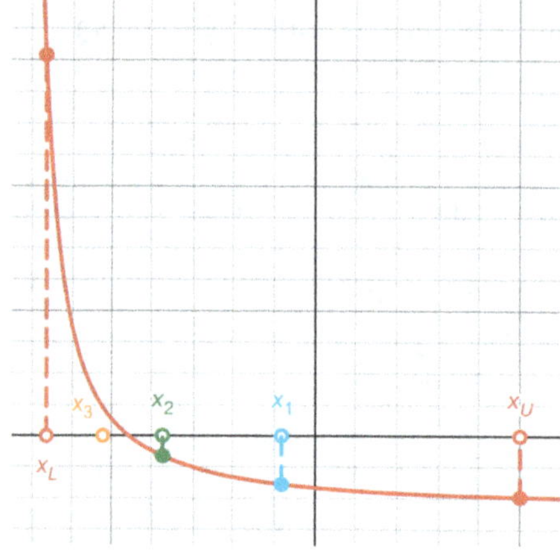

Figure 1.6: The open numbered circles represent the midpoint of the interval containing the zero estimate for the ith iteration. Compare how much closer to the true zero these x_2 and x_3 estimates using the bisection algorithm are than the corresponding x_2 and x_3 estimates using the false position algorithm of Figure 1.5.

the lower bound, but because of the initial upper bound and the way the function curves, the upper bound will be replaced repeatedly in small increments before getting close to the true zero. With the same number of iterations and starting interval, Figure 1.6 demonstrates the bisection algorithm would be a better choice for this function and interval.

Error analysis

Each iteration of the false position algorithm eliminates a variable length of the current interval, unlike the bisection algorithm, which always removes exactly half the interval. While removing more than half the interval is an improvement, Figure 1.4 illustrates even removing less than half could require fewer total iterations than halving the interval each time.

Conversely, Figure 1.5 depicts a case in which removing a small portion of the current interval is not an improvement over halving the interval. Furthermore, Figure 1.6 compares using the bisection algorithm on the same function with the same starting interval as Figure 1.5, more efficiently in this instance than applying the false position algorithm under the same conditions. Knowledge of function behavior can determine which algorithm works better; neither is optimal in all situations.

Theorem 1.1 established the estimation relative change error E_c, defined by Equation (1.2), as an upper bound for the true relative change error ε_c, defined by Equation (1.3), when used as a stopping criterion for the bisection algorithm. That is not the case for the false position algorithm, precisely because the length eliminated from the current interval varies and is not fixed at one half. Nevertheless, even without the guarantee of providing an upper bound, E_c is a reasonable stopping criterion for the false position algorithm.

1.3 Fixed-point algorithm

Recall any equation can be recast as $f(x) = 0$ by putting all terms on one side and defining the nonzero side as $f(x)$. Furthermore, if $f(x)$ can be rewritten in the form $f(x) = g(x) + kx$ where k is a constant, x can be isolated to produce an iterative fixed-point algorithm zero estimate at x_{i+1}.

$$f(x) = 0$$
$$g(x) + kx = 0$$
$$kx = -g(x)$$
$$x_{i+1} = -\frac{g(x_i)}{k}. \tag{1.5}$$

Equation (1.5) only potentially solves the initial equation but the following theorem guarantees Equation (1.5) converges to a solution if consecutive terms approach each other.

Theorem 1.2. *If* $\lim_{i \to \infty} x_{i+1} = x_i$, *Equation (1.5) converges to a solution.*

Proof. Taking the limit as $i \to \infty$ of both sides of Equation (1.5)

$$\lim_{i \to \infty} x_{i+1} = \lim_{i \to \infty} -\frac{g(x_i)}{k}.$$

If $\lim_{i \to \infty} x_{i+1} = x_i$, then

$$x_i = -\frac{g(x_i)}{k}$$
$$kx_i = -g(x_i)$$
$$g(x_i) + kx_i = 0.$$

Since $f(x) = g(x) + kx$ by definition, in particular,

$$f(x_i) = g(x_i) + kx_i = 0.$$

Thus, $f(x_i) = 0$ and x_i is a solution to the initial equation. $\qquad\square$

Example 1.4. Use the fixed-point algorithm to estimate a solution for $2x^3 - x^2 + 15x = 3$ with an initial estimate of $x_1 = 2.25$ and accurate to the thousandths place.

Solution. Define $f(x) = 2x^3 - x^2 + 15x - 3$ and solve $f(x) = 0$. Isolating for x produces the iterative step for this particular function per Equation (1.5).

$$f(x) = 2x^3 - x^2 + 15x - 3 = 0$$
$$15x = 3 + x^2 - 2x^3$$

$$x_{i+1} = \frac{3 + x_i^2 - 2x_i^3}{15}$$

If $x_1 = 2.25$, then

$$x_2 = \frac{3 + (2.25)^2 - 2(2.25)^3}{15} = -0.981$$

$$x_3 = \frac{3 + (-0.981)^2 - 2(-0.981)^3}{15} = 0.390$$

$$x_4 = \frac{3 + (0.390)^2 - 2(0.390)^3}{15} = 0.202$$

$$x_5 = \frac{3 + (0.202)^2 - 2(0.202)^3}{15} = 0.202.$$

Rounded to the thousandths place, the estimated zero is $x_4 = 0.202$; see the following note concerning subscripts. This solves $f(0.202) = 0$, which in turn solves the initial equation $2(0.202)^3 - (0.202)^2 + 15(0.202) = 3$.

Although five iterations were necessary to confirm the estimated zero rounded to the thousandths place in Example 1.4 above, the final estimation value 0.202 first occurred at the fourth iteration, hence the notation using x_4 in the solution.

The best laid plans...

Care must be taken when solving for x so extraneous solutions are not introduced or domain restrictions are not violated. One example from algebra is squaring both sides of an equation, which could produce extra solutions that do not satisfy the original equation. Another example is exponentiating a logarithmic function, which inadvertently removes the positive domain restriction on the argument of the log. Also, consider isolating x may create an unwieldy function containing pieces such as high degree roots or inverse trigonometric functions. An easy workaround to all these issues is recasting the equation as $f(x) = 0$ as before and then simply adding an x term to both sides of this equation. This produces the iterative step that again potentially solves the initial equation, provided the condition $\lim_{i \to \infty} x_{i+1} = x_i$ is met. This alternate fixed-point algorithm zero estimate at x_{i+1} is

$$f(x) = 0 \text{ or } 0 = f(x)$$

$$x = x + f(x)$$

$$x_{i+1} = x_i + f(x_i). \tag{1.6}$$

Theorem 1.3. *If* $\lim_{i \to \infty} x_{i+1} = x_i$, *Equation (1.6) converges to a solution.*

Proof. Taking the limit as $i \to \infty$ of both sides of Equation (1.6)

$$\lim_{i\to\infty} x_{i+1} = \lim_{i\to\infty} \left(x_i + f(x_i) \right)$$

If $\lim_{i\to\infty} x_{i+1} = x_i$, then

$$x_i = x_i + f(x_i)$$
$$0 = f(x_i).$$

Thus x_i is a solution to the initial equation recast as $f(x) = 0$. $\qquad\qquad\square$

Example 1.5. Use the fixed-point algorithm to estimate a solution for $\sqrt{x} = \cos x^2$ with and initial estimate of $x_1 = 0.5$ until the estimation relative change error $E_c < 1\%$.

Solution. Squaring both sides of the equation introduces potential extraneous solutions or negative values in the iterative process which are not part of the domain of the original equation. Instead, the root is moved to the right side and an x is added to both sides to form the iterative step for this particular function per Equation (1.6).

$$0 = \cos x^2 - \sqrt{x}$$
$$x = \cos x^2 - \sqrt{x} + x$$
$$x_{i+1} = \cos x_i^2 - \sqrt{x_i} + x_i$$

The iterations are summarized in Table 1.3. Checking $x_5 = 0.734$ verifies.

$$0.734 \approx \cos(0.734)^2 - \sqrt{0.734} + 0.734$$
$$0 \approx \cos(0.734)^2 - \sqrt{0.734}.$$

This is equivalent to the initial equation $\sqrt{0.734} \approx \cos(0.734)^2$.

Table 1.3: The fixed-point algorithm details of Example 1.5.

i	x_i	$x_{i+1} = \cos x_i^2 - \sqrt{x_i} + x_i$	$E_c, \%$
1	0.500	0.762	
2	0.762	0.725	34.37
3	0.725	0.738	5.04
4	0.738	0.734	1.79
5	0.734		0.60

The first three columns of Table 1.3 are duplicated in Table 1.4 to illustrate the condition necessary for convergence to a solution, $\lim_{i\to\infty} x_{i+1} = x_i$, is satisfied by showing the equivalent $\lim_{i\to\infty}(x_{i+1} - x_i) = 0$. The difference between row entries in the second and third columns, x_i and x_{i+1} respectively, is displayed in the fourth column $x_{i+1} - x_i$ and approaches zero. After a two more iterations $x_7 = 0.735$ and $x_8 = 0.735$, the true zero rounded to the thousandths place.

Table 1.4: Using the data of Example 1.5, the fourth column indicates $\lim_{i \to \infty}(x_{i+1} - x_i) = 0$, the condition equivalent to $\lim_{i \to \infty} x_{i+1} = x_i$ that is necessary for convergence to a solution and verifies $x_5 = 0.734$ is a valid estimation.

i	x_i	x_{i+1}	$x_{i+1} - x_i$
1	0.500	0.762	0.262
2	0.762	0.725	−0.037
3	0.725	0.738	0.013
4	0.738	0.734	−0.004
5	0.734	0.736	0.002
6	0.736	0.735	−0.001
7	0.735	0.735	0.000

Remember, solving for x and using Equation (1.5) or adding an x to both sides of an equation and using Equation (1.6) only potentially solves an initial equation. Theorems 1.2 and 1.3 each include the supposition $\lim_{i \to \infty} x_{i+1} = x_i$. While the condition is the same, it may not necessarily hold for both Equations (1.5) and (1.6) for the same function and starting interval. The convergence to a solution, or divergence away from one, does not imply the same result from the other. Therefore, if either Equation (1.5) or (1.6) fails there is merit in applying the other. Divergence manifests in estimates, which continue to increase in magnitude, estimation relative change errors which stop decreasing, or either estimates or estimation relative change errors that alternate infinitely.

If Equation (1.6) was used instead of Equation (1.5) in Example 1.4, divergence is imminent, as both the estimates and estimation relative change errors increase, as shown in Table 1.5.

Table 1.5: Using Equation (1.6) by adding an x to both sides of the equation in Example 1.4 diverges; both the x_i values and E_c values increase in magnitude.

i	x_i	$x_{i+1} = 2x_i^3 - x_i^2 + 16x_i - 3$	E_c, %
1	2.25	50.179	
2	50.179	259,173.081	95.56
3	259,173.081	34,817,600,270,656,100	99.98

Conversely if Equation (1.5) was used in lieu of Equation (1.6) in Example 1.5, x_i values alternate infinitely starting at the fifth and sixth iterations and, if more than three digits to the right of the decimal point are retained for the x_i values, estimation relative change errors begin alternating infinitely at the 13th and 14th iterations, as shown in Table 1.6. Note neither $x = 0.308$ nor $x = 0.991$ solves the initial equation $\sqrt{x} = \cos x^2$, since

$$\sqrt{0.308} = 0.555 \neq \cos (0.308)^2 = 0.996$$

$$\sqrt{0.991} = 0.995 \neq \cos (0.991)^2 = 0.555.$$

Table 1.6: Using Equation (1.5) by solving for x in Example 1.5 diverges; both the x_i values and the E_c values eventually infinitely alternate.

i	x_i	$x_{i+1} = \cos^2 x_i^2$	E_c, %
1	1.000	0.292	
2	0.292	0.993	242.55
3	0.993	0.305	70.59
4	0.305	0.991	225.34
5	0.991	0.308	69.22
6	0.308	0.991	222.18
7	0.991	0.308	68.95
8	0.308	0.991	221.53
⋮	⋮	⋮	⋮
12	0.308	0.991	221.36
13	0.991	0.308	68.88
14	0.308	0.991	221.35
15	0.991	0.308	68.88
16	0.308	0.991	221.35

A poor choice for the starting estimate also contributes to the failure of single-point algorithms like fixed-point and Newton–Raphson of Section 1.4. If the initial estimate is not close enough to the solution, single-point algorithms may diverge and iterate farther from the true zero and produce no solution. Bracketed algorithms, those that use an upper and lower bound to surround the zero like bisection and false position, are not subject to divergence and are guaranteed to converge to a solution, even if slowly. To reduce the possibility of divergence for single-point algorithms, a similar bracketed algorithm approach locates a small interval $[x_L, x_U]$ in which $f(x_L)$ and $f(x_U)$ differ in sign, and selects either x_L or x_U as the starting estimate.

Example 1.6 demonstrates that carefully considering a starting value for single-point algorithms can be crucial to their success.

Example 1.6. Show the fixed-point algorithm diverges from a zero of $f(x) = 2x^3 - 8x - 3$ using an initial estimate of $x_1 = 2.5$, but converges to a zero using an initial estimate of $x_1 = 0.25$.
Solution. The detailed iterations for both starting values are summarized in Tables 1.7 and 1.8. The divergence in Table 1.7 exhibits as the increasing magnitude of both x_{i+1} and the estimation relative change error E_c.

1.4 Newton–Raphson algorithm

Like the fixed-point algorithm, the Newton–Raphson algorithm requires a single starting estimate. The motivation is that a tangent line approximates a function for values near the point of tangency. This method uses the x-intercept of the tangent line at the current

Table 1.7: Using the fixed-point algorithm on $f(x) = 2x^3 - 8x - 3$ diverges when $x_1 = 2.5$.

i	x_i	$x_{i+1} = (2x_i{}^3 - 3)/8$	E_c, %
1	2.5	3.531	
2	3.531	10.633	29.20
3	10.633	300.205	66.79
4	300.205	6,763,852.049	94.46
5	6,763,852.049	7.736×10^{19}	100.00

Table 1.8: Using the fixed-point algorithm on $f(x) = 2x^3 - 8x - 3$ converges to a solution when $x_1 = 0.25$.

i	x_i	$x_{i+1} = (2x_i{}^3 - 3)/8$	E_c, %
1	0.250	−0.371	
2	−0.371	−0.388	167.37
3	−0.388	−0.390	4.30
4	−0.390	−0.390	0.46

estimate. If this value does not meet the stopping criterion, the derivative of the function at this new estimate is used to generate a new tangent line and x-intercept.

From calculus, the tangent line to $f(x)$ at $x = x_i$ in point-slope form is

$$y - f(x_i) = f'(x_i)(x - x_i).$$

The x-intercept of this line is the zero estimate, at $(x, 0)$. So setting $y = 0$ and solving for x

$$0 - f(x_i) = f'(x_i)(x - x_i)$$

$$-\frac{f(x_i)}{f'(x_i)} = x - x_i$$

$$x = x_i - \frac{f(x_i)}{f'(x_i)}$$

If x does not meet the stopping criterion it becomes the next estimate x_{i+1} and the process repeats. Thus the general formula for generating the Newton–Raphson algorithm zero estimate at x_{i+1} is

$$x_{i+1} = x_i - \frac{f(x_i)}{f'(x_i)}. \tag{1.7}$$

Two iterations of the Newton–Raphson algorithm are depicted in Figures 1.7 and 1.8. A tangent line is generated at the initial estimate x_1 using $f(x_1)$ and $f'(x_1)$, and the x-intercept of that line is the next estimate x_2 (Figure 1.7). The process repeats with a tangent line now generated at x_2 using $f(x_2)$ and $f'(x_2)$, and the x-intercept of this new

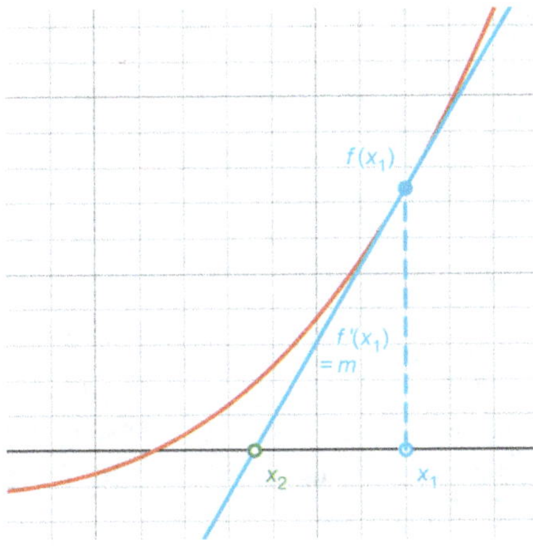

Figure 1.7: Using the initial estimate x_1, a tangent line there produces the next estimate at the x-intercept of x_2.

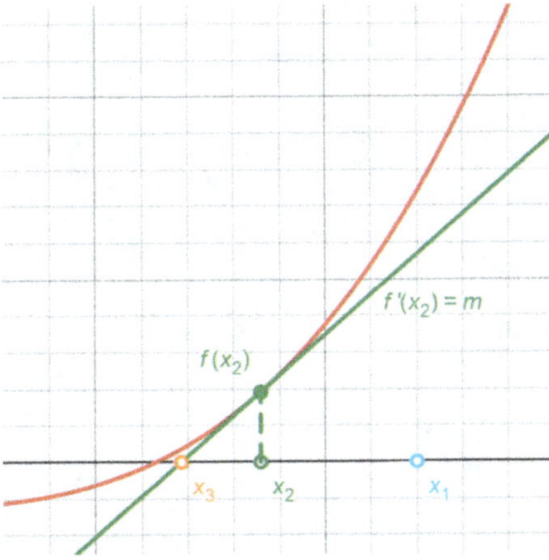

Figure 1.8: A new tangent line at the second estimate x_2 produces the next estimate at the x-intercept of x_3, relatively close to the true zero.

line is the next estimate x_3 (Figure 1.8). After just two iterations, the estimate at x_3 is reasonably close to the true zero.

As with previous methods, the stopping criterion can be the estimation relative change error E_c of Equation (1.2). However, if $f(x_i)$ is indeed approaching zero as i increases, or $\lim_{i \to \infty} f(x_i) = 0$, taking the limit as $i \to \infty$ of both sides of Equation (1.7) yields

$$\lim_{i \to \infty} x_{i+1} = \lim_{i \to \infty} \left(x_i - \frac{f(x_i)}{f'(x_i)} \right)$$

$$x_{i+1} = \left(x_i - \frac{0}{f'(x_i)} \right)$$

$$x_{i+1} = x_i$$

These equalities produce the double implication

$$f(x_i) = 0 \iff x_{i+1} = x_i.$$

Similarly for the approximations

$$f(x_i) \approx 0 \iff x_{i+1} \approx x_i$$
$$f(x_i) \approx 0 \iff x_{i+1} - x_i \approx 0.$$

Therefore, the stopping criterion can be based on how close the difference between the last two estimates are to zero. This mimics the alternate stopping criterion for the fixed-point algorithm of Section 1.3.

Example 1.7. Use the Newton–Raphson algorithm to estimate a zero for $f(x) = x^3 - 2x^2 - 7$ with an initial estimate of $x_1 = 3$, accurate to thousandths place.

Solution. Using Equation (1.7), $f(3) = 2$, $f'(x) = 3x^2 - 4x$, and $f'(3) = 15$, so the next estimate is

$$x_2 = x_1 - \frac{f(x_1)}{f'(x_1)} = 3 - \frac{f(3)}{f'(3)} = 3 - \frac{2}{15} = 2.867$$

All iterations are summarized in Table 1.9.

Table 1.9: The Newton–Raphson algorithm iteration details of Example 1.7.

i	x_i	$f(x_i)$	$f'(x_i)$	$E_c, \%$
1	3.000	2.000	15.000	
2	2.867	0.122	13.187	4.65
3	2.857	0.001	13.065	0.32
4	2.857	0.000	13.064	0.00

When the plan does not come together

The Newton–Raphson algorithm routinely converges quickly, but there are a few potential drawbacks. First, since it uses a tangent line, the derivative of the function must be known. This motivates the derivation of the secant and modified secant algorithms of Sections 1.5 and 1.6, which are in the same vein graphically without requiring the derivative. Second, a poor initial estimate can lead to one of three conditions that produces no solution: divergence away from a solution, an infinite alteration, or a domain violation.

1. Divergence away from a solution

Figures 1.9 through 1.12 illustrate the first phenomena that might result from a poor initial estimate, divergence away from a solution. Numerically, estimates will increase in magnitude or the estimation relative change errors will stop decreasing. This is akin to the divergence exhibited in Table 1.7 of Example 1.6 using the fixed-point algorithm.

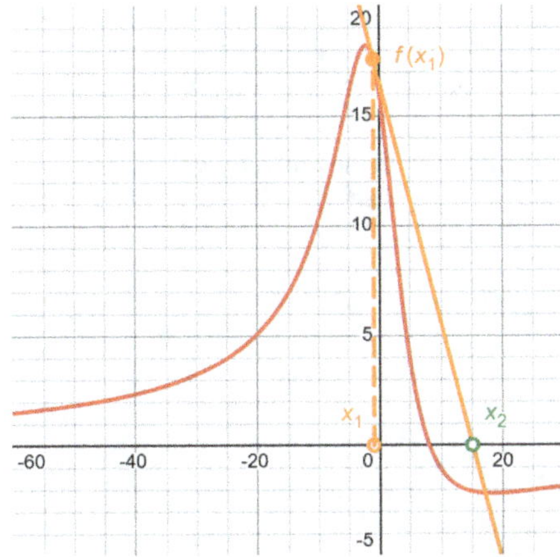

Figure 1.9: The initial estimate is the open circle at x_1, and the tangent line at the corresponding point on the function $(x_1, f(x_1))$ generates the next estimate, its x-intercept is the open circle at x_2 near 15.

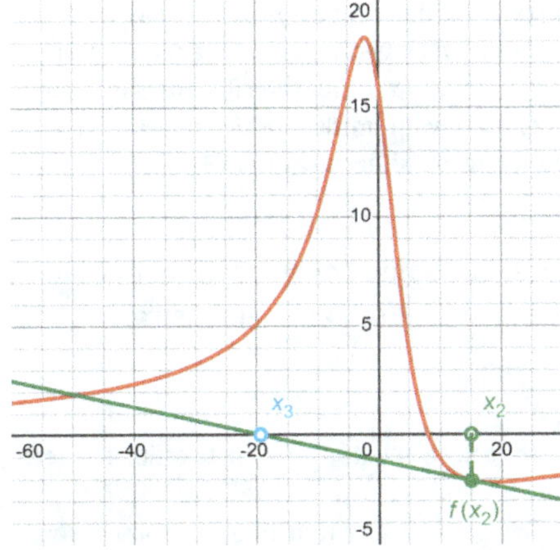

Figure 1.10: The tangent line at the closed circle $(x_2, f(x_2))$ intercepts the x-axis at the next estimate, the open circle at x_3 near -20. Note the true zero is around $x = 8$, so the estimate is already moving away from the true zero; the distance has changed from $|x_2 - 8| = |15 - 8| = 7$ to $|x_3 - 8| = |-20 - 8| = 28$.

2. Infinite alteration

The second pitfall of an ill-chosen initial estimate is an infinite alteration in which, starting at some iteration and for all iterations following, consecutive values for the zero

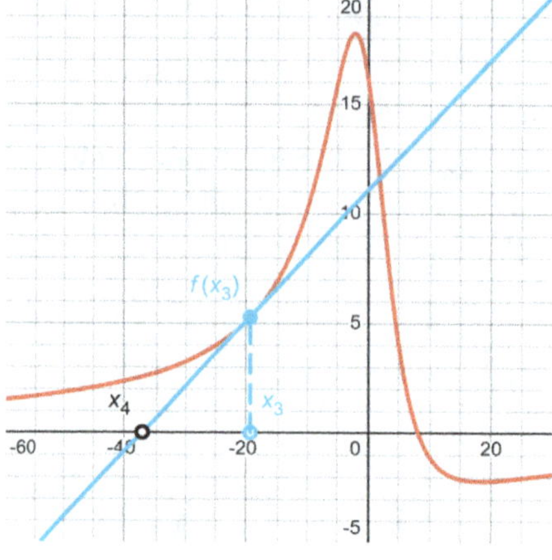

Figure 1.11: The tangent line at the closed circle $(x_3, f(x_3))$ intercepts the x-axis at the next estimate, the open circle at x_4 near -37. This estimate has moved farther from the true zero, with an increased distance of $|x_4 - 8| = |-37 - 8| = 45$.

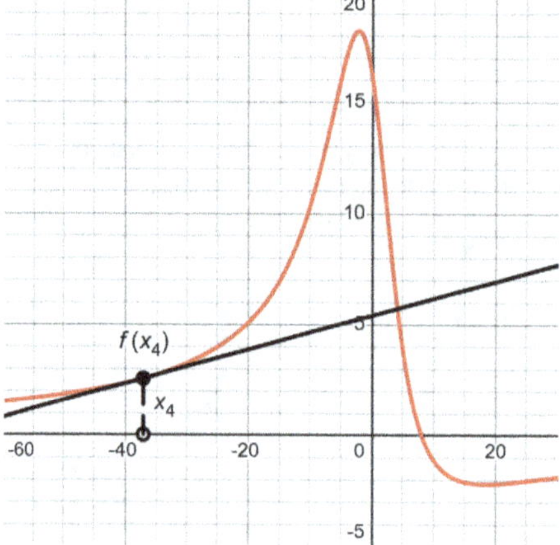

Figure 1.12: The tangent line at the closed circle $(x_4, f(x_4))$ will intercept the x-axis somewhere to the left of -60, and it becomes clear the situation will only get worse. As the function flattens, subsequent tangent lines with smaller positive slopes will move the x-intercept farther away from 8.

estimates alternate between the same two values. The same behavior in successive estimation relative change errors can be observed as well. Both are demonstrated in Example 1.8.

Example 1.8. Show the Newton–Raphson algorithm fails, due to an infinite alteration, to estimate a zero for $f(x) = 3x^{1/3} - x^2 + 1$ using an initial estimate of $x_1 = -1.5$.

Solution. The iterations are summarized in Table 1.10. Two patterns of alternating row entries which stabilize indicate an infinite alteration. First, comparing estimated zeros x_i in the second column, starting at $i = 14, 16, 18, \ldots$ $x_i = -0.556$ and $x_{i+1} = 0.129$, implying the algorithm will never converge to a single value or true zero for an initial estimate of $x_1 = -1.5$. In computer science parlance, the algorithm remains stuck in an infinite loop.

Secondly, and retaining more than three digits to the right of the decimal for the x_i values, comparing estimation relative change errors E_c in the last column, starting at $j = 16, 18, 20, \ldots$ $x_j = 1.23$ and $j_{j+1} =$

5.29, also implying the algorithm will keep generating these two alternating values and will not converge to a single value or true zero for an initial estimate of $x_1 = -0.15$.

Table 1.10: The Newton–Raphson algorithm may fail with an ill-chosen initial estimate, exhibited by an infinite alteration of either estimates x_i or estimation relative change errors E_c, which never converge to a single value.

i	x_i	$f(x_i)$	$f'(x_i)$	E_c, %
1	−1.500	−4.684	3.763	
2	−0.255	−0.968	2.996	4.88
3	0.068	2.220	5.870	4.76
⋮	⋮	⋮	⋮	⋮
13	0.130	2.501	3.647	5.28
14	−0.556	−1.777	2.591	1.23
15	0.129	2.501	3.649	5.30
16	−0.556	−1.776	2.591	1.23
17	0.129	2.501	3.648	5.29
18	−0.556	−1.776	2.591	1.23
19	0.129	2.501	3.649	5.29
20	−0.556	−1.776	2.591	1.23
21	0.129	2.501	3.648	5.29

3. Domain violation

The third kind of result that produces no solution because of a poor initial estimate is a domain violation, depicted in Figure 1.13. At $x_1 = 0.25$ the blue tangent line at $(0.25, -1.551)$ to the red function $f(x) = x^2 - \ln x - 3$ has slope $f'(0.25) = -3.5$ and intersects the x-axis at $x_2 = -0.193$. This immediately terminates the algorithm prematurely since the $\ln x$ term requires a positive argument. However, if $x_1 = 1.25$ the orange tangent line at $(1.25, -1.661)$ has slope $f'(1.25) = 1.7$ and intersects the x-axis at $x_2 = 2.227$. The algorithm then quickly converges to the true zero at $x_4 = 1.910$ accurate to the thousandths place.

Further analysis using calculus, specifically where the first derivative equals zero, indicates the minimum value of $f(x) = x^2 - \ln x - 3$ occurs at

$$f'(x) = 2x - \frac{1}{x} = 0$$

$$x^2 = \frac{1}{2}$$

$$x = \frac{1}{\sqrt{2}} = 0.707.$$

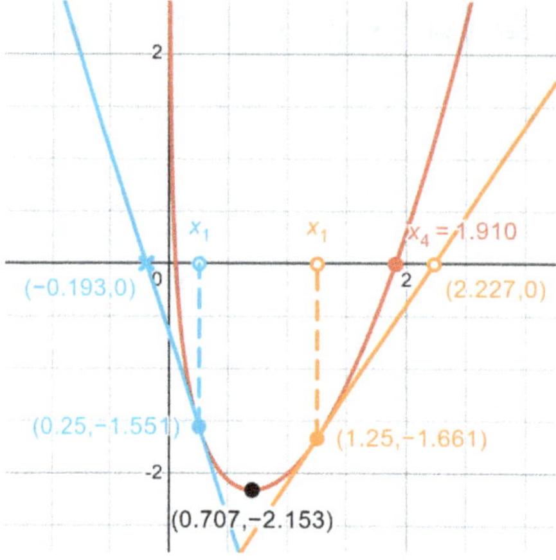

Figure 1.13: Tangent lines at points with x values sufficiently close to the left of $x = 0.707$ like the blue point at $x_1 = 0.25$ intersect the x-axis at a negative value; $x_2 = -0.193$ is not valid for the $\ln x$ term and stops the Newton–Raphson algorithm. Tangent lines at points with x values to the right of $x = 0.707$ like the orange point at $x_1 = 1.25$ intersect the x-axis at a positive value; $x_2 = 2.227$ and subsequent estimates quickly converge to the true zero at $x_4 = 1.910$.

Figure 1.13 further demonstrates for any $x > 0.707$, the tangent line slope at that point $(x, f(x))$ is $f'(x) > 0$, the tangent line intersects the positive x-axis, and the algorithm continues. Conversely for any $x < 0.707$ and sufficiently close to 0.707, the tangent line slope at that point $(x, f(x))$ is $f'(x) < 0$, the tangent line intersects the negative x-axis, and the algorithm halts prior to converging to a true zero. Note if $x < 0.707$ but sufficiently close to the other true zero at $x = 0.050$, the tangent line will intersect the *positive* x-axis and converge to that true zero at $x = 0.050$. Calculus verifies sufficiently close means $x \le 0.132$.

1.5 Secant algorithm

For relatively small intervals, the derivative at one point is approximately equal to the secant line through that point and another close point, or

$$f'(x_i) \approx \frac{f(x_i) - f(x_{i-1})}{x_i - x_{i-1}}. \tag{1.8}$$

Substituting Equation (1.8) into the iterative step of the Newton–Raphson algorithm of Equation (1.7) yields the iterative step for the secant algorithm zero estimate at x_{i+1} as follows.

$$x_{i+1} = x_i - \frac{f(x_i)}{f'(x_i)}$$

$$x_{i+1} = x_i - \frac{f(x_i)}{\frac{f(x_i) - f(x_{i-1})}{x_i - x_{i-1}}}$$

$$x_{i+1} = x_i - \frac{f(x_i)(x_i - x_{i-1})}{f(x_i) - f(x_{i-1})}. \tag{1.9}$$

Each iteration of the secant algorithm uses two points at x_{i-1} and x_i to generate a third point at x_{i+1}. In the next iteration x_{i-1} is discarded and replaced with x_i, x_i is replaced with x_{i+1}, and both x_i and x_{i+1} are used to produce x_{i+2}, and so forth. Table 1.11 expounds how x values are determined at each iteration.

Table 1.11: In consecutive iterations of the secant algorithm, the previous x_{i-1} value is removed, the previous x_i and x_{i+1} values shift one column left, and a new x_{i+1} is generated in the last column. The colors of the x values correspond to the those in Figures 1.14 through 1.16.

Iteration	Points used		Point generated
i	x_{i-1}	x_i	x_{i+1}
1	x_0	x_1	x_2
2	x_1	x_2	x_3
3	x_2	x_3	x_4

Figures 1.14 through 1.16 illustrate three iterations of the secant algorithm. While two points must initiate the secant algorithm, they are different in nature than the two points that start the bisection and false position algorithms, both of which require a bracketed interval *around* the zero. The secant algorithm only requires two relatively close points to simulate a tangent line and can be on the same side of the true zero. Figure 1.14 highlights this difference, as the two starting values at x_0 and x_1 are both to the left of the true zero and do not bracket it. It is possible during the secant algorithm to generate points that do bracket the zero, in which case the iteration is equivalent to the false position algorithm. Both x_3 and x_4 in Figures 1.15 and 1.16, respectively, are located between the two previous estimates.

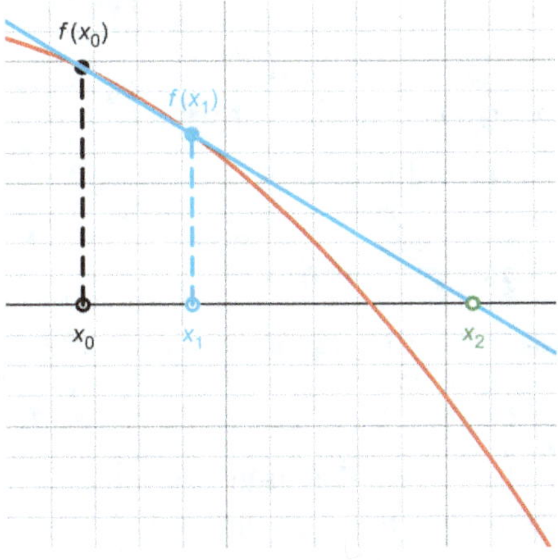

Figure 1.14: The secant algorithm starts with two relatively close points $(x_0, f(x_0))$ and $(x_1, f(x_1))$, commonly on the same side of the true zero. The secant line through those points intersects the x-axis at the next estimate at x_2.

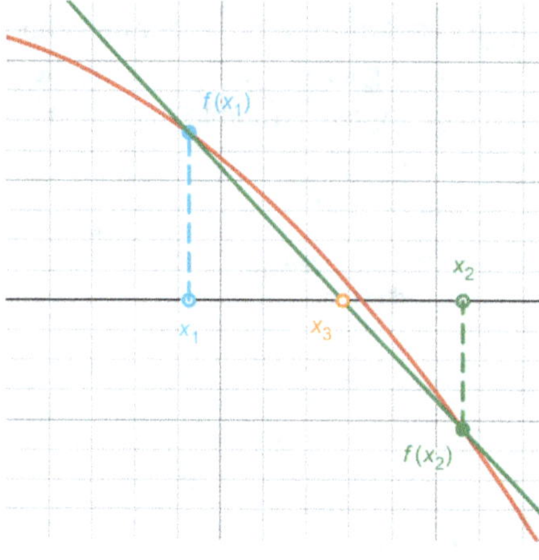

Figure 1.15: The point $(x_0, f(x_0))$ is discarded and the new secant line through $(x_1, f(x_1))$ and $(x_2, f(x_2))$ intersects the x-axis at the next estimate at x_3. In this iteration x_1 and x_2 are on opposite sides of the true zero, unlike the first iteration, in which x_0 and x_1 were both on the left side of the true zero.

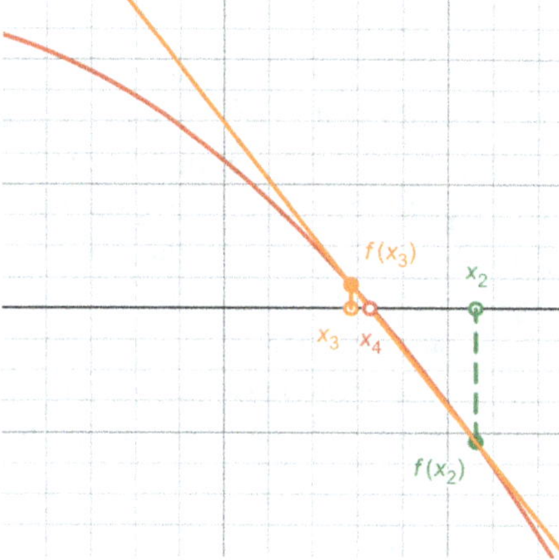

Figure 1.16: Now the point $(x_1, f(x_1))$ is removed and the next secant line through $(x_2, f(x_2))$ and $(x_3, f(x_3))$ intersects the x-axis at the new estimate at x_4.

Example 1.9. Use the secant algorithm to estimate a zero for $f(x) = x^{1/4} - x^3 + 5$ with initial values $x_0 = 1$ and $x_1 = 1.2$ until $E_c < 1\%$.

Solution. $f(1) = 5$ and $f(1.2) = 4.319$. Using Equation (1.9)

$$x_2 = x_1 - \frac{f(x_1)(x_1 - x_0)}{f(x_1) - f(x_0)} = 1.2 - \frac{4.319(1.2 - 1)}{4.319 - 5} = 2.468.$$

Finding x_3 then uses x_1 and x_2 with just one additional calculation of $f(x_2)$, since $f(x_1) = f(1.2) = 4.319$ has been found. The remaining iterations are summarized in Table 1.12.

Because the secant algorithm approximates the Newton–Raphson algorithm, albeit without using the derivative explicitly, it can fall prey to the same three pitfalls which produce no solution, namely, divergence away from a solution, an infinite alteration, or a domain violation; see Section 1.4.

Table 1.12: The secant algorithm iteration details of Example 1.9.

i	x_{i-1}	$f(x_{i-1})$	x_i	$f(x_i)$	x_{i+1}	E_c, %
1	1.000	5.000	1.200	4.319	2.468	
2	1.200	4.319	2.468	−8.733	1.618	52.50
3	2.468	−8.733	1.618	1.891	1.769	8.52
4	1.618	1.891	1.769	0.619	1.842	3.98
5	1.769	0.619	1.842	−0.087	1.833	0.49

1.6 Modified secant algorithm

The secant algorithm begins with two relatively close points at x_{i-1} and x_i to generate the next one at x_{i+1}. However, there is no guarantee x_{i+1} will be close to the x_i used in the calculation of x_{i+1}. This might cause the algorithm to converge slowly or even diverge. The modified secant algorithm prevents a large interval emerging between consecutive x values by using a small *constant* value h to generate x_{i+1} rather than relying on the previous point at x_i.

Recall the secant line through the points at x_{x-1} and x_i are used in the secant algorithm to approximate $f'(x_i)$. The modified secant algorithm also uses a secant line as a substitute for $f'(x)$, but now uses the points at x_i and $x_i + x_i h$, where h is a small fixed value.

$$f'(x_i) \approx \frac{f(x_i + x_i h) - f(x_i)}{x_i h}. \tag{1.10}$$

Substituting Equation (1.10) into the iterative step of the Newton–Raphson algorithm of Equation (1.7) produces the iterative step for the modified secant algorithm zero estimate at x_{i+1}.

$$x_{i+1} = x_i - \frac{f(x_i)}{f'(x_i)}$$

$$x_{i+1} = x_i - \frac{f(x_i)}{\frac{f(x_i + x_i h) - f(x_i)}{x_i h}}$$

$$x_{i+1} = x_i - \frac{x_i h f(x_i)}{f(x_i + x_i h) - f(x_i)}. \tag{1.11}$$

The first iteration of the modified secant algorithm is displayed in Figure 1.17; a secant line generates the next estimate at the x-intercept x_2. Figure 1.18 shows the second iteration, which uses only that x-intercept at x_2 from the previous iteration and neither of the two points at x_1 and $x_1 + x_1 h$ which generated the previous secant line. In the second iteration, the second point at $x_2 + x_2 h$ is produced from a perturbation to x_2 using the same small fixed value of h as in the calculation of $x_1 + x_1 h$.

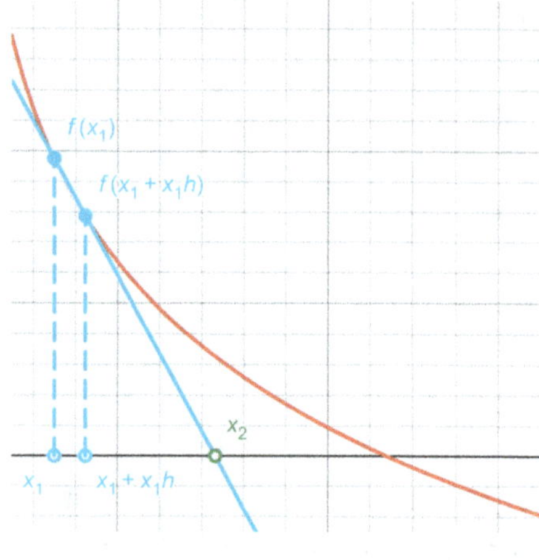

Figure 1.17: The modified secant algorithm uses the x-intercept at x_2 of the line through two close points. The second point at $x_1 + x_1 h$ is calculated using a small fixed perturbation h to the first point at x_1.

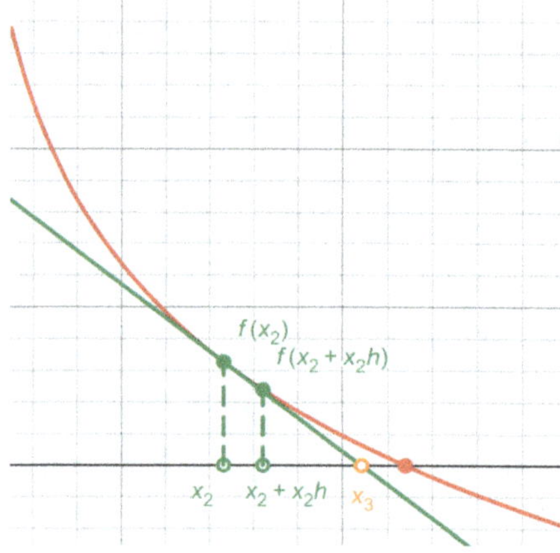

Figure 1.18: In the second iteration of the modified secant algorithm, both previous points at x_1 and $x_1 + x_1 h$ of Figure 1.17 are discarded and replaced by a new pair of points. The first point in this pair is located at the x-intercept from the prior iteration at x_2 and the second point at $x_2 + x_2 h$ is again found using the same h value. The x-intercept of the line through the new pair of points produces the next estimate at x_3.

The widths of the two intervals of the first two iterations, $[x_1, x_1 + x_1 h]$ in Figure 1.17 and $[x_2, x_2 + x_2 h]$ in Figure 1.18, are fairly consistent and that is precisely the advantage the modified secant algorithm may provide. In contrast, the intervals of the first two iterations of the secant algorithm, $[x_0, x_1]$ in Figure 1.14 and $[x_1, x_2]$ in Figure 1.15, show a much more pronounced difference.

A final graphical case in favor of the modified secant algorithm over the secant algorithm is demonstrated in Figure 1.19. If instead the secant algorithm is employed based on Figures 1.17 and 1.18, the second iteration would remove the point at x_1 but retain the point at $x_1 + x_1 h$. Along with x_2, a different second secant line forms, corresponding to a different estimate at the x-intercept x_3 in Figure 1.19. This x_3 not as close to the true zero at the red point as the x_3 in Figure 1.18 is to the red point, the latter of which was found using the modified secant algorithm.

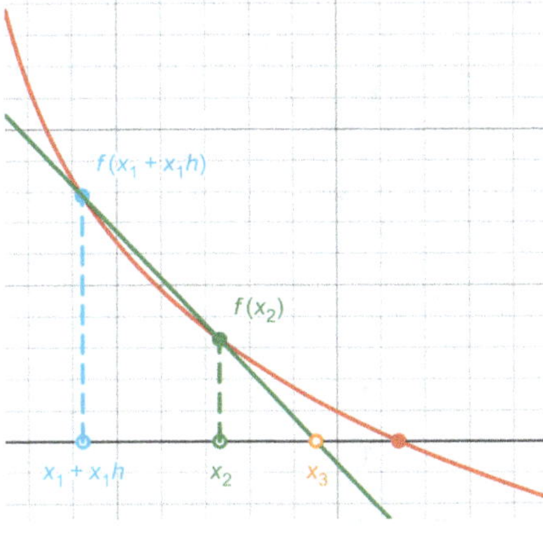

Figure 1.19: Using the first two points in Figure 1.17 and then the secant algorithm, the point at $x_1 + x_1 h$ is retained and used with x_2 to find the estimate at the x-intercept at x_3. The distance between this x_3 and the true zero at the red point is greater than the distance between the x_3 produced by the modified secant algorithm and the true zero at the red point in Figure 1.18.

Example 1.10. Use the modified secant algorithm to estimate a zero for $f(x) = 3 - x - 0.1x^5$ with an initial estimate $x_1 = 0.5$ and $h = 0.025$ until $E_c < 1\,\%$.

Solution. Using $x_1 + x_1 h = 0.5 + 0.5(0.025) = 0.513, f(0.5) = 2.497, f(0.513) = 2.484$, and Equation (1.11)

$$x_2 = x_1 - \frac{x_1 h f(x_1)}{f(x_1 + x_1 h) - f(x_1)} = 0.5 - \frac{(0.5)(0.025)(2.497)}{2.484 - 2.497} = 2.917.$$

The next estimate x_3 is found with the same small fixed value $h = 0.025$ applied to the previous x_2, such that

$$x_2 + x_2 h = 2.917 + 2.917(0.025) = 2.990.$$

The remaining iterations are summarized in Table 1.13, which includes the distance between the two points used for the ith secant line, $d_i = x_i h$, referenced in the comparison below.

Table 1.13: The modified secant algorithm iteration details of Example 1.10.

i	x_i	$f(x_i)$	$x_i + x_i h$	$f(x_i + x_i h)$	x_{i+1}	$d_i = x_i h$	E_c, %
1	0.500	2.497	0.513	2.484	2.917	0.013	
2	2.917	−21.053	2.990	−23.904	2.379	0.073	22.65
3	2.379	−6.995	2.438	−8.055	1.986	0.059	19.75
4	1.986	−2.079	2.036	−2.535	1.760	0.050	12.86
5	1.760	−0.449	1.804	−0.715	1.686	0.044	4.41
6	1.686	−0.047	1.728	−0.268	1.677	0.042	0.54

The conditions of Example 1.10 illustrate why the modified secant algorithm sometimes converges more quickly than the secant algorithm. Table 1.14 uses the secant algorithm for the same function and x_1 value used in Example 1.10, with the additional initial value $x_0 = 0.487$, so that the first distance between points, d_1, is the same for both algorithms.

Table 1.14: Using the secant algorithm for the conditions of Example 1.10 takes ten iterations, whereas the modified secant algorithm needed six iterations, as shown in Table 1.13.

i	x_{i-1}	$f(x_{i-1})$	x_i	$f(x_i)$	x_{i+1}	$d_i = x_i - x_{i-1}$	E_c, %
1	0.487	2.510	0.500	2.497	2.925	0.013	
2	0.500	2.497	2.925	−21.333	0.754	2.425	287.88
3	2.925	−21.333	0.754	2.222	0.959	−2.171	21.35
4	0.754	2.222	0.959	1.960	2.494	0.205	61.56
5	0.959	1.960	2.494	−9.145	1.230	1.535	102.81
6	2.494	−9.145	1.230	1.489	1.407	−1.264	12.58
7	1.230	1.489	1.407	1.042	1.820	0.177	22.69
8	1.407	1.042	1.820	−0.815	1.639	0.413	11.06
9	1.820	−0.815	1.639	0.181	1.671	−0.181	1.97
10	1.639	0.181	1.671	0.024	1.677	0.033	0.31

As the d_i columns highlight, subsequent secant algorithm distances in Table 1.14 can vary quite a lot in both magnitude and sign between iterations compared to the modified secant algorithm distances in Table 1.13 which remain relatively stable. Recall secant algorithm distances $d_i = x_i - x_{i-1}$ always rely on the previous point at x_{i-1} which may not be relatively close necessarily to the current point at x_i, while modified secant algorithm distances $d_i = x_i h$ are always forced to be small based on the constant h value. Using the secant algorithm meets the stopping criterion eventually, but requires almost twice the number of iterations as the modified secant algorithm; see Tables 1.14 and Table 1.13, respectively.

Since the modified secant algorithm is motivated by the theory of the Newton–Raphson algorithm without using the derivative, divergence away from a solution, an infinite alteration, or a domain violation might derail the process, as explained in Section 1.4.

Roots and zeros concepts summary

1. The bracketed bisection and false position algorithms guarantee a solution for $f(x) = 0$ on $[a, b]$ if $f(a)$ and $f(b)$ differ in sign and $f(x)$ is continuous.
2. The estimation relative change error E_c is an upper bound for the true relative change error ε_c when used with the bisection algorithm.
3. Once the bisection algorithm satisfies the E_c criterion at the ith interval, the estimate at x_j for $j = 1 \ldots i$ that produces the smallest absolute value difference with zero is the best estimate, sometimes not at x_i.
4. Using the fixed-point algorithm by solving for the independent variable may introduce extraneous solutions.
5. The Newton–Raphson algorithm requires calculation of the first derivative.
6. The secant and modified secant algorithms mimic the Newton–Raphson algorithm without relying on the first derivative.

7. The two points that iterate the secant and modified secant algorithms are closely spaced, not bracketed around the solution.
8. The modified secant algorithm removes the possibility of an increasing distance between the two points used for iteration that exists with the secant algorithm.
9. The fixed-point, Newton–Raphson, secant, and modified secant algorithms might fail and diverge from a solution for selected initial values.
10. Divergence from a solution is manifested by estimated solutions increasing in magnitude, E_c values which stop decreasing, or infinitely alternating estimated solutions or E_c values.
11. The bracketed bisection and false position algorithms tend to be less efficient than the Newton–Raphson algorithm, which iterates on a single point, and the secant and modified secant algorithms, which use two closely spaced points.
12. The specific function and initial values all contribute to the efficacy of all zero-finding algorithms.

Exercises

Section 1.1

Use the bisection algorithm to estimate either a zero for $f(x)$ or a solution to the equation on the starting interval $[a, b]$ until $E_c < p$.

1. $f(x) = x^5 - 0.7x^3 - 1.3x + 0.24$ $[1, 2]$ $p = 3\%$
2. $f(x) = 0.9 - 1.2x^2 + 1.5x^4 - x^5$ $[0.3, 2.1]$ $p = 2\%$
3. $f(x) = 5 + x - x^2 - 3 \sin x$ $[1, 4]$ $p = 2.5\%$
4. $f(x) = 2 \cos x + 2.7x - 0.21x^3$ $[2, 4]$ $p = 2.25\%$
5. $f(x) = e^{-x} + 0.5x^2 - 2.7x$ $[2.6, 6.8]$ $p = 5\%$
6. $f(x) = 2^x - 0.25x^3 - 3.1$ $[4.5, 6]$ $p = 1\%$
7. $f(x) = \ln x - x + 3$ $[1, 7]$ $p = 2.25\%$
8. $f(x) = 2 + 1.4x - x \log_2 x$ $[2.75, 5.25]$ $p = 1.25\%$
9. $\sqrt{x} + 2.15 = 0.48x^2$ $[2, 4]$ $p = 1.5\%$
10. $\sqrt[3]{x} = 0.128x^3 - 0.96$ $[-1.5, 3.5]$ $p = 1.75\%$
11. $0.2x^5 + \cos x = 3^x + 1.75$ $[2, 3]$ $p = 2.75\%$
12. $7.4 + \sin x = 0.22x^2 - \ln x$ $[6.1, 7.3]$ $p = 0.75\%$
13. $\log_3 x = 1.8x - x^{2.5}$ $[1, 2]$ $p = 2.5\%$
14. $1.32e^x - x^{3.5} = 1.28$ $[0, 2]$ $p = 2\%$

Section 1.2

Use the false position algorithm to estimate either a zero for $f(x)$ or a solution to the equation on the starting interval $[a, b]$ until $E_c < p$.

15. $f(x) = 0.05x^7 - x^2 + 0.85$ $[-0.6, 1.6]$ $p = 1.5\%$
16. $f(x) = 0.03x^4 - 0.4x^2 + 1$ $[0.1, 2.8]$ $p = 0.25\%$
17. $f(x) = 3 \cos x - 0.8x + 5$ $[6, 10]$ $p = 1\%$
18. $f(x) = x^3 - \sin 2x - 1.4$ $[0.5, 1.5]$ $p = 0.75\%$
19. $f(x) = 10^x - x^3 - 2$ $[-1.5, -0.5]$ $p = 1\%$

20. $f(x) = e^{x-6} - 0.135x^2 + 1.5$ $[0, 6.5]$ $p = 0.1\%$
21. $f(x) = 0.16x^3 - x^2 + 2\log_3 x + 2$ $[0.5, 5]$ $p = 0.15\%$
22. $f(x) = x^3 - 3.5x^2 + \ln x + 3.5$ $[0.4, 2.6]$ $p = 0.5\%$
23. $0.65x^2 = \sqrt[3]{x} + 1$ $[-2, 1]$ $p = 1.75\%$
24. $4x^2 - 1 = 3\sqrt{x} + x^3$ $[0.5, 2]$ $p = 0.5\%$
25. $0.05x^4 + 2 = \frac{2}{3-x}$ $[1.8, 2.8]$ $p = 1.25\%$
26. $3^{-x} = x^2 - x + 1.5$ $[-2, 0]$ $p = 1.5\%$

Sections 1.1 and 1.2

Compare the bisection and the false position algorithms by using both to estimate either a zero for $f(x)$ or a solution to the equation on the starting interval $[a, b]$ until $E_c < p$.

27. $x^4 - 3x^3 = 5 - 4x^2$ $[0.5, 2]$ $p = 0.5\%$
28. $0.1x^4 - x = 3$ $[1.5, 3.5]$ $p = 1\%$
29. $f(x) = 1.5 - \frac{1}{x} - 5e^{-2x}$ $[0.25, 2.25]$ $p = 3\%$
30. $f(x) = 5^x + 0.025x - 0.75$ $[-2, 1]$ $p = 10\%$
31. $f(x) = 1.75\sin x - 0.35x^2 - 2.4x + 1.6$ $[-10, -2]$ $p = 0.3\%$
32. $f(x) = 5.2 - 0.08x^3 - 1.25\cos x$ $[3.2, 5.6]$ $p = 0.15\%$

Section 1.3

Use the fixed-point algorithm to estimate either a zero for $f(x)$ or a solution to the equation with an initial estimate x_1 until $E_c < p$, or explain why it diverges.

33. $f(x) = 0.05x^4 - 0.2x^3 - 2x + 2.5$ $x_1 = 3$ $p = 3\%$
34. $f(x) = 4 - 3x + x^2 - 0.2x^3$ $x_1 = 3.5$ $p = 0.5\%$
35. $f(x) = 3\sin x - 5 + 4x - x^2$ $x_1 = 0.15$ $p = 3.5\%$
36. $f(x) = -x^3 - x^2 + 8x - \cos x$ $x_1 = 1$ $p = 0.75\%$
37. $e^x = 0.48x^4 - 0.3$ $x_1 = -1.25$ $p = 0.25\%$
38. $0.84x^3 = 3.7 - 2^{-x}$ $x_1 = 2$ $p = 2\%$
39. $f(x) = \ln x + x^3 - 3x^2 + 3$ $x_1 = 1$ $p = 1\%$
40. $f(x) = 1.5 - 0.036x^3 - \log(x^2 + 1)$ $x_1 = 0.25$ $p = 0.25\%$

Use the fixed-point algorithm to estimate either a zero for $f(x)$ or a solution to the equation with an initial estimate x_1 accurate to the thousandths place, or explain why it diverges.

41. $0.4x^3 - 1.5x^2 = 4x - 12$ $x_1 = 3.25$
42. $1.5x - x^3 - 0.5x^4 = -0.5$ $x_1 = -1$
43. $f(x) = x^4 - 5x^3 - 7x + 9$ $x_1 = 2$
44. $f(x) = 5 + 6x - 2x^3$ $x_1 = 2.1$
45. $f(x) = -0.05x^4 + 1.6 - 2\cos x$ $x_1 = -1.5$
46. $f(x) = -0.1x^2 + 1 + \sin x$ $x_1 = -3$

Section 1.4

Use the Newton–Raphson algorithm to estimate either a zero for $f(x)$ or a solution to the equation with an initial estimate x_1 until $E_c < p$, or explain why it is not possible.

47. $f(x) = x^5 - 3x^2 - 4$ $x_1 = 1.25$ $p = 2\%$
48. $f(x) = 2 - 5x^3 - 2x^4$ $x_1 = 0.25$ $p = 4\%$
49. $2\sin x + \sqrt{x} + x = 7$ $x_1 = 7.5$ $p = 0.5\%$
50. $0.75x^3 = 4x - \cos x$ $x_1 = -1$ $p = 1.25\%$
51. $2x^2 = 6x + \ln x$ $x_1 = 2$ $p = 2.5\%$
52. $2\log_3 x = 0.25x^4 - 12x + 15$ $x_1 = 2.5$ $p = 1.5\%$
53. $f(x) = 3\cos x - x^4 + 5x^2$ $x_1 = 0.5$ $p = 1\%$
54. $f(x) = 8\sin x - 6\sqrt[3]{x} + 3$ $x_1 = 5$ $p = 0.05\%$

Use the Newton–Raphson algorithm to estimate either a zero for $f(x)$ or a solution to the equation with an initial estimate x_1 accurate to the thousandths place, or explain why it is not possible.

55. $f(x) = 3x^3 - x^{2/3} - 10$ $x_1 = 1.25$
56. $f(x) = 2x^5 - 10x^3 - x + 8$ $x_1 = 1.5$
57. $x^6 + 3x = \frac{1}{x}$ $x_1 = -0.5$
58. $6\sqrt{x} + x = x^2 + 2$ $x_1 = 2.75$
59. $1.2x^2 - 0.25x^3 = \log_2 x$ $x_1 = 1.5$
60. $9x - \ln x = x^4 + 2$ $x_1 = 0.5$
61. $f(x) = 5\tan^{-1} x - 1.2x + 3.4$ $x_1 = 5$
62. $f(x) = 3\cos x + 0.5x - 2.9$ $x_1 = 5.5$

Section 1.5

Use the secant algorithm to estimate either a zero for $f(x)$ or a solution to the equation with initial values x_0 and x_1 until $E_c < p$, or explain why it is not possible.

63. $f(x) = 0.2x^5 - 1.1x^2 - 3x + 1.5$ $x_0 = 3$ $x_1 = 3.15$ $p = 1\%$
64. $f(x) = 4 + 7.3x^3 - 1.6x^5$ $x_0 = 0$ $x_1 = 1$ $p = 1\%$
65. $f(x) = 2\tan x - 4x^3 + 0.45$ $x_0 = -1$ $x_1 = -0.75$ $p = 1.5\%$
66. $f(x) = 1.25x^4 + \tan x - 0.6$ $x_0 = 0.75$ $x_1 = 1$ $p = 2\%$
67. $x^4 - 5x + 6\sqrt{x} = 4$ $x_0 = 0.8$ $x_1 = 1.2$ $p = 1.5\%$
68. $x^3 = 4x^2 - \frac{1}{x^2} - 5.5$ $x_0 = 0.5$ $x_1 = 0.7$ $p = 0.5\%$
69. $12\sqrt{x} = 0.225x^2 + 3.9$ $x_0 = 0.5$ $x_1 = 0.75$ $p = 1\%$
70. $0.1x^5 - 3.2x^2 + 6\sqrt[4]{x} = 2.4$ $x_0 = 0.2$ $x_1 = 0.25$ $p = 0.5\%$

Use the secant algorithm to estimate either a zero for $f(x)$ or a solution to the equation with initial values x_0 and x_1 accurate to the thousandths place, or explain why it is not possible.

71. $f(x) = 3.7 + 2.75x + 0.35x^2 - e^x$ $x_0 = 0$ $x_1 = 0.25$
72. $f(x) = 4^{-x} + 0.64x^2 - 1.8x - 3.2$ $x_0 = 2.5$ $x_1 = 3$
73. $f(x) = 10 + 0.75x^3 - 6\sqrt[5]{x}$ $x_0 = -0.83$ $x_1 = -0.79$

74. $f(x) = 2\sqrt[5]{x} - 0.8x^3 - 11$ $x_0 = 0.725$ $x_1 = 0.735$
75. $5\ln x + 0.12x^5 = 2.4x + 7$ $x_0 = 3$ $x_1 = 3.1$
76. $1 + 4x = 0.072x^3 + 3\log_2 x$ $x_0 = 6.5$ $x_1 = 7$
77. $\cos 0.3x = \sqrt[3]{x^2} - 0.36x$ $x_0 = 14.1$ $x_1 = 14.3$
78. $0.5x^3 - 1.5\sin x = 12$ $x_0 = 2$ $x_1 = 2.1$

Section 1.6

Use the modified secant algorithm to estimate either a zero for $f(x)$ or a solution to the equation with an initial estimate x_1 and h until $E_c < p$, or explain why it is not possible.

79. $f(x) = 3.4 - 1.9x - 0.24x^4$ $x_1 = 1$ $h = 0.25$ $p = 0.15\,\%$
80. $f(x) = 2x^4 - 0.3x^3 + 4x - 10$ $x_1 = 0.75$ $h = 0.05$ $p = 1\,\%$
81. $0.01x^7 = 0.1x^4 + x + 1.7$ $x_1 = 3$ $h = 0.05$ $p = 0.1\,\%$
82. $f(x) = x^6 - 3.1x^2 - 0.5x + 2.2$ $x_1 = 0.5$ $h = 0.1$ $p = 0.25\,\%$
83. $f(x) = 3.8 + 4.5\sqrt[3]{x} - 3.7\sin x$ $x_1 = 4$ $h = 0.5$ $p = 2.5\,\%$
84. $3x^4 - 4.5x^2 + 2.5\cos x = 5.5$ $x_1 = 0.6$ $h = 0.01$ $p = 1.5\,\%$
85. $5^{-x} + 8.3 = 5.5x - 0.18x^3$ $x_1 = 3.75$ $h = 0.1$ $p = 1\,\%$
86. $0.04x^5 - 3x - 0.25e^x = 1.36$ $x_1 = 2.5$ $h = 0.15$ $p = 1\,\%$

Use the modified secant algorithm to estimate either a zero for $f(x)$ or a solution to the equation with an initial estimate x_1 and h accurate to the thousandths place, or explain why it is not possible.

87. $f(x) = 0.08x^5 - 0.4x^4 + 7x - 7.8$ $x_1 = 2.6$ $h = 0.05$
88. $0.128x^5 + 0.6x^2 = 5.8$ $x_1 = 1.5$ $h = 0.1$
89. $x^5 + \ln x = 2.6x^2 - 2$ $x_1 = 0.75$ $h = 0.5$
90. $f(x) = 5x^4 - 2x^5 - \log_3 x$ $x_1 = 0.5$ $h = 0.25$
91. $f(x) = 0.054x^3 + x\log x - 4.65$ $x_1 = 2$ $h = 0.15$
92. $f(x) = \ln(x^2 + 3) - 0.1x^7 - 1.86$ $x_1 = -1.25$ $h = 0.1$
93. $\sin x = -\frac{3}{x} - 1.56$ $x_1 = -4$ $h = 0.01$
94. $\tan^{-1} 2x = 1 - 0.48x^2$ $x_1 = -3$ $h = 0.15$

Sections 1.5 and 1.6

Compare the secant algorithm using x_0 and x_1 and the modified secant algorithm using x_1 and h to estimate either a zero for $f(x)$ or a solution to the equation accurate to the thousandths place.

95. $f(x) = 0.28x^3 - 4.5x + 1.72$ $x_0 = 2.673$ $x_1 = 2.7$ $h = 0.01$
96. $f(x) = 0.16x^3 - 0.84x^2 + 0.35$ $x_0 = 3.8$ $x_1 = 4$ $h = 0.05$
97. $0.01x^4 + 0.38\log_2 x = 0.7\sqrt{x}$ $x_0 = 1.484$ $x_1 = 1.4$ $h = -0.06$
98. $0.29x^2 = 3.2\sqrt{x} + 3.5\ln x$ $x_0 = 3.92$ $x_1 = 4$ $h = 0.02$
99. $2.25 + 1.3x - 0.5x^3 = 2.6\cos 2x$ $x_0 = -2.268$ $x_1 = -2.4$ $h = 0.055$
100. $\sin x = x^6 - 2.2x^2 - 3.4$ $x_0 = 0.408$ $x_1 = 0.4$ $h = -0.02$

Sections 1.4 and 1.6

Compare the Newton–Raphson algorithm using x_1 and the modified secant algorithm using x_1 and h to estimate either a zero for $f(x)$ or a solution to the equation accurate to the thousandths place.

101. $f(x) = 0.03x^4 - 0.17x^3 + 1.1x - 2.57$ $\quad x_1 = 1.4 \quad\quad h = 0.01$

102. $f(x) = 7.4 - 4x + x^3 - 0.22x^4$ $\quad\quad\quad x_1 = 5 \quad\quad\quad h = 0.1$

103. $f(x) = 0.5x^5 + 1.2x - 2^x$ $\quad\quad\quad\quad x_1 = 1.5 \quad\quad h = -0.03$

104. $f(x) = 0.85e^x - 2.05x^2 + 4.3$ $\quad\quad\quad x_1 = -1.2 \quad h = -0.25$

105. $x^7 + 4\sqrt{x^2 + 3} = 6.6$ $\quad\quad\quad\quad\quad x_1 = -1.15 \quad h = 0.2$

106. $0.18x^3 = 6\sqrt{x} + 4.4$ $\quad\quad\quad\quad\quad\quad x_1 = 2.5 \quad\quad h = 0.5$

2 Extrema

Finding the maximum or minimum, collectively extrema, values of functions and where they occur is a common optimization application of calculus. Analytical means are straightforward: Take the first derivative, find the critical values where the first derivative equals zero or is undefined, and perform either a sign test or a second derivative test to determine if a critical value is a maximum or minimum. As demonstrated in the previous chapter, finding where a function equals zero is not always straightforward, and this also applies to derivative functions. The algorithms dissected in this chapter illustrate three unique approaches to the problem. The first uses a variation of a method described in the last chapter, the second uses the well-known golden ratio to optimize the number of calculations, and the third uses interpolation, a topic expounded upon and tied to the next chapter.

2.1 Newton–Raphson algorithm

Recall the Newton–Raphson algorithm in Section 1.4 used to find zeros of a function. Since non-endpoint extrema occur only where the first derivative equals zero (or is undefined), the same algorithm can be employed to find zeros of the first derivative. Finding a zero for $f(x)$ was iterated thusly using Equation (1.7).

$$x_{i+1} = x_i - \frac{f(x_i)}{f'(x_i)}.$$

The analogous iterative step using the Newton–Raphson algorithm zero estimate at x_{i+1} for $f'(x)$ is

$$x_{i+1} = x_i - \frac{f'(x_i)}{f''(x_i)}. \tag{2.1}$$

Certainly, this can be extended to find the zeros of any derivative, where the next zero estimate for the nth derivative is

$$x_{i+1} = x_i - \frac{f^{(n)}(x_i)}{f^{(n+1)}(x_i)}.$$

Note a particular zero x_p for the first derivative, where $f'(x_p) = 0$, still needs verification as a maximum, a minimum, or neither for the original function $f(x)$. The second derivative is known because it is used in the iterative step of Equation (2.1). From calculus, if $f''(x_p) < 0$, x_p is at a maximum and if $f''(x_p) > 0$, x_p is at a minimum. In the unfortunate case where $f''(x_p) = 0$, values around x_p must be tested in the first derivative for a sign change to confirm or deny the existence of an extremum at x_p.

https://doi.org/10.1515/9783112221051-002

> **Example 2.1.** Use the Newton–Raphson algorithm to estimate where an extremum occurs on $f(x) = 5x - 3x^2 - 0.01x^5$ with an initial estimate of $x_1 = 2$ until $E_c < 1\%$. Use the second derivative to classify the extremum as a maximum or minimum.
>
> **Solution.** The derivatives are
>
> $$f'(x) = 5 - 6x - 0.05x^4$$
> $$f''(x) = -6 - 0.2x^3$$
>
> The iterative steps are summarized in Table 2.1. The extremum occurs at $x = 0.829$ and is a maximum because $f''(0.829) = -6.114 < 0$. The maximum value of the function is $f(0.829) = 2.079$.

Table 2.1: The Newton–Raphson algorithm iteration details of Example 2.1.

x	$f'(x)$	$f''(x)$	$E_c, \%$
2.000	−7.800	−7.600	
0.974	−0.887	−6.185	105.41
0.830	−0.005	−6.114	17.28
0.829	0.000	−6.114	0.10

2.2 Golden ratio search

In streamlining the development of the golden ratio search (GRS), initial concessions are made that a single maximum exists in the initial open interval, with the goal of locating that maximum. Changes for a single minimum are straightforward and discussed in Section 2.2.4, after the case of a maximum is completely described, demonstrated (Example 2.2), and proved (Theorem 2.1) in its entirety. The implications of multiple or no extrema in the initial interval are briefly addressed in Section 2.2.5.

2.2.1 The value and properties of the golden ratio

Deriving the golden ratio illuminates its usefulness in the GRS. Let an interval length L equal the sum of two subinterval lengths such that $L = L_1 + L_2$, $L_1 < L_2$, and the ratio of the entire interval length L to the larger subinterval length L_2 is equal to the ratio of the larger subinterval length L_2 to the smaller subinterval length L_1.

$$\frac{L}{L_2} = \frac{L_2}{L_1}. \tag{2.2}$$

Equivalently, substituting $L = L_1 + L_2$ in Equation (2.2)

$$\frac{L_1 + L_2}{L_2} = \frac{L_2}{L_1}.$$

$$\frac{L_1}{L_2} + 1 = \frac{L_2}{L_1}.$$

$$\frac{L_1}{L_2} = \frac{L_2}{L_1} - 1. \tag{2.3}$$

Define the golden ratio as

$$\phi = \frac{L_1}{L_2}. \tag{2.4}$$

Combining Equations (2.3) and (2.4),

$$\phi = \frac{1}{\phi} - 1$$

$$\phi^2 = 1 - \phi$$

$$\phi^2 + \phi - 1 = 0.$$

Using the quadratic equation, the value of the golden ratio ϕ equals the positive root solution.

$$\phi = \frac{-1 + \sqrt{(1)^2 - 4(1)(-1)}}{2(1)}$$

$$\phi = \frac{\sqrt{5} - 1}{2} \approx 0.618.$$

2.2.2 How the GRS efficiently uses the golden ratio

The GRS is applied to an interval with a lower bound at x_L and an upper bound at x_U. Two interior points at x_1 and x_2 are calculated using ϕ, proportional to the length of the entire interval and defined as distance d.

$$d = \phi(x_U - x_L) \tag{2.5}$$

$$x_1 = x_L + d \tag{2.6}$$

$$x_2 = x_U - d. \tag{2.7}$$

Since $\phi > 0.5$, by Equation (2.5), distance d must be greater than half the length of the entire interval $x_U - x_L$. Consequently, adding d to the lower bound at x_1 and subtracting it from the upper bound at x_2 means the points are, in ascending order, at x_L, x_2, x_1, and x_U; see Figure 2.1.

Given an interval described above by the four ordered points, there are two overlapping intervals containing the ordered points at $[x_L, x_2, x_1]$ and at $[x_2, x_1, x_U]$. Note x_2 and x_1 are, respectively, the interior x values in each interval, and one of them represents where the next extremum estimate occurs. Whichever has the larger function value dictates which of the two overlapping intervals is retained for the next iteration and what part of the other overlapping interval can be discarded.

If $f(x_2) > f(x_1)$, the first of the overlapping intervals containing $[x_L, x_2, x_1]$ replaces the initial interval, eliminating the need for x values between $[x_1, x_U]$. Conversely if $f(x_2) < f(x_1)$, the second of the overlapping intervals containing $[x_2, x_1, x_U]$ replaces the initial interval, eliminating the need for x values between $[x_L, x_2]$. A new x value in the replacement interval is calculated and the process of narrowing the interval continues until the stopping criterion is met.

Theorem 2.1 verifies only one new x value needs to be calculated in the next iteration because the three ordered x values of the replacement interval chosen are already spaced according to the golden ratio; this is crux of the GRS efficiency. If the first of the overlapping intervals is used, only a new x_2 must be calculated and if the second of the overlapping intervals is used, only a new x_1 must be calculated. In either case, by extension, just one new function value must be computed as well.

Theorem 2.1. *Using the GRS given the values x_L, x_2, x_1, and x_U at any iteration, only x_1 or x_2 needs recalculation for the next iteration; eliminating x_L or x_U leaves three values already spaced according to the golden ratio ϕ.*

Proof. Without loss of generality, suppose $f(x_2) > f(x_1)$ so that the first overlapping interval $[x_L, x_1]$ containing the three ordered points at x_L, x_2, and x_1 is the replacement interval used in the next iteration. These three values are proved spaced according to the golden ratio ϕ.

Calculating the distance d of the golden ratio ϕ proportional to the entire interval by Equation (2.5), Equation (2.6) divides the initial interval into subintervals $[x_L, x_1]$ and $[x_1, x_U]$ so that the lengths $L_2 = x_1 - x_L$ and $L_1 = x_U - x_1$ are precisely the larger and smaller lengths of ϕ, respectively. By the definition of ϕ of Equation (2.4)

$$\phi = \frac{L_1}{L_2}$$

$$\phi = \frac{x_U - x_1}{x_1 - x_L}.$$

Similarly, using the proportional distance d of Equation (2.5) again and Equation (2.7) divides the initial interval into two different subintervals $[x_L, x_2]$ and $[x_2, x_U]$ in which the corresponding lengths are the smaller and larger lengths of ϕ, respectively. Because the same proportional distance d is used on the same initial interval $[x_L, x_U]$, the smaller length of $[x_L, x_2]$ is the same as the smaller length of $[x_1, x_U]$, so that

$$x_2 - x_L = x_U - x_1$$

$$x_2 - x_L = L_1.$$

Using this definition of L_1, the previous definition of L_2, and Equation (2.4)

$$\phi = \frac{L_1}{L_2}$$

$$\phi = \frac{x_2 - x_L}{x_1 - x_L}.$$

Thus, if the replacement interval used is $[x_L, x_1]$ containing x_2, the ratio of the length of the subinterval $[x_L, x_2]$ contained within the length of the entire replacement interval $[x_L, x_1]$ is equal to the golden ratio, constructed entirely by the three known values x_L, x_2, and x_1. $\qquad\square$

Figure 2.1 illuminates Theorem 2.1 by labeling the interval lengths referenced in the proof and highlighting which ones are equal. The blue length L_2 generated from x_L and marking x_1, and the adjacent orange length $L_1 = x_U - x_1$, are in the golden ratio. Similarly the green length L_2 generated from x_U and marking x_2, and the adjacent orange length $L_1 = x_2 - x_L$ are in the golden ratio. Since the blue and green lengths L_2 are equal, so are the orange lengths L_1. Thus, the blue length L_2 is in the golden ratio with the orange length *below* it, i. e., $L_1 = x_2 - x_L$. If the blue length L_2 is retained as the new entire interval for the next iteration, it contains x_2 from the orange length L_1 and therefore already marks one interior x value according to the golden ratio, where x_2 is reassigned as x_1 in the next iteration.

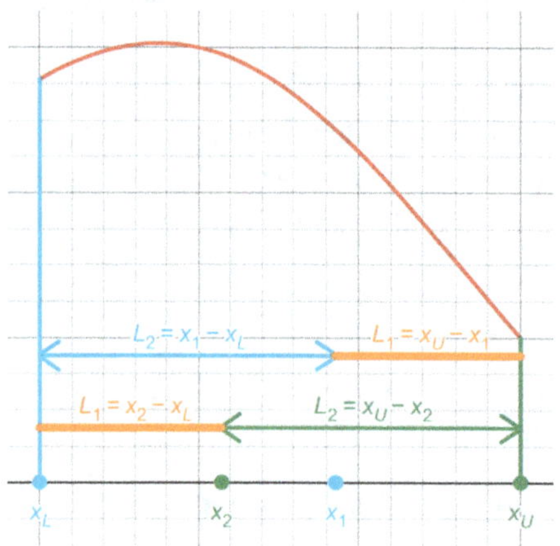

Figure 2.1: The blue and green lengths L_2 are equal and in the golden ratio with both equal orange lengths L_1.

When assigning the four x values of the next iteration of the GRS, first consider the case in which $f(x_2) > f(x_1)$ and the first overlapping interval is used as the start of the next iteration, as in Figure 2.1. The lower bound x_L stays the same, x_2 becomes the new x_1, and x_1 becomes the new x_U. The new x_2 is found as before, but now subtracting $d = \phi(x_U - x_L)$ from x_U, all of which are based on the current values of x_L and x_U.

For the case in which $f(x_1) > f(x_2)$ the second overlapping interval is used to start the next iteration. Instead of the lower bound remaining the same, it is the upper bound x_U that stays the same, with x_2 becoming the new x_L and x_1 becoming the new x_2. The new x_1 is found by adding $d = \phi(x_U - x_L)$ to x_L, again based on the new values of x_L and x_U.

Table 2.2 summarizes the two cases and is a quick reference guide for implementing the algorithm for locating a maximum; finding a minimum is simply a matter of reversing inequalities and will be detailed after Example 2.2. The complimentary Tables 2.3 and 2.4 each elaborate one column of Table 2.2 by illustrating how the x values shift for the next iteration and which x value is recalculated. Note if either x_L or x_U does not remain the same, $d = \phi(x_U - x_L)$ uses their current values, the old x_2 for x_L or the old x_1 for x_U.

Table 2.2: GRS x value reassignment for locating a maximum.

To find	if $f(x_2) > f(x_1)$ use	if $f(x_1) > f(x_2)$ use
new x_L:	same	old x_2
new x_2:	current $x_U - d$	old x_1
new x_1:	old x_2	current $x_L + d$
new x_U:	old x_1	same

Table 2.3: GRS reassignment for locating a maximum if $f(x_2) > f(x_1)$. Keep x_L, shift right, and recalculate x_2.

x_L	x_2		x_1		x_U
$\downarrow$	new $x_2 =$	$\searrow$		$\searrow$	
x_L	current $x_U - d$		old x_2		old x_1
	$=$ old $x_1 - d$				

Table 2.4: GRS reassignment for locating a maximum if $f(x_1) > f(x_2)$. Keep x_U, shift left, and recalculate x_1.

x_L	x_2		x_1	x_U
	$\swarrow$	$\swarrow$	new $x_1 =$	$\downarrow$
old x_2	old x_1		current $x_L + d$	x_U
			$=$ old $x_2 + d$	

Figures 2.2 and 2.3 graphically depict an iteration of the GRS in which the maximum value occurs in the second overlapping interval $[x_2, x_U]$ and how part of the first overlapping interval, the subinterval $[x_L, x_2]$, is eliminated. Figure 2.4 shows the three retained x values and how they shift and are reassigned. Figure 2.5 illustrates how only a new x_2 needs to be calculated, and subsequently how part of the second overlapping interval, the subinterval $[x_1, x_U]$, is removed when the maximum value occurs in the first overlapping interval.

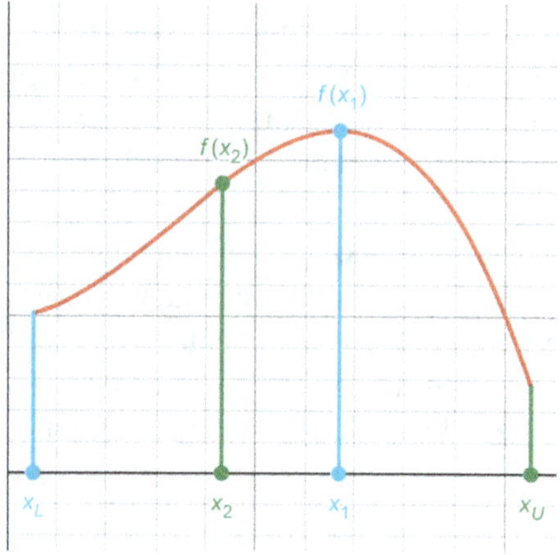

Figure 2.2: Since $f(x_1) > f(x_2)$, the second overlapping interval contains the maximum and the next iteration will retain the current points at x_2, x_1, and x_U and discard the one at x_L.

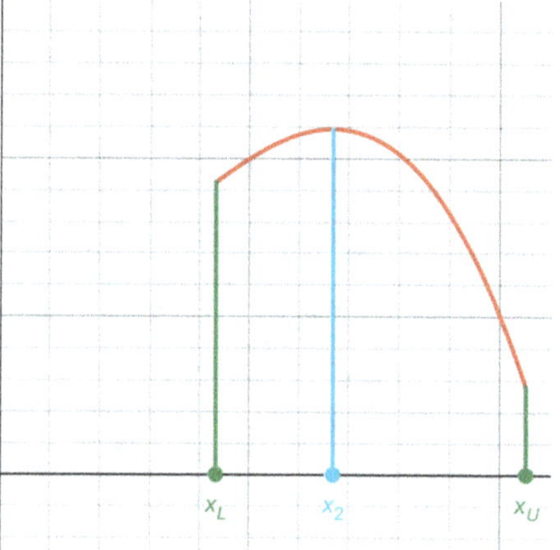

Figure 2.3: The points between the old x_L and the old x_2 in Figure 2.2 have been eliminated because the second overlapping interval was used. The new x_L is the old x_2, the new x_2 is the old x_1, and x_U remains the same.

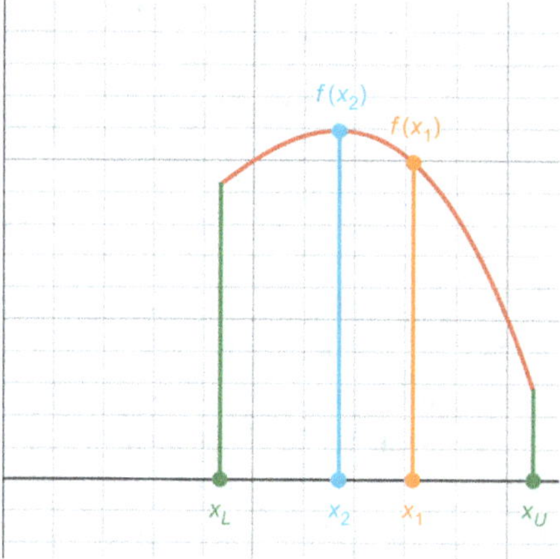

Figure 2.4: The only new point needed is at x_1, which is found with the current value of $d = \phi(x_U - x_L)$ added to the current x_L. This time, $f(x_2) > f(x_1)$, so the first overlapping interval will be used in the next iteration, using points at x_L, x_2, and x_1, while discarding x_U.

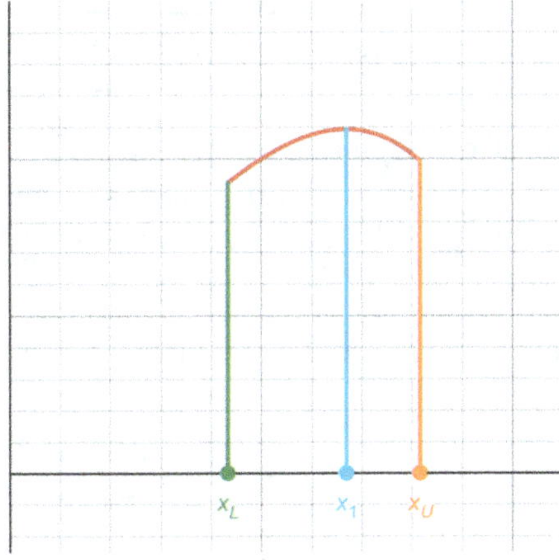

Figure 2.5: The points between the old x_1 and the old x_U in Figure 2.4 have been eliminated because the first overlapping interval was used. This time, x_L remains the same, the new x_1 is the old x_2, and the new x_U is the old x_1. The next point found would be at a new x_2 between x_L and x_1, found by subtracting the current value of $d = \phi(x_U - x_L)$ from the current x_U.

2.2.3 Error analysis

An estimation relative change error can evaluate the GRS, and like the E_c for the bisection algorithm in Section 1.1, an upper bound can be found with two modifications. Recall the upper bound for E_c in conjunction with the bisection algorithm is predicated on the estimate occurring, at maximum, halfway between the endpoints. Equation (1.2) defined $E_c = |\frac{x_i - x_{i-1}}{x_i}| \times 100\,\%$, where x_i is the current estimate of the zero and x_{i-1} is the previous estimate, and the previous estimate always becomes one of the current endpoints x_L or x_U of the interval in which x_i lies. If x_i is exactly halfway between the endpoints x_L and x_U, one of which is the previous estimate x_{i-1}, the distance between x_i and x_{i-1} is equivalent to half the length of the entire interval, or $|x_i - x_{i-1}| = 0.5|x_U - x_L|$. Substituting the right side of this equation into Equation (1.2) produces an equivalent E_c, the upper bound E_c for ε_c using the bisection algorithm.

$$E_c = \left| \frac{x_i - x_{i-1}}{x_i} \right| \times 100\,\%$$

$$= \text{Bisection Upper } E_c = 0.5 \left| \frac{x_U - x_L}{x_i} \right| \times 100\,\%. \qquad (2.8)$$

In contrast, the upper bound E_c in conjunction with the GRS must be considered for the estimate x_i occurring at a maximum distance from *each* endpoint x_L and x_U. These two distances are different because the estimate x_i is at either x_1 or x_2, neither of which is halfway between the endpoints but at a point that divides the length of the interval into a golden ratio. Thus the first modification to Equation (2.8) is replacing the factor 0.5, found in the following Theorem 2.2. The second modification is determining whether the divisor is x_1 or x_2; each GRS iteration produces both, and either could be the current estimate.

Theorem 2.2. *The replacement factor for 0.5 in Equation (2.8) for the upper bound of E_c for the GRS is 0.382.*

Proof. Without loss of generality, assume the second overlapping interval is used for the next iteration in finding a maximum, so that the estimated maximum occurs at x_1 and the two endpoints are x_2 on the left and x_U on the right. The distance between x_1 and each endpoint is considered.

Case 1 (Maximum at the left endpoint). Let distance $d_L = x_1 - x_2$. Substituting according to Equations (2.6) and (2.7)

$$
\begin{aligned}
d_L &= x_1 - x_2 \\
&= x_L + d - (x_U - d) \\
&= x_L - x_U + 2d.
\end{aligned}
$$

Substituting Equation (2.5) for d

$$
\begin{aligned}
d_L &= x_L - x_U + 2\phi(x_U - x_L) \\
&= (x_U - x_L)(2\phi - 1) \\
&= (x_U - x_L)\left(2\left(\frac{\sqrt{5}-1}{2}\right) - 1\right) \\
&= (x_U - x_L)(\sqrt{5} - 2).
\end{aligned}
$$

Thus, if the actual maximum is at the left endpoint x_2 instead of the estimate at x_1, the maximum error is the length of the entire interval $[x_L, x_U]$ multiplied by $\sqrt{5} - 2 \approx 0.236$.

Case 2 (Maximum at the right endpoint). Now, let distance $d_R = x_U - x_1$. Substituting using Equations (2.6) and (2.5)

$$
\begin{aligned}
d_R &= x_U - x_1 \\
&= x_U - (x_L + d) \\
&= x_U - x_L - \phi(x_U - x_L) \\
&= (x_U - x_L)(1 - \phi) \\
&= (x_U - x_L)\left(1 - \left(\frac{\sqrt{5}-1}{2}\right)\right) \\
&= (x_U - x_L)\left(\frac{3 - \sqrt{5}}{2}\right)
\end{aligned}
$$

Here if the actual maximum is at the right endpoint x_U rather than the estimate at x_1, the maximum error is the length of the entire interval $[x_L, x_U]$ multiplied by $\frac{3-\sqrt{5}}{2} \approx 0.382$.

The multiple of the entire interval in Case 2 is greater than Case 1, $0.382 > 0.236$, so the overall replacement factor for 0.5 in Equation (2.8) for the upper bound of E_c for the GRS is 0.382. $\qquad\square$

For each iteration, the estimate for the maximum is at either x_1 or x_2, whichever has the larger function value. Therefore, define

$$x_M = \begin{cases} x_1, & f(x_1) > f(x_2) \\ x_2, & f(x_2) > f(x_1) \end{cases} \tag{2.9}$$

The upper bound $E_c(x_M)$ for a maximum using the GRS is now defined by substituting 0.382 for 0.5 according to Theorem 2.2, and x_M of Equation (2.9) for x_i, into Equation (2.8).

$$\text{Bisection Upper } E_c = 0.5 \left| \frac{x_U - x_L}{x_i} \right| \times 100\,\%$$

$$E_c(x_M) = 0.382 \left| \frac{x_U - x_L}{x_M} \right| \times 100\,\%. \tag{2.10}$$

Example 2.2. Use the GRS to estimate the maximum value of $f(x) = 5\ln x - 0.12x^2$ and where it occurs on the interval $[4, 6]$ until $E_c(x_M) < 2\,\%$.

Solution. The iterations are summarized in Table 2.5. In the first two iterations $f(x_2) > f(x_1)$, so the first of the overlapping intervals determines the next configuration of x values. In those cases, the lower bound remains the same while x_2 and x_1 shift right, to x_1 and x_U, respectively, and a new x_2 is calculated, all according to Table 2.3. In the third and fourth iterations $f(x_1) > f(x_2)$, so the second of the overlapping intervals determines the next configuration of x values. In those cases the upper bound remains the same but now x_2 and x_1 shift left, to x_L and x_2, respectively, and a new x_1 is calculated, all using Table 2.4. $E_c(x_M)$ uses the divisor x_2 in the first, second, and fifth iterations, and x_1 in the third, fourth, and sixth iterations, according to which function value is larger by Equation (2.9), because the estimation is locating a maximum.

Table 2.5: GRS iteration details of Example 2.2. At each step, the larger of $f(x_2)$ and $f(x_1)$, shown in green, determines how x_L, x_2, x_1, and x_U are assigned in the next iteration. *At $x_2 = 4.541$, $f(x_2) = 5.09125$, and at $x_1 = 4.584$, $f(x_1) = 5.09130$.

i	x_L	x_2	$f(x_2)$	x_1	$f(x_1)$	x_U	d	$E_c(x_M)(\%)$
1	4.000	4.764	5.082	5.236	4.988	6.000	1.236	16.04
2	4.000	4.472	5.089	4.764	5.082	5.236	0.764	10.56
3	4.000	4.292	5.073	4.472	5.089	4.764	0.472	6.53
4	4.292	4.472	5.089	4.584	5.091	4.764	0.292	3.93
5	4.472	4.584	5.091	4.652	5.090	4.764	0.180	2.43
6	4.472	4.541	5.091*	4.584	5.091*	4.652	0.111	1.50

The function of Example 2.2 was chosen because it can be solved analytically and compared to the estimate. From calculus, equate the first derivative to zero and solve for the location of the critical value.

$$f'(x) = \frac{5}{x} - 0.24x = 0$$

$$\frac{125}{6} = x^2$$

$$x = \sqrt{\frac{125}{6}} = 4.564.$$

The negative root is not in the domain of $f(x)$ because of the $5 \ln x$ term and is discarded. The second derivative confirms the critical value is at a maximum.

$$f''(x) = \left(-\frac{5}{x^2} - 0.24 \right)\Big|_{x=4.564} = -0.480 < 0$$

The analytic maximum value is then $f(4.564) = 5.091$. The identical estimated maximum value of 5.091 in Table 2.5 indicates accuracy to the thousandths place.

The upper bound $E_c(x_M)$ for the GRS represents another worst-case scenario, just like E_c for the bisection algorithm of Section 1.1. Similarly, the maximum value can be achieved at a particular $E_c(x_M)$, with additional iterations confirming that maximum value satisfies a smaller $E_c(x_M)$ threshold. In Table 2.5 the maximum value 5.091 of $f(x)$ was found during the fourth iteration with $E_c(x_M) = 3.93\,\%$. The next two iterations also produced 5.091, but with smaller $E_c(x_M)$ values of 2.43 % and 1.5 %, validating 5.091 as the maximum value found earlier, only with less certainty for $i = 4$.

Unlike the bisection algorithm, a better estimate in a previous GRS iteration is impossible. $f(x_1)$ and $f(x_2)$ always converge to the maximum value from values less than the maximum value, and neither can ever exceed the maximum value. When locating minima discussed next, $f(x_1)$ and $f(x_2)$ always converge to a minimum value from values greater than the minimum value, and each are never smaller than the minimum value.

2.2.4 Adjusting the GRS for minima

To locate a minimum with the GRS, the only difference is retaining the overlapping interval containing the smaller function value rather than the larger. Simply reversing the inequalities of Tables 2.2 through 2.4 produces Tables 2.6 through 2.8 and the template for finding a minimum value using the GRS.

Table 2.6: GRS x value reassignment for locating a minimum.

To find	if $f(x_2) < f(x_1)$ use	if $f(x_1) < f(x_2)$ use
new x_L:	same	old x_2
new x_2:	current $x_U - d$	old x_1
new x_1:	old x_2	current $x_L + d$
new x_U:	old x_1	same

Table 2.7: GRS reassignment for locating a minimum if $f(x_2) < f(x_1)$. **Keep x_L, shift right, and recalculate x_2.**

x_L	x_2	x_1	x_U
$\downarrow$	new $x_2 =$	$\searrow$	$\searrow$
x_L	current $x_U - d$	old x_2	old x_1
	$=$ old $x_1 - d$		

Table 2.8: GRS reassignment for locating a minimum if $f(x_1) < f(x_2)$. **Keep x_U, shift left, and recalculate x_1.**

x_L	x_2	x_1	x_U
$\swarrow$	$\swarrow$	new $x_1 =$	$\downarrow$
old x_2	old x_1	current $x_L + d$	x_U
		$=$ old $x_2 + d$	

Analogous to estimating a maximum, the minimum is at either x_1 or x_2 for each iteration, whichever has the smaller function value. Correspondingly adjusting Equation (2.9), define

$$x_m = \begin{cases} x_1, & f(x_1) < f(x_2) \\ x_2, & f(x_2) < f(x_1) \end{cases} \tag{2.11}$$

Accordingly, the upper bound $E_c(x_m)$ for a minimum using the GRS is defined by altering Equation (2.10) and substituting x_m for x_M.

$$E_c(x_m) = 0.382 \left| \frac{x_U - x_L}{x_m} \right| \times 100\,\% \tag{2.12}$$

Example 2.3. Use the GRS to estimate the minimum value of $f(x) = 0.06x^3 - 2\sqrt{x}$ and where it occurs on the interval $[0.5, 2.5]$ until $E_c(x_m) < 3\,\%$.

Solution. The iterations are summarized in Table 2.9. In the first, second, fourth, and seventh iterations $f(x_1) < f(x_2)$, so the second of the overlapping intervals determines the next configuration of x values. In those cases the upper bound remains the same while x_2 and x_1 shift left, to x_L and x_2, respectively, and a new x_1 is calculated, all using Table 2.8. In the third, fifth, and sixth iterations $f(x_2) < f(x_1)$, so the first of the overlapping intervals determines the next configuration of x values. In these cases, the lower bound remains the same, but now x_2 and x_1 shift right, to x_1 and x_U, respectively, and a new x_2 is calculated, all according to Table 2.7. Correspondingly $E_c(x_m)$ uses the divisor x_1 in the first, second, fourth, and seventh iterations, and x_2 in the third, fifth, and sixth iterations, according to which function value is smaller by Equation (2.11), because the estimation is locating a minimum.

Table 2.9: GRS iteration details of Example 2.3. At each step the smaller of $f(x_2)$ and $f(x_1)$, shown in yellow, determines how x_L, x_2, x_1, and x_U are assigned for the next iteration.

i	x_L	x_2	$f(x_2)$	x_1	$f(x_1)$	x_U	d	$E_c(x_m)$, %
1	0.500	1.264	−2.127	1.736	−2.321	2.500	1.236	44.01
2	1.264	1.736	−2.321	2.028	−2.348	2.500	0.764	23.28
3	1.736	2.028	−2.348	2.208	−2.326	2.500	0.472	14.39
4	1.736	1.916	−2.346	2.028	−2.348	2.208	0.292	8.89
5	1.916	2.028	−2.348	2.097	−2.343	2.208	0.180	5.50
6	1.916	1.985	−2.349	2.028	−2.348	2.097	0.111	3.47
7	1.916	1.959	−2.348	1.985	−2.349	2.028	0.069	2.14

Like Example 2.2, the true minimum of Example 2.3 can be found analytically for comparison purposes. Setting the first derivative equal to zero, there is a critical value at

$$f'(x) = 0.18x^2 - \frac{1}{\sqrt{x}} = 0$$

$$x^{5/2} = \frac{100}{18}$$

$$x = \left(\frac{50}{9}\right)^{2/5} = 1.986.$$

The second derivative verifies this is a minimum.

$$f''(x) = \left(0.36x + \frac{1}{2x^{3/2}}\right)\Bigg|_{x=1.986} = 0.894 > 0.$$

The analytic minimum value is $f(1.986) = 0.06(1.986)^3 - 2\sqrt{1.986} = -2.349$, the same as the estimated minimum value in Table 2.9 and accurate to the thousandths place.

2.2.5 Multiple or no extrema

The examples and exercises in this chapter all contain a single extremum in the initial open interval. Ensuring this condition is a subproblem, not germane to demonstrating the GRS theoretically. Nevertheless, considering how the GRS behaves in non-ideal situations is warranted.

Suppose the GRS is used locate a maximum on an interval containing multiple maxima. Depending on the starting endpoints and characteristics of the function, *any* of the maxima could be found. If finding a particular maximum is the goal, such as the absolute maximum, there is no guarantee the GRS will locate it in lieu of another relative maximum.

Figure 2.6 depicts this scenario. Since $f(x_1) > f(x_2)$, the shaded left subinterval $[x_L, x_2]$ is removed after the first iteration, eliminating the absolute maximum. The GRS converges to the relative maximum near x_2 after additional iterations, which perhaps has merit, but the intended goal of finding the particular, absolute maximum is not satisfied. Alternatively, if the initial x_U is placed at the initial x_1, the absolute maximum *is* found; the problem lies not with the GRS itself, but in choosing the starting interval.

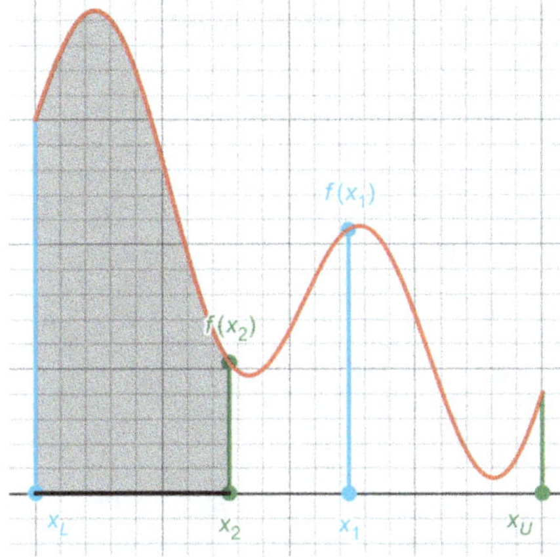

Figure 2.6: The GRS eliminates the shaded subinterval containing the absolute maximum and instead finds the relative maximum near x_2, a possible result when the starting interval contains multiple maxima.

A less apparent but erroneous conclusion results from a hasty application of the GRS on a monotonic function, i. e., strictly increasing or strictly decreasing, on an open interval with no extrema. In Figure 2.7, a GRS for a maximum on the monotonically increasing function discards x_2 and the shaded left subinterval $[x_L, x_2]$; all subsequent iterations proceed likewise. Convergence terminates at the right endpoint at x_U, not a true extremum but an arbitrary point used to initiate the search. Observe by initially moving x_U to the right at $x_{U'}$, the GRS behaves as before, continually discarding x_2 and the left subinterval until it converges, this time at $x_{U'}$. Clearly, no maximum existed at x_U, but no extremum exists at $x_{U'}$ either for the same reason; both resulted merely because each was chosen as the right endpoint.

Figure 2.8 illustrates drawing a related false conclusion. The interval contains a single maximum, but suppose the GRS is tasked with finding a minimum instead. Per the first iteration, the shaded left subinterval $[x_L, x_2]$ is removed because $f(x_1) < f(x_2)$. Similarly, all following left subintervals are removed, and convergence is at the right endpoint at x_U. If chosen arbitrarily to initialize the GRS, this is not a true minimum.

Note if the functions $f(x)$ of Figures 2.7 and 2.8 had a bounded domain with endpoints at x_L and x_U which were not arbitrary, x_L and x_U *would* represent locations of

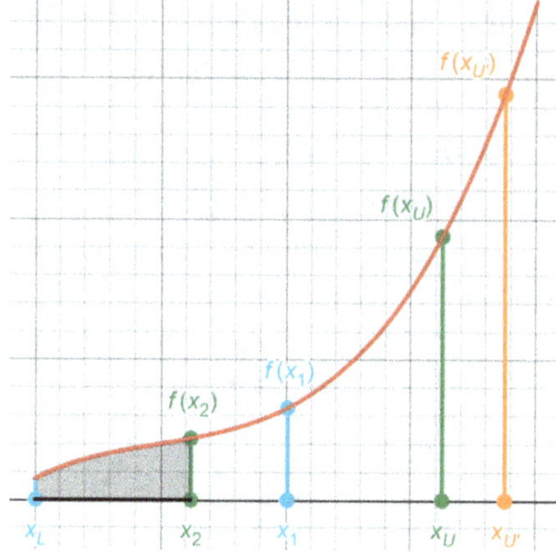

Figure 2.7: Finding an extremum on a monotonic function terminates at an endpoint. If arbitrarily selected, the endpoint is not a true extremum. Here the GRS indicates x_U or $x_{U'}$ is the maximum depending on which is the initial right endpoint; neither is the maximum.

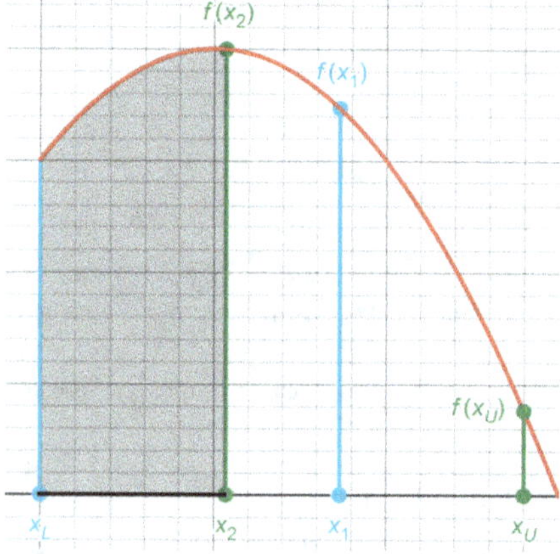

Figure 2.8: The interval $[x_L, x_U]$ has a single maximum, but if the goal is finding a minimum, the GRS will converge to the arbitrarily chosen right endpoint at x_U.

extrema. However, extrema at the endpoints and extrema in the open interval, where the first derivative equals zero or is undefined, are often distinguished because they are mathematically distinct.

2.3 Parabolic interpolation

A parabola, or second-degree polynomial, can approximate a nonlinear function over a relatively small interval, an improvement over using a line or first-degree polynomial. A line on a closed interval has extrema only at the endpoints. However, a parabola on a closed interval can have an additional extremum if the interval contains the vertex of the parabola, a better fit for estimating extrema of a nonlinear function. Example 2.4 and Figures 2.9 through 2.12 step through parabolic interpolation.

Three points define a unique parabola $f(x) = ax^2 + bx + c$ for parameters a, b, and c. An initial set of points $(x_1, f(x_1))$, $(x_2, f(x_2))$, and $(x_3, f(x_3))$ fit to such a parabola must simultaneously satisfy

$$ax_1^2 + bx_1 + c = f(x_1) \tag{2.13}$$

$$ax_2^2 + bx_2 + c = f(x_2) \tag{2.14}$$

$$ax_3^2 + bx_3 + c = f(x_3). \tag{2.15}$$

If this parabola is used to estimate where a maximum or minimum occurs, the derivative must equal zero at that x value x_4, so it must also satisfy the condition

$$f'(x) = 2ax + b$$

$$f'(x_4) = 2ax_4 + b = 0$$

$$b = -2ax_4. \tag{2.16}$$

The ultimate goal is not finding parameters a, b, and c but solving x_4 using the three known points. The algebraic steps involve first eliminating c by subtracting one of Equations (2.13) through (2.15) from the other two. For instance, subtracting Equation (2.15) from Equations (2.13) and (2.14) leaves

$$ax_1^2 - ax_3^2 + bx_1 - bx_3 = f(x_1) - f(x_3) \tag{2.17}$$

$$ax_2^2 - ax_3^2 + bx_2 - bx_3 = f(x_2) - f(x_3). \tag{2.18}$$

Second, eliminate b by substituting Equation (2.16) into Equations (2.17) and (2.18).

$$ax_1^2 - ax_3^2 - 2ax_4x_1 + 2ax_4x_3 = f(x_1) - f(x_3) \tag{2.19}$$

$$ax_2^2 - ax_3^2 - 2ax_4x_2 + 2ax_4x_3 = f(x_2) - f(x_3) \tag{2.20}$$

Third, eliminate a by solving Equations (2.19) and (2.20) for a and setting them equal to each other.

$$a\left(x_1^2 - x_3^2 - 2x_4x_1 + 2x_4x_3\right) = f(x_1) - f(x_3)$$

$$a = \frac{f(x_1) - f(x_3)}{x_1^2 - x_3^2 - 2x_4x_1 + 2x_4x_3}$$

$$a\left(x_2^2 - x_3^2 - 2x_4x_2 + 2x_4x_3\right) = f(x_2) - f(x_3)$$

$$a = \frac{f(x_2) - f(x_3)}{x_2^2 - x_3^2 - 2x_4x_2 + 2x_4x_3}$$

$$\frac{f(x_1) - f(x_3)}{x_1^2 - x_3^2 - 2x_4x_1 + 2x_4x_3} = \frac{f(x_2) - f(x_3)}{x_2^2 - x_3^2 - 2x_4x_2 + 2x_4x_3}$$

$$\frac{x_1^2 - x_3^2 - 2x_4x_1 + 2x_4x_3}{f(x_1) - f(x_3)} = \frac{x_2^2 - x_3^2 - 2x_4x_2 + 2x_4x_3}{f(x_2) - f(x_3)}$$

Lastly, the variable x_4 can be isolated by grouping and factoring.

$$\frac{x_1^2 - x_3^2}{f(x_1) - f(x_3)} - \frac{2x_4x_1 - 2x_4x_3}{f(x_1) - f(x_3)} = \frac{x_2^2 - x_3^2}{f(x_2) - f(x_3)} - \frac{2x_4x_2 - 2x_4x_3}{f(x_2) - f(x_3)}$$

$$\frac{x_1^2 - x_3^2}{f(x_1) - f(x_3)} - 2x_4\frac{x_1 - x_3}{f(x_1) - f(x_3)} = \frac{x_2^2 - x_3^2}{f(x_2) - f(x_3)} - 2x_4\frac{x_2 - x_3}{f(x_2) - f(x_3)}$$

$$-2x_4\frac{x_1 - x_3}{f(x_1) - f(x_3)} + 2x_4\frac{x_2 - x_3}{f(x_2) - f(x_3)} = \frac{x_2^2 - x_3^2}{f(x_2) - f(x_3)} - \frac{x_1^2 - x_3^2}{f(x_1) - f(x_3)}$$

$$2x_4\left(\frac{x_2 - x_3}{f(x_2) - f(x_3)} - \frac{x_1 - x_3}{f(x_1) - f(x_3)}\right) = \frac{x_2^2 - x_3^2}{f(x_2) - f(x_3)} - \frac{x_1^2 - x_3^2}{f(x_1) - f(x_3)} \tag{2.21}$$

Rewriting the left side of Equation (2.21) with a common denominator,

$$2x_4\frac{(x_2 - x_3)(f(x_1) - f(x_3)) - (x_1 - x_3)(f(x_2) - f(x_3))}{(f(x_2) - f(x_3))(f(x_1) - f(x_3))} \tag{2.22}$$

Likewise, using a common denominator to rewrite the right side of Equation (2.21),

$$\frac{(x_2^2 - x_3^2)(f(x_1) - f(x_3)) - (x_1^2 - x_3^2)(f(x_2) - f(x_3))}{(f(x_2) - f(x_3))(f(x_1) - f(x_3))} \tag{2.23}$$

The alternate forms of Equations (2.22) and (2.23) of the left and right sides of Equation (2.21), respectively, are equal and have the same denominators. Therefore, x_4 is simply the numerator of Equation (2.23) divided by the twice the numerator of Equation (2.22).

$$x_4 = \frac{(x_2^2 - x_3^2)(f(x_1) - f(x_3)) - (x_1^2 - x_3^2)(f(x_2) - f(x_3))}{2(x_2 - x_3)(f(x_1) - f(x_3)) - 2(x_1 - x_3)(f(x_2) - f(x_3))} \tag{2.24}$$

Distributing terms in the numerator of Equation (2.24) in the first two following steps, removing two opposite-signed terms in the third step, and regrouping yields Equation (2.25).

$$(x_2^2 - x_3^2)f(x_1) - (x_2^2 - x_3^2)f(x_3) - (x_1^2 - x_3^2)f(x_2) + (x_1^2 - x_3^2)f(x_3)$$

$$(x_2^2 - x_3^2)f(x_1) - x_2^2 f(x_3) + x_3^2 f(x_3) - (x_1^2 - x_3^2)f(x_2) + x_1^2 f(x_3) - x_3^2 f(x_3)$$

$$(x_2^2 - x_3^2)f(x_1) - (x_1^2 - x_3^2)f(x_2) + x_1^2 f(x_3) - x_2^2 f(x_3)$$

$$f(x_1)(x_2^2 - x_3^2) + f(x_2)(x_3^2 - x_1^2) + f(x_3)(x_1^2 - x_2^2) \tag{2.25}$$

Repeating the same four operations on the terms in the denominator of Equation (2.24) produces

$$2(x_2 - x_3)f(x_1) - 2(x_2 - x_3)f(x_3) - 2(x_1 - x_3)f(x_2) + 2(x_1 - x_3)f(x_3)$$

$$2(x_2 - x_3)f(x_1) - 2x_2 f(x_3) + 2x_3 f(x_3) - 2(x_1 - x_3)f(x_2) + 2x_1 f(x_3) - 2x_3 f(x_3)$$

$$2(x_2 - x_3)f(x_1) - 2(x_1 - x_3)f(x_2) + 2x_1f(x_3) - 2x_2f(x_3)$$
$$2f(x_1)(x_2 - x_3) + 2f(x_2)(x_3 - x_1) + 2f(x_3)(x_1 - x_2) \qquad (2.26)$$

Substituting the alternate forms of Equations (2.25) and (2.26) of the numerator and denominator, respectively, of Equation (2.24) produces the iterative step for the parabolic interpolation estimate at x_4 using three previous points.

$$x_4 = \frac{f(x_1)(x_2{}^2 - x_3{}^2) + f(x_2)(x_3{}^2 - x_1{}^2) + f(x_3)(x_1{}^2 - x_2{}^2)}{2f(x_1)(x_2 - x_3) + 2f(x_2)(x_3 - x_1) + 2f(x_3)(x_1 - x_2)} \qquad (2.27)$$

Similar to the bisection algorithm, the location of the estimate at x_4 relative to the current three ordered points at $x_1 < x_2 < x_3$ dictates which of those three is eliminated for the next iteration. The strategy is to include the new x_4 value and retain the two closest x values, and then renumber them in order. Practically, if x_4 is the smallest or second smallest of the four, x_3 is dropped. If x_4 is the second largest or the largest of the four, x_1 is dropped. Whatever x values remain are renumbered x_1, x_2, and x_3 in order from smallest to largest for the next iteration. This procedure can be used for finding a maximum *or* minimum because the fitting parabola will be oriented appropriately to estimate either extremum. Table 2.10 summarizes the entire renumbering procedure and Tables 2.11 through 2.14 each illustrate one of the four possible shifting scenarios.

Table 2.10: Parabolic interpolation x value reassignment.

To find	if $x_4 < x_1$	if $x_1 < x_4 < x_2$	if $x_2 < x_4 < x_3$	if $x_3 < x_4$
new x_1:	x_4	same	old x_2	old x_2
new x_2:	old x_1	x_4	x_4	old x_3
new x_3:	old x_2	old x_2	same	x_4

Table 2.11: Parabolic interpolation reassignment if $x_4 < x_1$. The first value is x_4, x_1 and x_2 are shifted right, and x_3 is removed.

x_1	x_2	x_3
current x_4 = new x_1	old x_1 = new x_2	old x_2 = new x_3

Table 2.12: Parabolic interpolation reassignment if $x_1 < x_4 < x_2$. x_4 is inserted between x_1 and x_2, x_1 remains the same, x_2 is shifted right, and x_3 is removed.

x_1	x_2	x_3
old x_1 = new x_1	current x_4 = new x_2	old x_2 = new x_3

Table 2.13: Parabolic interpolation reassignment if $x_2 < x_4 < x_3$. x_4 is inserted between x_2 and x_3, x_2 is shifted left, x_3 remains the same, and x_1 is removed.

x_1	x_2	x_3
↙		↓
old x_2	current x_4	old x_3
= new x_1	= new x_2	= new x_3

Table 2.14: Parabolic interpolation reassignment if $x_3 < x_4$. The last value is x_4, x_2 and x_3 are shifted left, and x_1 is removed.

x_1	x_2	x_3
↙	↙	
old x_2	old x_3	current x_4
= new x_1	= new x_2	= new x_3

Example 2.4. Use parabolic interpolation to estimate the maximum value of $f(x) = 3x^2 - 0.5x^5$ and where it occurs using initial values $x_1 = 0.4$, $x_2 = 0.6$, and $x_3 = 1.7$ until $E_c < 1\,\%$.

Solution. Using Equation (2.27) and the x value reassignment techniques described in Tables 2.10 through 2.14, the five required iterations are summarized in Table 2.15. Figures 2.9 through 2.12 illustrate the first four iterations and are detailed below.

Table 2.15: Parabolic interpolation algorithm iteration details of Example 2.4. Each x_4 generated is assigned a different cell color and inserted in order in the next iteration. The two x values retained from the previous iteration keep their cell color and are shifted accordingly.

x_1	$f(x_1)$	x_2	$f(x_2)$	x_3	$f(x_3)$	x_4	$f(x_4)$	E_c, %
0.400	0.475	0.600	1.041	1.700	1.571	1.283	3.200	
0.600	1.041	1.283	3.200	1.700	1.571	1.187	3.050	8.06
0.600	1.041	1.187	3.050	1.283	3.200	1.526	2.847	22.20
1.187	3.050	1.283	3.200	1.526	2.847	1.323	3.224	15.33
1.283	3.200	1.323	3.224	1.526	2.847	1.333	3.226	0.73

Figure 2.9 shows the initial configuration of $x_1 = 0.4$, $x_2 = 0.6$, and $x_3 = 1.7$ of Example 2.4. The black parabola is fit through the corresponding points on the red function at those x values, and the maximum function value on the black parabola determines where the new estimate of the maximum value occurs, at $x_4 = 1.283$. Since $x_4 = 1.283$ is between $x_2 = 0.6$ and $x_3 = 1.7$, $x_1 = 0.4$ is discarded and the remaining x values are renumbered in order from left to right, with x_2 becoming $x_1 = 0.6$, x_4 becoming $x_2 = 1.283$ and $x_3 = 1.7$ remaining the same, according to Table 2.13. This new configuration begins the second iteration in Figure 2.10.

Using the new assignment of x values, Figure 2.10 displays the blue parabola fit through those corresponding points on the red function at those x values. The maxi-

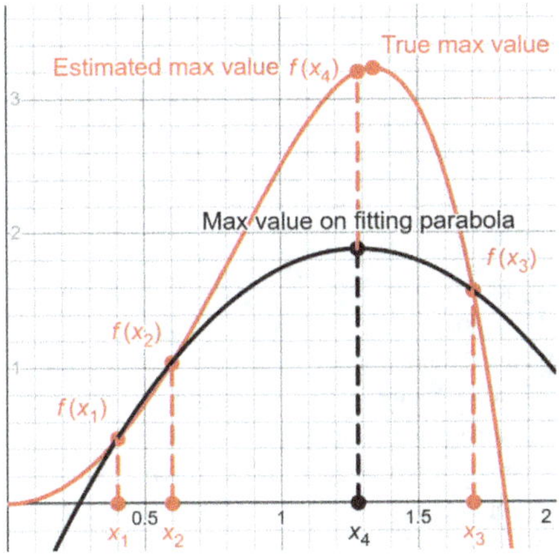

Figure 2.9: The first iteration of Example 2.4. The black parabola fit through the red function points at $x_1 = 0.4$, $x_2 = 0.6$, and $x_3 = 1.7$ has a maximum function value at $x_4 = 1.283$, so the two adjacent points at x_2 and x_3 are retained, while x_1 is dropped. The value at $f(x_4)$ is the estimated maximum value of the red function $f(x)$, already close to the true maximum value.

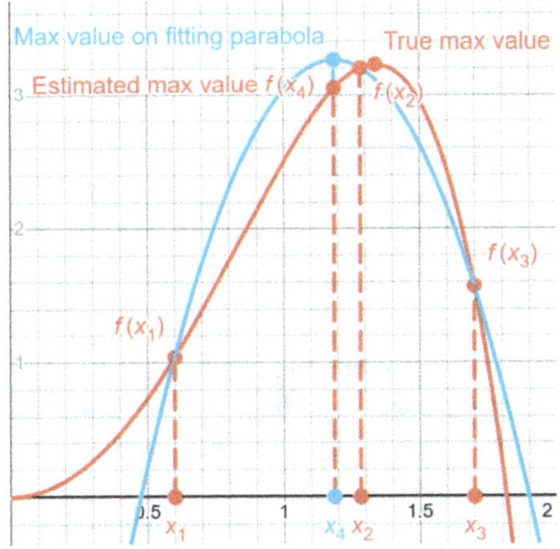

Figure 2.10: The second iteration of Example 2.4. The blue parabola fit through the reassigned red function points at $x_1 = 0.6$, $x_2 = 1.283$, and $x_3 = 1.7$ has a maximum function value at $x_4 = 1.187$, so the two adjacent points at x_1 and x_2 are retained while x_3 is dropped. The value at $f(x_4)$ is the estimated maximum value of the red function $f(x)$, not quite as accurate as the first iteration, but eventual convergence is still possible.

mum function value on the blue parabola is at $x_4 = 1.187$, which is between $x_1 = 0.6$ and $x_2 = 1.283$. Thus $x_3 = 1.7$ is discarded this time, and the remaining x values of x_1, x_4, and x_2 are renumbered left to right as $x_1 = 0.6$, $x_2 = 1.187$, and $x_3 = 1.283$, respectively, according to Table 2.12. The new assignment begins the third iteration in Figure 2.11.

Using the current numbering of x values, Figure 2.11 depicts the green parabola fit through those corresponding points on the red function at those x values. The maximum function value on the green parabola is at $x_4 = 1.526$, larger than all the currently numbered x values since $x_3 = 1.283$. According to Table 2.14, x_1 is removed, x_2 and x_3 shift left, and x_4 is inserted as the last x value. For the fourth iteration the remaining points are renumbered left to right such that $x_1 = 1.187$, $x_2 = 1.283$, and $x_3 = 1.526$. Figure 2.12 depicts this renumbering.

The updated arrangement of x values in Figure 2.12 demonstrates the orange parabola fit through those corresponding points on the red function at those x values.

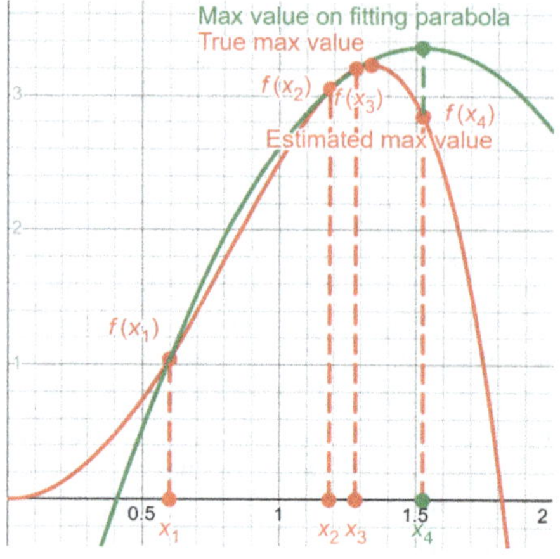

Figure 2.11: The third iteration of Example 2.4. The green parabola fit through the renumbered red function points at $x_1 = 0.6$, $x_2 = 1.187$, and $x_3 = 1.283$ has a maximum function value at $x_4 = 1.526$, so x_2 and x_3 are retained and x_1 is dropped. The estimated maximum value of the red function at $f(x_4)$ is less accurate than the two previous iterations, but this does not indicate divergence as the next iteration demonstrates.

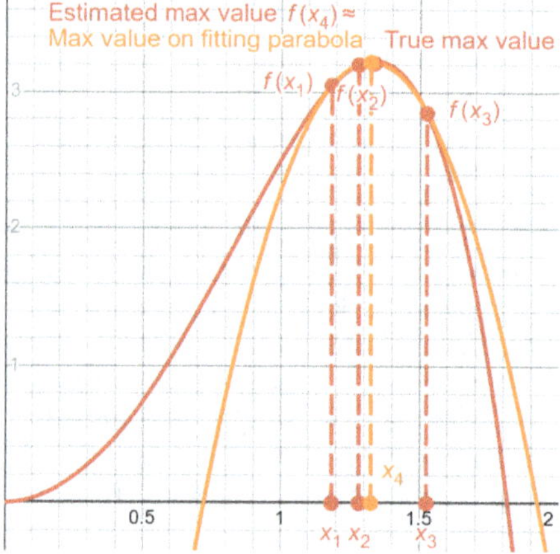

Figure 2.12: The fourth iteration of Example 2.4. The orange parabola fit through the updated red function points at $x_1 = 1.187$, $x_2 = 1.283$, and $x_3 = 1.526$ has a maximum function value at $x_4 = 1.323$, very close to both the estimated maximum value of the red function $f(x_4)$ and the true maximum value.

The maximum function value on the orange parabola is at $x_4 = 1.323$ between $x_2 = 1.283$ and $x_3 = 1.526$, so x_4 is inserted between those two x values and x_1 is discarded. However, unlike the previous iteration in which x_1 was removed, this x_4 is inserted between x_2 and x_3 rather than as the rightmost x value so that the renumbering for the next iteration is different. Here ordering left to right, x_2 becomes $x_1 = 1.283$, x_4 becomes $x_2 = 1.323$, and $x_3 = 1.526$ remains the same, according to Table 2.13.

Figures 2.9 through 2.12 and Table 2.15 additionally illustrate how the new x value at x_4 not only propels the parabolic interpolation algorithm forward, but is used to estimate the true maximum value of $f(x)$, i.e. $f(x_4) \approx$ the maximum function value of $f(x)$. Figure 2.9 shows the maximum function value on the estimating black parabola at $x_4 = 1.283$, but from that point the red dashed vertical line up to $f(x_4)$ is the estimated maximum value of $f(x)$, already within a small proximity to the true maximum value. According to Table 2.15, this value is $f(1.283) = 3.200$, which will later support this graphical evidence.

In Figure 2.10, the point marking the maximum function value on the blue parabola at $x_4 = 1.187$ can be connected down by the red dashed vertical line to the estimated maximum value $f(x_4)$. The function value $f(x_4) = f(1.187) = 3.050$ from Table 2.15 and the graph both indicate the distance between the location of the true maximum value and the location of the current estimate of the maximum value has actually increased, which is possible before eventual convergence.

The maximum function value at the point on the green parabola at $x_4 = 1.526$ in Figure 2.11 likewise can be connected down by the red dashed vertical line to the estimated maximum value $f(x_4)$. Graphically, the distance between the location of this estimate and the location of the true maximum value has increased again. Table 2.15 confirms the inferior estimate with $f(1.526) = 2.847$. The estimation relative change error $E_c = 22.20\,\%$ has increased as well.

In Figure 2.12, the point on the orange parabola of the fourth iteration indicating its maximum function value at $x_4 = 1.323$ is so close to estimated maximum value on the red function at $f(x_4)$ that the red dashed vertical line between them is indiscernible.

Four specific values in Table 2.15 support $f(1.323)$ as a valid estimate of the maximum value. The first and weakest is the $E_c = 15.33\,\%$ of the fourth iteration, indicating at least a better estimate than the third iteration. The second and third are more compelling, x_4 and E_c of the fifth iteration, in which the estimate of $x_4 = 1.333$ changed slightly from $x_4 = 1.323$ of the fourth iteration, quantified by an $E_c = 0.73\,\%$. The fourth is the estimated maximum value $f(1.333) = 3.226$ of the fifth iteration compared to $f(1.323) = 3.224$ of the fourth iteration, an even smaller change than the x_4 values of the corresponding iterations.

Extrema concepts summary

1. Solving $f'(x) = 0$ with the Newton–Raphson algorithm of Section 1.4 locates a possible extremum at $x = c$; the sign of $f''(c)$ determines a maximum or minimum.
2. The bracketed Golden Ratio Search (GRS) uses four merging x values per iteration, efficiently reusing three values from the previous iteration.
3. The estimation relative change errors $E_c(x_M)$ and $E_c(x_m)$ provide upper bounds for the GRS, constructed similarly to the upper bound E_c for the bisection algorithm of Section 1.1.
4. The GRS is not guaranteed to find a specific extremum in a starting interval containing multiple extrema.
5. The GRS can erroneously identify endpoints as extrema on a monotonic function or if used to find a maximum or minimum on an open interval containing none.
6. The inherent maximum or minimum of a quadratic function fit through three points makes parabolic interpolation suitable for estimating extrema.
7. Parabolic interpolation locates either a maximum or minimum in an interval without distinction; inequalities in the algorithm require no adjustments as in the GRS.

Exercises

Section 2.1

Use the Newton–Raphson algorithm to estimate where an extremum occurs on $f(x)$ with an initial estimate x_1 until $E_c < p$. Use the analytic second derivative to classify the extremum as a maximum or minimum.

1. $f(x) = x^5 - 2x^4 - 3x + 6$ $\qquad$ $x_1 = 1.5$ $\qquad$ $p = 1\%$
2. $f(x) = 0.8 + 0.3x + 2.1x^2 - 0.25x^4$ $\qquad$ $x_1 = 3$ $\qquad$ $p = 2\%$
3. $f(x) = 0.16x^2 + 5\sin x - 2$ $\qquad$ $x_1 = -5.75$ $\quad$ $p = 0.5\%$
4. $f(x) = \cos x - 0.14x^3 + 0.7x - 3$ $\qquad$ $x_1 = -6$ $\qquad$ $p = 0.25\%$
5. $f(x) = e^{-x} - 0.125x^4 + 4.5x + 0.29$ $\quad$ $x_1 = -1.5$ $\quad$ $p = 1\%$
6. $f(x) = 3 - 3x + x^3 - 3^x$ $\qquad$ $x_1 = -0.5$ $\qquad$ $p = 1\%$
7. $f(x) = \ln x - 0.04x^5 + 9x - 15$ $\qquad$ $x_1 = 1.75$ $\quad$ $p = 0.25\%$
8. $f(x) = 4 + x - 2\sqrt{x} - \log_3 x$ $\qquad$ $x_1 = 1.3$ $\qquad$ $p = 0.2\%$

Use the Newton–Raphson algorithm to estimate where an extremum occurs on $f(x)$ with an initial estimate x_1 accurate to the thousandths place. Use the analytic second derivative to classify the extremum as a maximum or minimum.

9. $f(x) = 1.5 + x + 0.5x^2 - 0.015x^4$ $\qquad$ $x_1 = 3.5$
10. $f(x) = x^5 - 6x^2 + 4x - 3$ $\qquad$ $x_1 = -0.5$
11. $f(x) = 0.32x^2 - 1.18x - 4\sqrt{x} - \frac{2}{x} + 2$ $\qquad$ $x_1 = 8$
12. $f(x) = 0.2x^3 - 4.6x - \frac{3}{x}$ $\qquad$ $x_1 = -8$
13. $f(x) = 2\tan^{-1} x + 0.02x^4 + 0.08x$ $\qquad$ $x_1 = 3$
14. $f(x) = 2x + 0.15x^2 - 3\tan^{-1} x$ $\qquad$ $x_1 = -3.5$
15. $f(x) = 3.2 + 0.66x + \log_2(x^2 + 1) - 2^x$ $\quad$ $x_1 = -2$
16. $f(x) = 0.01e^x - 0.9\ln x + 4.12$ $\qquad$ $x_1 = 5$

Section 2.2

Use the golden ratio search (GRS) to estimate the maximum value and where it occurs on $f(x)$ on the interval $[x_L, x_U]$ until $E_c(x_M) < p$.

17. $f(x) = 1 + 2x + 0.8x^2 - 0.2x^5$ $\qquad$ $[0, 2]$ $\qquad$ $p = 5\%$
18. $f(x) = 0.3x^4 - x^3 + 2.15x + 0.9$ $\qquad$ $[-1, 1.5]$ $\qquad$ $p = 10\%$
19. $f(x) = 1.28x - 0.07x^3 - 4\cos x$ $\qquad$ $[-5, -1]$ $\qquad$ $p = 3\%$
20. $f(x) = \sin x - 0.06x^4 + 0.62x^2 + 3.7$ $\quad$ $[-4, -1]$ $\qquad$ $p = 3\%$
21. $f(x) = 3^x - 0.18x^5 + 0.82x^3 - 12$ $\qquad$ $[1.75, 2.5]$ $\qquad$ $p = 2\%$
22. $f(x) = x^4 - x^6 - e^x$ $\qquad$ $[-1.25, 0.25]$ $\quad$ $p = 7.5\%$
23. $f(x) = 0.25x^3 - x - \log(2 - x) - 1.45$ $\quad$ $[-1.5, 0.25]$ $\quad$ $p = 15\%$
24. $f(x) = \ln(e^x + 1) - 0.03x^5 - 2.67$ $\qquad$ $[0.75, 2.25]$ $\qquad$ $p = 8\%$

Use the golden ratio search (GRS) to estimate the minimum value and where it occurs on $f(x)$ on the interval $[x_L, x_U]$ until $E_c(x_m) < p$.

25. $f(x) = 0.35x^6 - 0.56x^5 - 0.32x + 1.79$ $[-1.5, 1.5]$ $p = 3\%$
26. $f(x) = 0.064x^5 - 0.32x^2 - 2x$ $[1, 3]$ $p = 2.5\%$
27. $f(x) = \sin 2.8x + \sqrt{12 - x} + \frac{3}{x}$ $[-6, -4]$ $p = 1.5\%$
28. $f(x) = 0.4 - \frac{1}{x} - \cos x$ $[-7, -4]$ $p = 1.75\%$
29. $f(x) = e^{2x} - x^2 - 6x$ $[0.25, 0.75]$ $p = 5\%$
30. $f(x) = x^4 - 4^{-x} + 4$ $[-2, -1]$ $p = 5\%$
31. $f(x) = x - 0.021x^3 - \ln(1 - x)$ $[-6, -1]$ $p = 4.5\%$
32. $f(x) = 1.5x - 0.15x^{1.5} - \log_2 x$ $[0.5, 2.5]$ $p = 10\%$

Section 2.3

Use parabolic interpolation to estimate the maximum value and where it occurs on $f(x)$ for initial values $\{x_1, x_2, x_3\}$ until $E_c < p$.

33. $f(x) = -0.2x^3 + 1.2x^2 + 0.09x - 5$ $\{0, 4, 8\}$ $p = 0.5\%$
34. $f(x) = x^3 - 6.7x^2 + 8.4x$ $\{-2, 0, 3\}$ $p = 4.5\%$
35. $f(x) = 7 - 0.015x^4 - 5^x - 10^x$ $\{-5, 0, 2\}$ $p = 1\%$
36. $f(x) = e^{4x} + 0.004x^5 - 2.5x$ $\{-4, -1, 0\}$ $p = 0.5\%$
37. $f(x) = 6.1 \cos x + \sqrt{x^2 + 2} + 2.3x + 4.5$ $\{-5.5, -3.75, 0\}$ $p = 1.5\%$
38. $f(x) = \sqrt{5 - x} - 3 \sin x - 0.48x^2 - 5.7$ $\{-3, -1.5, 2\}$ $p = 0.5\%$
39. $f(x) = 2x^7 - 5x^2 + x - 2$ $\{0.5, 1.3, 2.1\}$ $p = 0.75\%$
40. $f(x) = 1 + 0.2x + 0.3x^3 - 0.123x^4$ $\{0, 2, 3\}$ $p = 1\%$

Use parabolic interpolation to estimate the minimum value and where it occurs on $f(x)$ for initial values $\{x_1, x_2, x_3\}$ until $E_c < p$.

41. $f(x) = 2 - 3\sqrt[3]{x} + x \cos x$ $\{-1, 4.25, 7\}$ $p = 0.075\%$
42. $f(x) = 5 \sin x + 0.35x^3 + 1.7x^2 - 1.57$ $\{-4.5, 1, 5.5\}$ $p = 0.5\%$
43. $f(x) = 18 - 4x \tan^{-1} x - 0.05x^3$ $\{-12, -8, -2\}$ $p = 0.5\%$
44. $f(x) = 15.3 \tan^{-1} x + 0.13x^2 - 3x - 6$ $\{0, 10, 25\}$ $p = 0.25\%$
45. $f(x) = \ln(x^2 + 4) - 0.075x^5 + 1.25x^3 + 15$ $\{-4.1, -3, -1.9\}$ $p = 0.25\%$
46. $f(x) = \log x + \frac{5}{x} + 0.21x^2 - 2.5$ $\{0.15, 1.85, 4.15\}$ $p = 1.5\%$
47. $f(x) = \frac{4}{x^2 + 1} + 0.065x^2 - e^{-2x}$ $\{0.4, 2.4, 4\}$ $p = 0.5\%$
48. $f(x) = 2^x - 0.011x^5 + 0.88$ $\{-3, -1, 1.75\}$ $p = 1\%$

3 Interpolation

A function may be defined by a set of points or data rather than an explicit expression. Measurements collected from an experiment at certain times is a common example. It may be necessary to either evaluate such a function at points not observed or to fit an explicit function using the known data. Interpolation solves each of these problems. Furthermore, in the case of finding a well-fitting function, interpolation can be a preliminary step to answering questions posed in the first two chapters or those considered in later chapters. For instance, the Newton–Raphson algorithm works from a known function $f(x)$, which can be approximated first by interpolation.

This chapter describes three common methods of interpolation either for estimating values at unknown points or fitting an entire function to a given set of points. The first two methods, proved algebraically equivalent, fit a single nth-degree polynomial through $n + 1$ points. The third method defines a piecewise function by small degree polynomials between each pair of consecutive points. The chapter concludes considering the question of inverse interpolation.

It should be noted interpolation estimates dependent variable values between the smallest and largest independent variable values of the known data. Extrapolation involves estimating values outside these bounds and is a much dicier proposition; it assumes the pattern of the known data automatically applies indefinitely in either or both directions beyond the bounds of the given interval. Additional information beyond the data should exist to support such an assumption.

3.1 Divided differences

3.1.1 A first-degree DD polynomial function $f_1(x)$

Newton's method of divided differences (DD) fits an nth-degree DD polynomial function $f_n(x)$ exactly through $n + 1$ points and in terms of those known points, $(x_i, f(x_i))$ for $i = 0, \ldots, n$. A unique first-degree polynomial function can be written using the point-slope form of a line.

$$f(x) = c_0 + c_1(x - x_0) \tag{3.1}$$

The goal is to write the unknown coefficients c_0 and c_1 in terms of two known points $(x_0, f(x_0))$ and $(x_1, f(x_1))$. Substituting $x = x_0$ in Equation (3.1)

$$f(x_0) = c_0 + c_1(x_0 - x_0)$$
$$c_0 = f(x_0) \tag{3.2}$$

Substituting $x = x_1$ and Equation (3.2) in Equation (3.1)

https://doi.org/10.1515/9783112221051-003

$$f(x_1) = c_0 + c_1(x_1 - x_0)$$
$$c_1(x_1 - x_0) = f(x_1) - f(x_0)$$
$$c_1 = \frac{f(x_1) - f(x_0)}{x_1 - x_0} \tag{3.3}$$

The coefficient c_1 is defined and generalized as a 1st DD $f[x_i, x_j]$.

$$f[x_i, x_j] = \frac{f(x_i) - f(x_j)}{x_i - x_j} \tag{3.4}$$

Combining Equations (3.3) and (3.4), c_1 written in DD notation is

$$c_1 = f[x_1, x_0] \tag{3.5}$$

Now, expressed in DD notation and defined by the two known points, the c_0 and c_1 of Equations (3.2) and (3.5), respectively, are substituted in Equation (3.1) and produce the first-degree DD polynomial function $f_1(x)$.

$$f_1(x) = f(x_0) + f[x_1, x_0](x - x_0) \tag{3.6}$$

The following subsections extend and generalize writing nth-degree DD polynomial functions $f_n(x)$ using this format and notation.

3.1.2 A second-degree DD polynomial function $f_2(x)$

A unique second-degree polynomial function requires an additional point $(x_2, f(x_2))$ for a total of three points. The general form of such a function is

$$f(x) = ax^2 + bx + c. \tag{3.7}$$

But like Equation (3.6), which defined the first-degree DD polynomial function $f_1(x)$ in terms of the known points and using DD notation, the goal is to write the second-degree DD polynomial function $f_2(x)$ in the same manner. To this end, it is convenient to use an alternate form of Equation (3.7) for a second-degree polynomial function, which extends the idea of Equation (3.1).

$$f(x) = c_0 + c_1(x - x_0) + c_2(x - x_0)(x - x_1) \tag{3.8}$$

It is easy to verify this is equivalent to the general form of Equation (3.7) by expanding and grouping the same powers of x as follows.

$$f(x) = c_0 + c_1 x - c_1 x_0 + c_2(x^2 - x_1 x - x_0 x + x_0 x_1)$$

$$f(x) = c_0 + c_1 x - c_1 x_0 + c_2 x^2 - c_2 x_1 x - c_2 x_0 x + c_2 x_0 x_1$$

$$f(x) = c_0 - c_1 x_0 + c_2 x_0 x_1 + (c_1 - c_2 x_1 - c_2 x_0)x + c_2 x^2$$

Equating the three constant terms, the three terms in parentheses comprising the coefficient of x, and the coefficient of x^2 with the corresponding same-degree terms of Equation (3.7) means

$$c = c_0 - c_1 x_0 + c_2 x_0 x_1$$

$$b = c_1 - c_2 x_1 - c_2 x_0$$

$$a = c_2$$

Thus, Equations (3.7) and (3.8) are equivalent, but the second-degree polynomial function of Equation (3.8) is used for two reasons. First, since it extends the first-degree polynomial function of Equation (3.1), c_0 and c_1 are already defined in DD notation by the first two points $(x_0, f(x_0))$ and $(x_1, f(x_1))$ in Equation (3.6). Confirming, if $x = x_0$ is substituted in Equation (3.8), then

$$f(x_0) = c_0 + c_1(x_0 - x_0) + c_2(x_0 - x_0)(x_0 - x_1)$$

$$f(x_0) = c_0.$$

This is exactly the result of Equation (3.2). Likewise, if $x = x_1$ and Equation (3.2) are substituted in Equation (3.8),

$$f(x_1) = c_0 + c_1(x_1 - x_0) + c_2(x_1 - x_0)(x_1 - x_1)$$

$$c_1 = \frac{f(x_1) - f(x_0)}{x_1 - x_0}.$$

This is precisely Equation (3.3), and with Equation (3.4), was defined by Equation (3.5) as $c_1 = f[x_1, x_0]$.

The second advantage of using Equation (3.8) is by substituting the additional point $(x_2, f(x_2))$ the coefficient c_2 can be found, written in DD notation, and used to write the second-degree DD polynomial function $f_2(x)$, analogous to $f_1(x)$ of Equation (3.6). Letting $x = x_2$ and substituting Equations (3.2) and (3.3) in the second step below

$$f(x_2) = c_0 + c_1(x_2 - x_0) + c_2(x_2 - x_0)(x_2 - x_1)$$

$$f(x_2) = f(x_0) + \frac{f(x_1) - f(x_0)}{x_1 - x_0}(x_2 - x_0) + c_2(x_2 - x_0)(x_2 - x_1)$$

$$c_2 = \frac{f(x_2) - f(x_0) - \frac{f(x_1) - f(x_0)}{x_1 - x_0}(x_2 - x_0)}{(x_2 - x_0)(x_2 - x_1)}$$

This form makes it possible to find c_2 using the three known points, but is further manipulated into a recursive DD term below.

A recursive term in general is defined by one or more preceding terms and a specified starting term or terms. Fibonacci terms $1, 1, 2, 3, 5, 8, \ldots$ are an example; each term a_i is defined by adding the previous two, i.e., $a_i = a_{i-1} + a_{i-2}$ with $a_1 = 1$ and $a_2 = 1$ specified. Writing DD terms recursively is useful for extending to any c_n where $n > 2$, n integer, shown in Section 3.1.3.

Using a common denominator in the numerator to combine and remove terms

$$c_2 = \frac{\frac{f(x_2)-f(x_0)}{x_1-x_0}(x_1 - x_0) - \frac{f(x_1)-f(x_0)}{x_1-x_0}(x_2 - x_0)}{(x_2 - x_0)(x_2 - x_1)}$$

$$= \frac{\frac{x_1 f(x_2)-x_0 f(x_2)-x_1 f(x_0)+x_0 f(x_0)}{x_1-x_0} - \frac{x_2 f(x_1)-x_0 f(x_1)-x_2 f(x_0)+x_0 f(x_0)}{x_1-x_0}}{(x_2 - x_0)(x_2 - x_1)}$$

$$= \frac{x_1 f(x_2) - x_0 f(x_2) - x_1 f(x_0) - x_2 f(x_1) + x_0 f(x_1) + x_2 f(x_0)}{(x_1 - x_0)(x_2 - x_0)(x_2 - x_1)}$$

In the first of the last three steps, each fraction in the numerator had a pair of binomials multiplied, generating four terms each in the next step. Two terms, $x_0 f(x_0)$ and $-x_0 f(x_0)$, reduced to zero leaving six terms in the last step. To regroup and rewrite as a pair of two new binomial factors, two terms must be introduced to replace the reduced ones and restore the total to eight terms. A common algebraic technique is employed, that of adding and subtracting the same term, which is a net addition of zero but necessary for the regrouping/rewriting process. In this case, $x_1 f(x_1)$ is added and subtracted in the numerator and terms are then grouped by the x values of x_0, x_1, and x_2.

$$c_2 = \frac{x_1 f(x_2) - x_0 f(x_2) - x_1 f(x_0) - x_2 f(x_1) + x_0 f(x_1) + x_2 f(x_0) + x_1 f(x_1) - x_1 f(x_1)}{(x_1 - x_0)(x_2 - x_0)(x_2 - x_1)}$$

$$= \frac{-x_0 f(x_2)+x_0 f(x_1)+x_1 f(x_2)-x_1 f(x_1)+x_1 f(x_1)-x_1 f(x_0)-x_2 f(x_1)+x_2 f(x_0)}{(x_1 - x_0)(x_2 - x_0)(x_2 - x_1)}$$

Common x values are factored out of consecutive terms (four times), followed by factoring out a difference of two function values (two times).

$$c_2 = \frac{-x_0(f(x_2) - f(x_1))+x_1(f(x_2) - f(x_1))+x_1(f(x_1) - f(x_0))-x_2(f(x_1) - f(x_0))}{(x_1 - x_0)(x_2 - x_0)(x_2 - x_1)}$$

$$= \frac{(f(x_2) - f(x_1))(-x_0+x_1) + (f(x_1) - f(x_0))(x_1-x_2)}{(x_1 - x_0)(x_2 - x_0)(x_2 - x_1)}$$

Rewriting as a complex fraction, each numerator term is divided by the two denominator factors $(x_1 - x_0)(x_2 - x_1)$, while the third term, $x_2 - x_0$, remains in the denominator.

$$c_2 = \frac{\frac{(f(x_2)-f(x_1))(-x_0+x_1)}{(x_1-x_0)(x_2-x_1)} + \frac{(f(x_1)-f(x_0))(x_1-x_2)}{(x_1-x_0)(x_2-x_1)}}{x_2 - x_0}$$

$$c_2 = \frac{\frac{f(x_2)-f(x_1)}{x_2-x_1} - \frac{f(x_1)-f(x_0)}{x_1-x_0}}{x_2 - x_0} \tag{3.9}$$

Observe each fraction in the numerator of Equation (3.9) can be written as a 1st DD according to Equation (3.4).

$$c_2 = \frac{f[x_2, x_1] - f[x_1, x_0]}{x_2 - x_0} \tag{3.10}$$

The coefficient c_2 is defined and generalized as a 2nd DD $f[x_i, x_j, x_k]$.

$$f[x_i, x_j, x_k] = \frac{f[x_i, x_j] - f[x_j, x_k]}{x_i - x_k} \tag{3.11}$$

Combining Equations (3.10) and (3.11), c_2 written in DD notation is

$$c_2 = f[x_2, x_1, x_0] \tag{3.12}$$

The second-degree DD polynomial function $f_2(x)$ can now be written, analogous to $f_1(x)$ of Equation (3.6). Using Equation (3.8) and substituting for the coefficients c_0, c_1, and c_2 in terms of the three known points, expressed in DD notation by Equations (3.2), (3.5), and (3.12), respectively

$$f(x) = c_0 + c_1(x - x_0) + c_2(x - x_0)(x - x_1)$$
$$f_2(x) = f(x_0) + f[x_1, x_0](x - x_0) + f[x_2, x_1, x_0](x - x_0)(x - x_1) \tag{3.13}$$

3.1.3 An *n*th-degree DD polynomial function $f_n(x)$

A final particular case reveals the general form clearly. Extending Equation (3.8), a third-degree polynomial function through four points with the additional point $(x_3, f(x_3))$ is

$$f(x) = c_0 + c_1(x - x_0) + c_2(x - x_0)(x - x_1) + c_3(x - x_0)(x - x_1)(x - x_2) \tag{3.14}$$

Recall the c_0 and c_1 defined during the construction of $f_1(x)$ were the same for $f_2(x)$ so that only c_2 was newly calculated for $f_2(x)$. Similarly, the third DD polynomial function $f_3(x)$ uses these same c_0, c_1, and c_2 definitions, with c_3 the only new coefficient needed for $f_3(x)$. By similar, albeit more algebraically intense manipulations that produced Equations (3.10) through (3.12), the corresponding Equations (3.15) through (3.17) respectively define c_3, a generalized 3rd DD $f[x_3, x_2, x_1, x_0]$, and c_3 in DD notation.

$$c_3 = \frac{f[x_3, x_2, x_1] - f[x_2, x_1, x_0]}{x_3 - x_0} \tag{3.15}$$

$$f[x_i, x_j, x_k, x_l] = \frac{f[x_i, x_j, x_k] - f[x_j, x_k, x_l]}{x_i - x_l} \tag{3.16}$$

$$c_3 = f[x_3, x_2, x_1, x_0] \tag{3.17}$$

The corresponding third-degree DD polynomial function $f_3(x)$, substituting c_3 of Equation (3.17) into Equation (3.14) and using the same definitions of c_0, c_1, and c_2 of $f_2(x)$ in Equation (3.13), is

$$f_3(x) = f(x_0) + f[x_1, x_0](x - x_0) + f[x_2, x_1, x_0](x - x_0)(x - x_1)$$
$$+ f[x_3, x_2, x_1, x_0](x - x_0)(x - x_1)(x - x_2) \tag{3.18}$$

The recursive nature of a 3rd DD written in terms of two 2nd DD values, and a 2nd DD written in terms of two 1st DD values motivates an algebraic confirmation that the nth DD $f[x_n, x_{n-1}, x_{n-2}, \ldots, x_1, x_0]$ can be generalized in terms of two $(n-1)$th DD values, following the same pattern as Equations (3.11) and (3.16).

$$f[x_n, x_{n-1}, x_{n-2}, \ldots, x_1, x_0] = \frac{f[x_n, x_{n-1}, \ldots, x_1] - f[x_{n-1}, x_{n-2}, \ldots, x_0]}{x_n - x_0} \tag{3.19}$$

The coefficient c_n can be expressed in the DD notation of Equation (3.19), analogous to c_2 of Equation (3.12) and c_3 of Equation (3.17).

$$c_n = f[x_n, x_{n-1}, x_{n-2}, \ldots, x_1, x_0] \tag{3.20}$$

Lastly, extending the third-degree polynomial function form of Equation (3.14) to an nth-degree polynomial function, the nth-degree DD polynomial function $f_n(x)$ in DD notation is defined

$$f(x) = c_0 + c_1(x - x_0) + c_2(x - x_0)(x - x_1) + \cdots + c_n(x - x_0)(x - x_1) \ldots (x - x_{n-1})$$
$$f_n(x) = f(x_0) + f[x_1, x_0](x - x_0) + f[x_2, x_1, x_0](x - x_0)(x - x_1) + \cdots$$
$$+ f[x_n, x_{n-1}, x_{n-2}, \ldots, x_1, x_0](x - x_0)(x - x_1) \ldots (x - x_{n-1}) \tag{3.21}$$

Table 3.1 demonstrates how the three pairs of consecutive points is used to calculate three 1st DD values, the two pairs of consecutive 1st DD values is used to calculate two 2nd DD values, and the one pair of 2nd DD values is used to calculate one 3rd DD

Table 3.1: A template for the recursive calculation of DD values. Values in red signify the c_0, c_1, c_2, and c_3 coefficients of a third-degree DD polynomial function $f_3(x)$.

	1st DD	2nd DD	3rd DD
$(x_0, f(x_0))$			
$(x_1, f(x_1))$	$f[x_1, x_0]$		
$(x_2, f(x_2))$	$f[x_2, x_1]$	$f[x_2, x_1, x_0]$	
$(x_3, f(x_3))$	$f[x_3, x_2]$	$f[x_3, x_2, x_1]$	$f[x_3, x_2, x_1, x_0]$

value. Furthermore, the values in red along the diagonal from left to right are precisely the definitions the coefficients c_0, c_1, c_2, and c_3 by Equations (3.2), (3.5), (3.12), and (3.17), respectively, used in the third-degree DD polynomial function $f_3(x)$ of Equation (3.18).

Example 3.1. Construct a third-degree DD polynomial function $f_3(x)$ through the known points.

i	x_i	$f(x_i)$
0	1	3
1	4	9
2	6	12
3	10	21

Solution. Coefficients c_0, c_1, c_2, and c_3 are needed for Equation (3.18), requiring DD values up to the 3rd DD. Using Equation (3.4) and the Table 3.1 template, three 1st DD values are found.

$$f[x_1, x_0] = \frac{f(x_1) - f(x_0)}{x_1 - x_0} = \frac{f(4) - f(1)}{4 - 1} = \frac{9 - 3}{3} = 2$$

$$f[x_2, x_1] = \frac{f(x_2) - f(x_1)}{x_2 - x_1} = \frac{f(6) - f(4)}{6 - 4} = \frac{12 - 9}{2} = 1.5$$

$$f[x_3, x_2] = \frac{f(x_3) - f(x_2)}{x_3 - x_2} = \frac{f(10) - f(6)}{10 - 6} = \frac{21 - 12}{4} = 2.25$$

Populating the Table 3.1 template with these values gives

	1st DD	2nd DD	3rd DD
(1, 3)			
(4, 9)	$f[x_1, x_0] = 2$		
(6, 12)	$f[x_2, x_1] = 1.5$		
(10, 21)	$f[x_3, x_2] = 2.25$		

Using Equation (3.11) and the Table 3.1 template, two 2nd DD values are calculated using the 1st DD values from the previous table.

$$f[x_2, x_1, x_0] = \frac{f[x_2, x_1] - f[x_1, x_0]}{x_2 - x_0} = \frac{1.5 - 2}{6 - 1} = -0.1$$

$$f[x_3, x_2, x_1] = \frac{f[x_3, x_2] - f[x_2, x_1]}{x_3 - x_1} = \frac{2.25 - 1.5}{10 - 4} = 0.125$$

The current state of the solution is

	1st DD	2nd DD	3rd DD
(1, 3)			
(4, 9)	$f[x_1, x_0] = 2$		
(6, 12)	$f[x_2, x_1] = 1.5$	$f[x_2, x_1, x_0] = -0.1$	
(10, 21)	$f[x_3, x_2] = 2.25$	$f[x_3, x_2, x_1] = 0.125$	

Only one 3rd DD is needed according to the Table 3.1 template. Using Equation (3.16) and the two 2nd DD values from the previous table

$$f[x_3, x_2, x_1, x_0] = \frac{f[x_3, x_2, x_1] - f[x_2, x_1, x_0]}{x_3 - x_0} = \frac{0.125 - (-0.1)}{10 - 1} = 0.025$$

The completed template of Table 3.2 highlights the coefficients c_0, c_1, c_2, and c_3 in red along the diagonal from left to right, respectively, and the values x_0, x_1, and x_2 in blue in the first column.

Substituting all the red and blue highlighted values from Table 3.2 into Equation (3.18) produces the third-degree DD polynomial function $f_3(x)$.

$$f_3(x) = 3 + 2(x - 1) - 0.1(x - 1)(x - 4) + 0.025(x - 1)(x - 4)(x - 6)$$

The graph of $y = f_3(x)$ in Figure 3.1 supports the cubic nature of the DD polynomial function and that it contains the four known points.

Table 3.2: All the red and blue highlighted values needed for the third-degree DD polynomial function $f_3(x)$ of Example 3.1.

	1st DD	2nd DD	3rd DD
(1, 3)			
(4, 9)	$f[x_1, x_0] = 2$		
(6, 12)	$f[x_2, x_1] = 1.5$	$f[x_2, x_1, x_0] = -0.1$	
(10, 21)	$f[x_3, x_2] = 2.25$	$f[x_3, x_2, x_1] = 0.125$	$f[x_3, x_2, x_1, x_0] = 0.025$

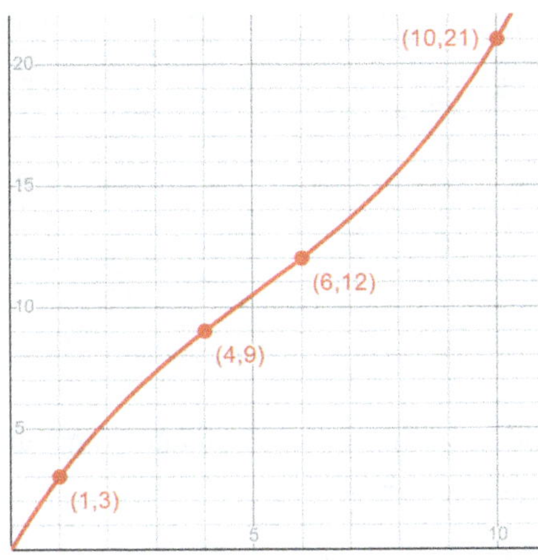

Figure 3.1: The DD polynomial function $y = f_3(x)$ of Example 3.1, which passes through all the given points and has a cubic shape.

3.1.4 Error analysis

Approximating the truncation error bound for an nth-degree DD polynomial function $f_n(x)$ stems from calculus. Representing $f(x)$ by an nth-degree Taylor polynomial $T_n(x)$ and the Lagrange form of the remainder term $R_n(x)$ of the associated Taylor series

$$f(x) = T_n(x) + R_n(x) \tag{3.22}$$

For the center of the series at $x = a$, and $x = c$ is a value between x and a

$$R_n(x) = \frac{f^{(n+1)}(c)}{(n+1)!}(x - a)^{n+1} \tag{3.23}$$

Solving for the truncation error bound $R_n(x)$ requires finding a bound on the function value $f^{(n+1)}(c)$; where it occurs is unimportant and thus neither is the specific value of c. Nevertheless, since the DD polynomial function $f_n(x)$ is an approximation of an unknown function $f(x)$, all derivatives of $f(x)$, including $f^{(n+1)}(x)$, are also unknown. Thus an approximated truncation error E_t for $R_n(x)$ is needed, and comparing the terms of Equation (3.22) to the DD polynomial function terms provides such an approximation.

The DD polynomial function $f_{n+1}(x)$ has degree $n + 1$ and by Equation (3.21) is

$$\begin{aligned}
f_{n+1}(x) = {} & f(x_0) + f[x_1, x_0](x - x_0) + f[x_2, x_1, x_0](x - x_0)(x - x_1) + \cdots \\
& + f[x_n, x_{n-1}, \ldots, x_1, x_0](x - x_0)(x - x_1) \ldots (x - x_{n-1}) \\
& + f[x_{n+1}, x_n, , \ldots, x_1, x_0](x - x_0)(x - x_1) \ldots (x - x_n)
\end{aligned} \tag{3.24}$$

Removing the last term equals the DD polynomial function $f_n(x)$ of Equation (3.21), so Equation (3.24) can be expressed alternately as

$$f_{n+1}(x) = f_n(x) + f[x_{n+1}, x_n, , \ldots, x_1, x_0](x - x_0)(x - x_1) \ldots (x - x_n) \tag{3.25}$$

Equation (3.25) is an approximation of $f(x)$, so combining with Equation (3.22)

$$f_{n+1}(x) \approx f(x) = T_n(x) + R_n(x).$$

Substituting the right side of Equation (3.25) for $f_{n+1}(x)$.

$$f_n(x) + f[x_{n+1}, x_n, , \ldots, x_1, x_0](x - x_0)(x - x_1) \ldots (x - x_n) \approx T_n(x) + R_n(x) \tag{3.26}$$

The DD polynomial function $f_n(x)$ on the left side and the Taylor polynomial function $T_n(x)$ on the right side are both polynomials of degree n, so that

$$f_n(x) \approx T_n(x). \tag{3.27}$$

The approximate and not equal relation is because $T_n(x)$ is centered around a single $x = a$ satisfying $T_n(x) = f(x)$. In contrast, $f_n(x)$ needs to satisfy *all* known points representing $f(x)$. Note each nth-degree term of $T_n(x)$ for $x = a$ satisfies

$$(x - a)^n = (x - a)(x - a) \cdots (x - a) = 0$$

The corresponding approximated nth-degree term of $f_n(x)$ for $x = x_0, x = x_1, \ldots, x = x_n$ must therefore satisfy

$$(x - x_0)(x - x_1) \cdots (x - x_n) = 0$$

Hence the approximation of Equation (3.27) hinges on

$$(x - a)^n \approx (x - x_0)(x - x_1) \cdots (x - x_n)$$

Subtracting each nth-degree polynomial function of Equation (3.27) from its respective side of Equation (3.26) leaves an $(n + 1)$th-degree term on each side.

$$f[x_{n+1}, x_n, \ldots, x_1, x_0](x - x_0)(x - x_1) \ldots (x - x_n) \approx R_n(x)$$

The left side defines the estimated truncation error E_t, an approximation of $R_n(x)$.

$$E_t = f[x_{n+1}, x_n, \ldots, x_1, x_0](x - x_0)(x - x_1) \ldots (x - x_n) \approx R_n(x) \tag{3.28}$$

Note since $R_n(x)$ is the truncation error bound for the nth-degree Taylor polynomial function $T_n(x)$, E_t is the estimated truncation error for the nth-degree DD polynomial function $f_n(x)$, according to Equation (3.26). E_t approximates the difference between an estimated value $f_n(k)$ and the true value $f(k)$ in either direction. The true value $f(k)$ can be expressed as lying in the interval

$$\left[f_n(k) - E_t, f_n(k) + E_t\right]. \tag{3.29}$$

By Equation (3.19), Equation (3.28) uses an $(n + 1)$th DD due to the presence of $n + 1$ points at $x_0, x_1, \ldots, x_n, x_{n+1}$. Therefore, to find the truncation error E_t for an nth DD polynomial function $f_n(x)$, an additional point not used in the construction of $f_n(x)$ is needed, and an additional row and column of DD values are required.

Since E_t depends on the magnitude of the function values, it is an incomplete evaluation of the estimation. As mentioned in Chapter 1, a relative error which scales according to magnitude is preferred, here according to the truncation error E_t. For the approximated value $f_n(k)$ at $x = k$ and the estimated truncation error E_t, define the estimated relative truncation error $E_{t,r}$ as the percentage

$$E_{t,r} = \left| \frac{E_t}{f_n(k)} \right| \times 100\,\% \tag{3.30}$$

The following Example 3.2 demonstrates all the error analysis concepts of this section by extending Example 3.1.

Example 3.2. Estimate $f(5)$ using $f_3(x)$ of Example 3.1. Use the additional fifth point $(x_4, f(x_4)) = (11, 30)$ to solve the associated truncation error E_t and express the range in which the true value $f(5)$ lies as an interval. Calculate $E_{t,r}$ to evaluate the quality of the estimated value $f(5)$.

Solution. The estimated value $f(5)$ using $x = 5$ and $f_3(5)$ is

$$f_3(5) = 3 + 2(5 - 1) - 0.1(5 - 1)(5 - 4) + 0.025(5 - 1)(5 - 4)(5 - 6) = 10.5$$

Finding E_t of Equation (3.28) necessitates calculating $f[x_4, x_3, x_2, x_1, x_0]$. This 4th DD requires an entire new row of DD calculations, the use of the last row of Table 3.2, and $x_2 = 6$, $x_1 = 4$, and $x_0 = 1$ of Example 3.1.

	1st DD	2nd DD	3rd DD	4th DD
$(10, 21)$	$f[x_3, x_2] = 2.25$	$f[x_3, x_2, x_1] = 0.125$	$f[x_3, x_2, x_1, x_0] = 0.025$	
$(11, 30)$	$f[x_4, x_3]$	$f[x_4, x_3, x_2]$	$f[x_4, x_3, x_2, x_1]$	$f[x_4, x_3, x_2, x_1, x_0]$

Each new 1st DD, 2nd DD, 3rd DD, and 4th DD, respectively, is calculated.

$$f[x_4, x_3] = \frac{f(x_4) - f(x_3)}{x_4 - x_3} = \frac{f(11) - f(10)}{11 - 10} = 30 - 21 = 9$$

$$f[x_4, x_3, x_2] = \frac{f[x_4, x_3] - f[x_3, x_2]}{x_4 - x_2} = \frac{9 - 2.25}{11 - 6} = 1.35$$

$$f[x_4, x_3, x_2, x_1] = \frac{f[x_4, x_3, x_2] - f[x_3, x_2, x_1]}{x_4 - x_1} = \frac{1.35 - 0.125}{11 - 4} = 0.175$$

$$f[x_4, x_3, x_2, x_1, x_0] = \frac{f[x_4, x_3, x_2, x_1] - f[x_3, x_2, x_1, x_0]}{x_4 - x_0} = \frac{0.175 - 0.025}{11 - 1} = 0.015$$

Substituting the above value for $f[x_4, x_3, x_2, x_1, x_0]$, $x = 5$, and the x values of the first four points of Example 3.1 in Equation (3.28), the estimated truncation error E_t is

$$E_t = f[x_4, x_3, x_2, x_1, x_0](x - x_0)(x - x_1)(x - x_2)(x - x_3)$$

$$E_t = 0.015(5 - 1)(5 - 4)(5 - 6)(5 - 10) = 0.3.$$

According to Equation (3.29), the true value of $f(5)$ lies in the interval

$$\left[f_3(5) - E_t, f_3(5) + E_t\right] = [10.5 - 0.3, 10.5 + 0.3] = [10.2, 10.8].$$

Lastly, by Equation (3.30), the estimated relative truncation error $E_{t,r}$ is

$$E_{t,r} = \left|\frac{0.3}{10.5}\right| \times 100\,\% = 2.86\,\%.$$

Figure 3.2 illustrates the estimated value $f_3(5) = 10.5$ and interval containing it $[10.2, 10.8]$.

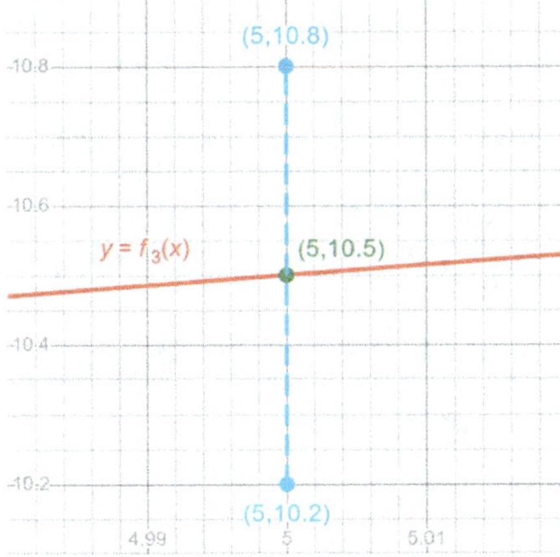

Figure 3.2: The solution of Example 3.2. The estimated value is $f_3(5) = 10.5$, but the true value $f(5)$ lies between 10.2 and 10.8.

3.1.5 Exact DD polynomial functions

Theorem 3.1. *Data generated from an unknown function $f(x)$, which itself is an nth-degree polynomial function can be written exactly as a DD polynomial function of the same degree.* 

Proof. Let $f(x)$ be an unknown nth-degree polynomial function. Select any $n + 1$ set of points N which satisfies $f(x)$ to construct the nth-degree DD polynomial function $f_n(x)$. By definition, any point in N must satisfy $f_n(x)$. Since $n + 1$ points defines a unique polynomial function, and the set of points in N satisfies both $f(x)$ *and* $f_n(x)$, so $f_n(x) = f(x)$. ☐

The following Corollary 3.1 identifies exact nth-degree polynomial functions during construction of the DD polynomial function $f_n(x)$.

Corollary 3.1. *If the nth column of DD values has at least two values and all values equal the same nonzero constant, the corresponding DD polynomial function $f_n(x)$ defines $f(x)$ exactly.*

Proof. By the recursive definition of DD values of Equation (3.19), every DD value in the $(n+1)$th column has a numerator consisting of the difference of a consecutive pair of DD values in the nth column. Since every DD value in the nth column is equal, the difference of all consecutive pairs equals zero. Thus, every DD value in the $(n + 1)$th column, and every DD value in subsequent columns greater than $n + 1$, all equal zero.

The last nonzero DD values occur in the nth column, the first of which is $f[x_n, \ldots, x_0]$. By Equations (3.20) and (3.21), this represents the coefficient c_n of nth term of the DD polynomial function $f_n(x)$.

Since $f_n(x)$ terminates at exactly the nth-degree term and contains all the points that $f(x)$ does, by Theorem 3.1, $f_n(x) = f(x)$. ☐

Example 3.3. Calculate a table of DD values and determine the degree of the DD polynomial function $f_n(x)$ which models $f(x)$ exactly. Construct $f_n(x)$ using the minimum number of points and verify it contains all point(s) not used in its construction.

Solution. Table 3.3 contains five points and up to the 3rd DD values, which are both zero, so no additional columns of DD values are necessary.

Since the last nonzero DD values are 2nd DD values, by Corollary 3.1, the data comes from a second-degree polynomial function $f(x)$, defined uniquely by any set of three of the five given points. Using the first three points and Equation (3.13), the second-degree DD polynomial function $f_2(x)$ is

$$f_2(x) = 1.48 - 0.4(x - 0.2) + 0.5(x - 0.2)(x - 1.4)$$

Evaluating this $f_2(x)$ at the two x values not used in the construction of $f_2(x)$

$$f_2(3.0) = 1.48 - 0.4(3.0 - 0.2) + 0.5(3.0 - 0.2)(3.0 - 1.4) = 2.6$$
$$f_2(4.4) = 1.48 - 0.4(4.4 - 0.2) + 0.5(4.4 - 0.2)(4.4 - 1.4) = 6.1$$

Both $f_2(x)$ values are exactly those given in Table 3.3.

Table 3.3: The points and DD values of Example 3.3. Since all three 2nd DD values are equal, the data is from a second-degree polynomial function $f(x)$ exactly and can be written as a second-degree DD polynomial function $f_2(x)$.

i	x_i	$f(x_i)$	1st DD	2nd DD	3rd DD
0	0.20	1.480			
1	1.40	1.000	−0.40		
2	2.50	1.825	0.75	0.5	
3	3.00	2.600	1.55	0.5	0
4	4.40	6.100	2.50	0.5	0

3.2 Lagrange polynomials

The Lagrange polynomial function is an equivalent form of Newton's DD polynomial function without using the DD coefficients. The *nth*-degree Lagrange polynomial function $\mathcal{L}_n(x)$ is defined as

$$\mathcal{L}_n(x) = \sum_{i=0}^{n} \left(\prod_{\substack{j=0 \\ j \neq i}}^{n} \frac{x - x_j}{x_i - x_j} \right) f(x_i) \tag{3.31}$$

The motivation behind this form becomes clear after examining the specific cases of $\mathcal{L}_1(x)$ and $\mathcal{L}_2(x)$.

Using Equation (3.31) with $n = 1$

$$\mathcal{L}_1(x) = \sum_{i=0}^{1} \left(\prod_{\substack{j=0 \\ j \neq i}}^{1} \frac{x - x_j}{x_i - x_j} \right) f(x_i)$$

The outside summation produces two terms for $i = 0, 1$. When $i = 0$, the product in parentheses produces just one fraction because while j runs from 0 to 1, there is an exception when $i = j$, in this case for $j = 0$. So for $i = 0, j = 1$ only. Likewise when $i = 1$, only $j = 0$ is relevant to the product in parentheses because $j \neq 1$ when $i = 1$. Writing both terms explicitly, the first-degree Lagrange polynomial function $\mathcal{L}_1(x)$ is

$$\mathcal{L}_1(x) = \frac{x - x_1}{x_0 - x_1} f(x_0) + \frac{x - x_0}{x_1 - x_0} f(x_1). \tag{3.32}$$

Equation (3.32) is guaranteed to satisfy the two known points. If $x = x_0$, then

$$\mathcal{L}_1(x_0) = \frac{x_0 - x_1}{x_0 - x_1} f(x_0) + \frac{x_0 - x_0}{x_1 - x_0} f(x_1)$$
$$\mathcal{L}_1(x_0) = 1 \cdot f(x_0) + 0 \cdot f(x_1)$$
$$\mathcal{L}_1(x_0) = f(x_0).$$

Similarly, if $x = x_1$, the opposite happens.

$$\mathcal{L}_1(x_1) = \frac{x_1 - x_1}{x_0 - x_1} f(x_0) + \frac{x_1 - x_0}{x_1 - x_0} f(x_1)$$
$$\mathcal{L}_1(x_1) = 0 \cdot f(x_0) + 1 \cdot f(x_1)$$
$$\mathcal{L}_1(x_1) = f(x_1)$$

Using Equation (3.31) with $n = 2$, $\mathcal{L}_2(x)$ has three terms because the outside summation runs from $i = 0$ to $i = 2$. In this case, each of these terms consists of two factors of fractions inside the parentheses; j runs from 0 to 2 but $j = i$ once for each of the three iterations of i. Excluding the fraction when $j = i$ is necessary because the definition of the denominator includes a factor of $x_i - x_j$ which equals zero when $j = i$.

The two factors of fractions for each value of i are

$$i = 0: \prod_{\substack{j=0 \\ j \neq 0}}^{2} \frac{x - x_j}{x_i - x_j} = \frac{x - x_1}{x_0 - x_1} \cdot \frac{x - x_2}{x_0 - x_2}$$

$$i = 1: \prod_{\substack{j=0 \\ j \neq 1}}^{2} \frac{x - x_j}{x_i - x_j} = \frac{x - x_0}{x_1 - x_0} \cdot \frac{x - x_2}{x_1 - x_2}$$

$$i = 2: \prod_{\substack{j=0 \\ j \neq 2}}^{2} \frac{x - x_j}{x_i - x_j} = \frac{x - x_0}{x_2 - x_0} \cdot \frac{x - x_1}{x_2 - x_1}$$

Substituting each of the above products in Equation (3.31) and multiplying each by $f(x_0)$, $f(x_1)$, and $f(x_2)$, respectively, completes the second-degree Lagrange polynomial function $\mathcal{L}_2(x)$.

$$\mathcal{L}_2(x) = \frac{(x - x_1)(x - x_2)}{(x_0 - x_1)(x_0 - x_2)}f(x_0) + \frac{(x - x_0)(x - x_2)}{(x_1 - x_0)(x_1 - x_2)}f(x_1) + \frac{(x - x_0)(x - x_1)}{(x_2 - x_0)(x_2 - x_1)}f(x_2)$$

$$(3.33)$$

Like $\mathcal{L}_1(x)$, the Lagrange polynomial function $\mathcal{L}_2(x)$ satisfies each known point.

$$\mathcal{L}_2(x_0) = (1)(1)f(x_0) + 0 \cdot f(x_1) + 0 \cdot f(x_2) = f(x_0)$$
$$\mathcal{L}_2(x_1) = 0 \cdot f(x_0) + (1)(1)f(x_1) + 0 \cdot f(x_2) = f(x_1)$$
$$\mathcal{L}_2(x_2) = 0 \cdot f(x_0) + 0 \cdot f(x_1) + (1)(1)f(x_2) = f(x_2)$$

These three verifications demonstrate the rationale behind Equation (3.31), how the product of factors is constructed, how the product affects each of the function values $f(x_i)$, and how it can be extended to the general nth-degree case. For each i value the product of factors multiplied by $f(x_i)$ is reduced to one when substituting x_i and reduced to zero when multiplied by all other known function values, assuring that each of the known points satisfies the Lagrange polynomial function $\mathcal{L}_2(x)$.

Proving the nth-degree Lagrange polynomial function $\mathcal{L}_n(x)$ and the nth-degree DD polynomial functions $f_n(x)$ are equal is algebraically intense. The particular case for $n = 1$ is shown as a template for higher-degree polynomials functions. Using Equation (3.6) with the 1st DD $f[x_1, x_0] = c_1$ of Equation (3.5), and substituting c_1 of Equation (3.3)

$$\begin{aligned}
f_1(x) &= f(x_0) + \frac{f(x_1) - f(x_0)}{x_1 - x_0}(x - x_0) \\
&= \frac{f(x_0)(x_1 - x_0)}{x_1 - x_0} + \frac{(f(x_1) - f(x_0))(x - x_0)}{x_1 - x_0} \\
&= \frac{x_1 f(x_0) - x_0 f(x_0) + x f(x_1) - x_0 f(x_1) - x f(x_0) + x_0 f(x_0)}{x_1 - x_0} \\
&= \frac{x_1 f(x_0) - x f(x_0) + x f(x_1) - x_0 f(x_1)}{x_1 - x_0} \\
&= \frac{(x_1 - x)f(x_0) + (x - x_0)f(x_1)}{x_1 - x_0} \\
&= \frac{x - x_1}{x_0 - x_1}f(x_0) + \frac{x - x_0}{x_1 - x_0}f(x_1)
\end{aligned}$$

This is exactly the Lagrange polynomial function $\mathcal{L}_1(x)$ of Equation (3.32).

Example 3.4. Construct a second-degree Lagrange polynomial function $\mathcal{L}_2(x)$ using the first three points of Example 3.1. Show it is equivalent to the second-degree DD polynomial function $f_2(x)$.

Solution. Although a third-degree DD polynomial function $f_3(x)$ was found in Example 3.1, the second-degree DD polynomial function $f_2(x)$ simply removes the 3rd DD term so that

$$f_2(x) = 3 + 2(x - 1) - 0.1(x - 1)(x - 4)$$

$$= 3 + 2x - 2 - \frac{1}{10}(x^2 - 5x + 4)$$

$$= \frac{15}{5} + \frac{4}{2}x - \frac{10}{5} - \frac{1}{10}x^2 + \frac{1}{2}x - \frac{2}{5}$$

$$= -\frac{1}{10}x^2 + \frac{5}{2}x + \frac{3}{5}$$

Using Equation (3.33) and the points (1, 3), (4, 9), and (6, 12) for the Lagrange polynomial function,

$$\mathcal{L}_2(x) = \frac{(x-4)(x-6)}{(1-4)(1-6)}(3) + \frac{(x-1)(x-6)}{(4-1)(4-6)}(9) + \frac{(x-1)(x-4)}{(6-1)(6-4)}(12)$$

$$= \frac{x^2 - 10x + 24}{5} - \frac{3(x^2 - 7x + 6)}{2} + \frac{6(x^2 - 5x + 4)}{5}$$

$$= \frac{2x^2 - 20x + 48}{10} - \frac{15(x^2 - 7x + 6)}{10} + \frac{12(x^2 - 5x + 4)}{10}$$

$$= \frac{2x^2 - 20x + 48 - 15x^2 + 105x - 90 + 12x^2 - 60x + 48}{10}$$

$$= \frac{-x^2 + 25x + 6}{10}$$

$$= -\frac{1}{10}x^2 + \frac{5}{2}x + \frac{3}{5}$$

Lagrange polynomial functions are advantageous because they are constructed directly from the known points, and the intermediate DD values do not need to be calculated. Also, they can be used for approximating derivatives of unknown functions at unknown points, discussed in Section 6.7.

Since the Lagrange polynomial function is equivalent to the DD polynomial function of the same degree, an identical error analysis using Equation (3.28) can be employed. However, that truncation error E_t uses $f[x_{n+1}, \ldots, x_0]$, so all DD values up to the $(n+1)$th DD would need to be calculated anyway, possibly diminishing the advantage of using a Lagrange polynomial function. Nevertheless, if the data is generated from an unknown but exact nth-degree polynomial function, the Lagrange polynomial function will model it exactly, just like DD polynomial functions. And if there is evidence that the data exactly fits an nth-degree polynomial function, determining the truncation error E_t may not be crucial.

3.3 Splines

Although it is always possible to fit an nth-degree polynomial function through $n + 1$ points, high-degree polynomials often poorly approximate the underlying function and consequently unknown points. Such polynomial functions tend to oscillate more as the degree increases and overestimate extreme values. The fifth-degree DD polynomial function $f_5(x)$ in Figure 3.3, fit exactly to the six points in Table 3.4, exhibits this phenomenon.

Furthermore, small changes in the data may exacerbate the fit of a high-degree polynomial. Suppose the point (5, 6.9) in Figure 3.3 is shifted right to (5.75, 6.9). In Figure 3.4, the corresponding change to the fifth-degree DD polynomial function in blue is com-

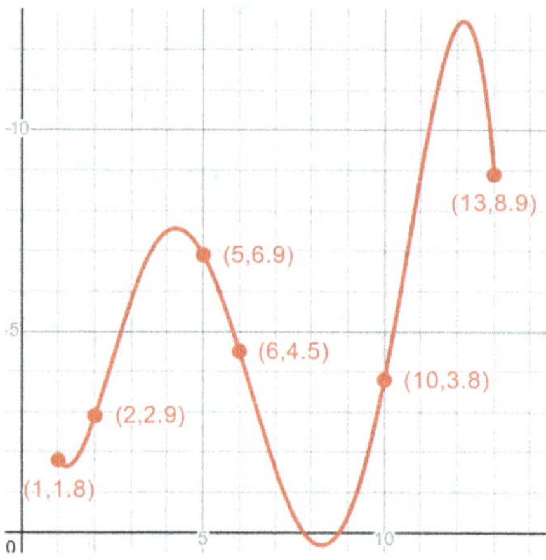

Figure 3.3: A fifth-degree DD polynomial function $f_5(x)$ fit exactly to the data in Table 3.4. High-degree polynomial functions tend to oscillate and overestimate extreme values.

Table 3.4: Six points used to fit a fifth-degree DD polynomial function $f_5(x)$ exactly in Figure 3.3. This data is used in comparison to the linear, quadratic, and cubic splines piecewise function examples that follow.

x	1	2	5	6	10	13
$f(x)$	1.8	2.9	6.9	4.5	3.8	8.9

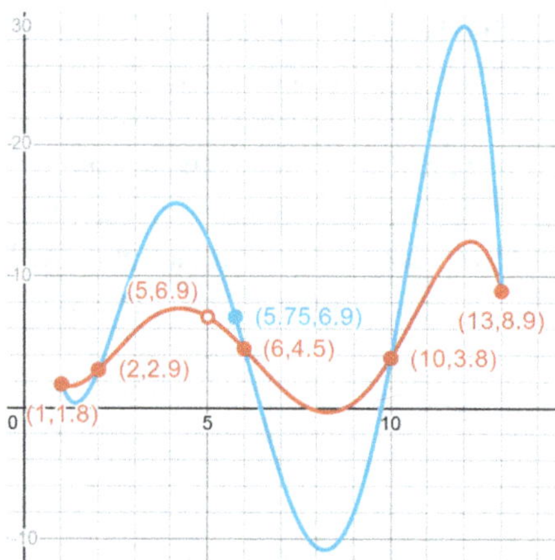

Figure 3.4: The red and blue fifth-degree DD polynomial functions fit to the same data, except for the red open point $(5, 6.9)$ and the blue closed point $(5.75, 6.9)$. The red function already oscillates between the points; the change from the red open point to the blue closed point magnifies the oscillations. Correspondingly, the maximum and minimum values are more extreme on the blue function.

pared to the original in red of Figure 3.3. The oscillations are more extreme and the two functions are quite different considering the similarity of the data, identical save for one shifted point.

An alternative to fitting one function through all known points is using a piecewise function by fitting an nth-degree spline $s_i(x)$ between each pair of points x_i and x_{i+1}. A particular x_p is then estimated using the specific spline $s_i(x)$ where $x_i < x_p < x_{i+1}$.

Spline is a holdover term that originally defined a flexible piece of metal. Using a board with pegs representing points, drafters smoothly bent splines around the pegs.

3.3.1 Linear splines

Each linear spline is determined by the line through consecutive known points. Thus, the first linear spline is defined by the line through the given points $(x_1, f(x_1))$ and $(x_2, f(x_2))$, and where the y value is called the estimated spline value $s_1(x)$. Using the point-slope form of a line

$$y - f(x_1) = m(x - x_1)$$

$$s_1(x) = f(x_1) + \frac{f(x_2) - f(x_1)}{x_2 - x_1}(x - x_1).$$

The general form of the linear spline $s_i(x)$ through consecutive points $(x_i, f(x_i))$ and $(x_{i+1}, f(x_{i+1}))$ follows as

$$s_i(x) = f(x_i) + \frac{f(x_{i+1}) - f(x_i)}{x_{i+1} - x_i}(x - x_i) \tag{3.34}$$

For n known points, the linear splines piecewise function $p_1(x)$ is written

$$p_1(x) = \begin{cases} s_1(x), & x_1 \le x < x_2 \\ s_2(x), & x_2 \le x < x_3 \\ \vdots & \vdots \\ s_{n-2}(x), & x_{n-2} \le x < x_{n-1} \\ s_{n-1}(x), & x_{n-1} \le x \le x_n \end{cases} \tag{3.35}$$

Example 3.5. Construct a linear splines piecewise function $p_1(x)$ using the points in Table 3.4. Use $p_1(x)$ to estimate function values at $x = 4$ and $x = 7$.

Solution. Repeated use of Equation (3.34) yields

$$s_1(x) = f(1) + \frac{f(2) - f(1)}{2 - 1}(x - 1) = 1.8 + 1.1(x - 1)$$

$$s_2(x) = f(2) + \frac{f(5) - f(2)}{5 - 2}(x - 2) = 2.9 + 1.333(x - 2)$$

$$s_3(x) = f(5) + \frac{f(6) - f(5)}{6 - 5}(x - 5) = 6.9 - 2.4(x - 5)$$

$$s_4(x) = f(6) + \frac{f(10) - f(6)}{10 - 6}(x - 6) = 4.5 - 0.175(x - 6)$$

$$s_5(x) = f(10) + \frac{f(13) - f(10)}{13 - 10}(x - 10) = 3.8 + 1.7(x - 10)$$

The linear piecewise function $p_1(x)$ according to Equation (3.35) is

$$p_1(x) = \begin{cases} 1.1x + 0.7, & 1 \leq x < 2 \\ 1.333x + 0.233, & 2 \leq x < 5 \\ -2.4x + 18.9, & 5 \leq x < 6 \\ -0.175x + 5.55, & 6 \leq x < 10 \\ 1.7x - 13.2, & 10 \leq x \leq 13 \end{cases}$$

For $x = 4$, since $2 \leq 4 < 5$, the second spline $s_2(x)$ is used to estimate $p_1(4)$.

$$p_1(4) = s_2(4) = 1.333(4) + 0.233 = 5.565$$

Because $6 \leq 7 < 10$ for $x = 7$, the fourth spline $s_4(x)$ is used to estimate $p_1(7)$.

$$p_1(7) = s_4(7) - 0.175(7) + 5.55 = 4.325$$

Figure 3.5 depicts the linear splines piecewise function $p_1(x)$ of Example 3.5 with the fifth-degree DD polynomial function $f_5(x)$ of Figure 3.3. The extreme values and range of $f_5(x)$ are noticeably reduced by using $p_1(x)$. Linear splines will never introduce values beyond the extreme values in the known data, but the exact fit DD polynomial function could, possibly significantly and erroneously. Figure 3.5 shows the range of the linear splines piecewise function $p_1(x)$ is $[1.8, 8.9]$; the range of the fifth-degree DD polynomial function $f_5(x)$ is roughly 1.8 times larger at $[-0.3, 12.7]$.

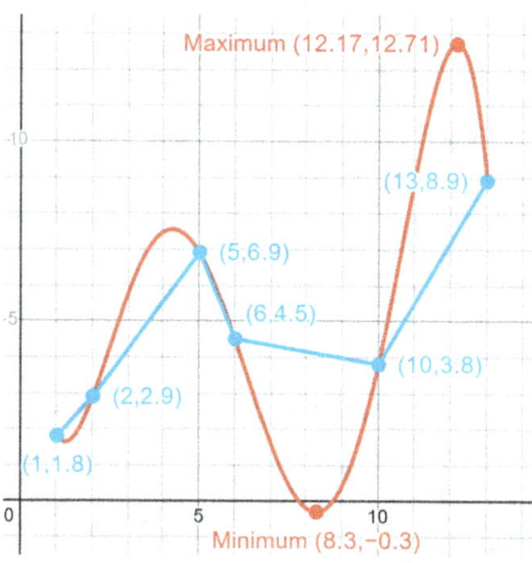

Figure 3.5: The linear splines piecewise function $p_1(x)$ for the data in Table 3.4 of Example 3.5 in blue, with the exact-fit fifth-degree DD polynomial function $f_5(x)$ and its extrema in red.

3.3.2 Quadratic splines

Linear splines are easily calculated but the trade-off is a function which is not smooth and, from calculus, not differentiable at the corners where the splines connect, sometimes called the knots. Quadratic splines solve this problem by adding the condition that the derivatives of each spline must equal at the knots, in addition to the function values equaling so that the function is continuous.

For n points $(x_i, f(x_i))$, $i = 1, \ldots, n$, $n - 1$ quadratic splines $s_i(x)$ defined between x_i and x_{i+1} are required. Each quadratic spline $s_i(x)$ is defined

$$s_i(x) = a_i + b_i(x - x_i) + c_i(x - x_i)^2 \tag{3.36}$$

An expression for each of the three parameters a_i, b_i, and c_i using the known points must be determined, and then used for each quadratic spline $s_i(x)$.

1. An expression for parameter a_i

Each point $(x_i, f(x_i))$ must satisfy its corresponding quadratic spline $s_i(x)$ so that $s_i(x_i) = f(x_i)$. By substituting x_i in Equation (3.36)

$$s_i(x_i) = a_i + b_i(x_i - x_i) + c_i(x_i - x_i)^2$$
$$s_i(x_i) = a_i + 0 + 0$$
$$s_i(x_i) = a_i$$

Since $s_i(x_i) = f(x_i)$ as noted above

$$a_i = f(x_i). \tag{3.37}$$

Thus, each parameter a_i is exactly the function value $f(x_i)$, which coincides with the start of the interval at x_i.

2. An expression for parameter c_i

Consecutive splines $s_i(x)$ and $s_{i+1}(x)$ are equal at the x value, where they connect, at x_{i+1}.

$$s_i(x_{i+1}) = s_{i+1}(x_{i+1})$$

Using Equation (3.36) for each of the above splines,

$$a_i + b_i(x_{i+1} - x_i) + c_i(x_{i+1} - x_i)^2 = a_{i+1} + b_{i+1}(x_{i+1} - x_{i+1}) + c_{i+1}(x_{i+1} - x_{i+1})^2$$
$$a_i + b_i(x_{i+1} - x_i) + c_i(x_{i+1} - x_i)^2 = a_{i+1} + 0 + 0.$$

Subtracting a_i from each side and substituting for both it and a_{i+1} according to Equation (3.37),

$$b_i(x_{i+1} - x_i) + c_i(x_{i+1} - x_i)^2 = f(x_{i+1}) - f(x_i). \tag{3.38}$$

Solving Equation (3.38) for parameter c_i defines each in terms of parameter b_i, determined in the following step.

$$c_i = \frac{f(x_{i+1}) - f(x_i) - b_i(x_{i+1} - x_i)}{(x_{i+1} - x_i)^2} \tag{3.39}$$

3. An expression for parameter b_i

Consider the derivative of each spline of Equation (3.36).

$$s_i{}'(x) = b_i + 2c_i(x - x_i) \tag{3.40}$$

Setting the derivatives of consecutive splines equal not only achieves the goal of a smooth connection but also produces a second equation with parameters b_i and c_i. If the derivatives of consecutive splines at the connecting x value x_{i+1} are equal, then

$$s_i{}'(x_{i+1}) = s_{i+1}{}'(x_{i+1}).$$

Writing each according to Equation (3.40),

$$b_i + 2c_i(x_{i+1} - x_i) = b_{i+1} + 2c_{i+1}(x_{i+1} - x_{i+1})$$
$$b_{i+1} = b_i + 2c_i(x_{i+1} - x_i). \tag{3.41}$$

Substituting Equation (3.39) into Equation (3.41) produces a recursive method for finding b_{i+1} using b_i only.

$$b_{i+1} = b_i + 2\left(\frac{f(x_{i+1}) - f(x_i) - b_i(x_{i+1} - x_i)}{(x_{i+1} - x_i)^2}\right)(x_{i+1} - x_i)$$
$$b_{i+1} = b_i + \frac{2(f(x_{i+1}) - f(x_i)) - 2b_i(x_{i+1} - x_i)}{x_{i+1} - x_i}$$
$$b_{i+1} = b_i + \frac{2(f(x_{i+1}) - f(x_i))}{x_{i+1} - x_i} - 2b_i$$
$$b_{i+1} = 2\left(\frac{f(x_{i+1}) - f(x_i)}{x_{i+1} - x_i}\right) - b_i \tag{3.42}$$

Equations (3.37), (3.42) and (3.39) provide the methods for finding parameters a_i, b_i, and c_i, respectively, for all quadratic splines $s_i(x)$ based on the known points, with one exception. Since c_i is defined by b_i and b_i is defined recursively, an initial b_1 must be defined so that all other b_i parameters, and subsequently all c_i parameters, can be found. A simple compromise defines the first spline $s_1(x)$ linearly while still smoothly transitioning to the second quadratic spline $s_2(x)$.

Letting $c_1 = 0$ for $s_1(x)$ of Equation (3.36) eliminates the second-degree term $c_1(x - x_1)^2$ and reduces $s_1(x)$ to a linear spline. Substituting $c_1 = 0$ in Equation (3.39) and finding b_1

$$0 = \frac{f(x_2) - f(x_1) - b_1(x_2 - x_1)}{(x_2 - x_1)^2}$$

$$b_1(x_2 - x_1) = f(x_2) - f(x_1)$$

$$b_1 = \frac{f(x_2) - f(x_1)}{x_2 - x_1} \tag{3.43}$$

Note that substituting Equation (3.43) into Equation (3.42) for $i = 1$ proves the first two b_i parameters, b_1 and b_2, will always be the same.

$$b_2 = 2\left(\frac{f(x_2) - f(x_1)}{x_2 - x_1}\right) - b_1$$

$$b_2 = 2b_1 - b_1$$

$$b_2 = b_1$$

Therefore, the recursion for determining parameters b_i starts with $i = 2$ for b_3.

The process of finding parameters a_i, b_i, and c_i for each quadratic spline $s_i(x)$ for $i = 1, \ldots, n - 1$ is summarized in Table 3.5.

Table 3.5: Determining parameters a_i, b_i, and c_i for each quadratic spline $s_i(x)$.

1.	Let every $a_i = f(x_i)$	Equation (3.37)
2.	Solve for b_1 and let $b_2 = b_1$	Equation (3.43)
3.	Solve for $b_3, \ldots, b_{n-1}$ recursively	Equation (3.42)
4.	Let $c_1 = 0$ and solve for each c_i using b_i (Steps 2 and 3) for $i = 2, \ldots, n - 1$	Equation (3.39)

The construction of the quadratic splines piecewise function $p_2(x)$ is analogous to the linear splines piecewise function $p_1(x)$ of Equation (3.35). Use the splines $s_i(x)$ defined by Equation (3.36) with parameters a_i, b_i, and c_i determined by the steps in Table 3.5.

$$p_2(x) = \begin{cases} s_1(x), & x_1 \leq x < x_2 \\ s_2(x), & x_2 \leq x < x_3 \\ \vdots & \vdots \\ s_{n-2}(x), & x_{n-2} \leq x < x_{n-1} \\ s_{n-1}(x), & x_{n-1} \leq x \leq x_n \end{cases} \tag{3.44}$$

Example 3.6. Construct a quadratic splines piecewise function $p_2(x)$ using the points in Table 3.4. Use $p_2(x)$ to estimate function values at $x = 4$ and $x = 7$.

Solution. According to Step 1 of Table 3.5, using Equation (3.37),

$$a_1 = f(x_1) = 1.8$$
$$a_2 = f(x_2) = 2.9$$
$$a_3 = f(x_3) = 6.9$$
$$a_4 = f(x_4) = 4.5$$
$$a_5 = f(x_5) = 3.8.$$

Following Step 2 of Table 3.5 and Equation (3.43)

$$b_1 = \frac{f(x_2) - f(x_1)}{x_2 - x_1} = \frac{2.9 - 1.8}{2 - 1} = 1.1$$
$$b_2 = b_1 = 1.1.$$

By Step 3 of Table 3.5 using Equation (3.42) recursively and starting with b_3,

$$b_3 = 2\left(\frac{f(x_3) - f(x_2)}{x_3 - x_2}\right) - b_2 = 2\left(\frac{6.9 - 2.9}{5 - 2}\right) - 1.1 = 1.567$$

$$b_4 = 2\left(\frac{f(x_4) - f(x_3)}{x_4 - x_3}\right) - b_3 = 2\left(\frac{4.5 - 6.9}{6 - 5}\right) - 1.567 = -6.367$$

$$b_5 = 2\left(\frac{f(x_5) - f(x_4)}{x_5 - x_4}\right) - b_4 = 2\left(\frac{3.8 - 4.5}{10 - 6}\right) - (-6.367) = 6.017.$$

Per Step 4 of Table 3.5 and using Equation (3.39) repeatedly starting with c_2,

$$c_1 = 0$$
$$c_2 = \frac{f(x_3) - f(x_2) - b_2(x_3 - x_2)}{(x_3 - x_2)^2} = \frac{6.9 - 2.9 - 1.1(5 - 2)}{(5 - 2)^2} = 0.078$$
$$c_3 = \frac{f(x_4) - f(x_3) - b_3(x_4 - x_3)}{(x_4 - x_3)^2} = \frac{4.5 - 6.9 - 1.567(6 - 5)}{(6 - 5)^2} = -3.967$$
$$c_4 = \frac{f(x_5) - f(x_4) - b_4(x_5 - x_4)}{(x_5 - x_4)^2} = \frac{3.8 - 4.5 - (-6.367)(10 - 6)}{(10 - 6)^2} = 1.548$$
$$c_5 = \frac{f(x_6) - f(x_5) - b_5(x_6 - x_5)}{(x_6 - x_5)^2} = \frac{8.9 - 3.8 - 6.017(13 - 10)}{(13 - 10)^2} = -1.439.$$

Using all the above values and Equation (3.44), the quadratic splines piecewise function $p_2(x)$, including the domain of each spline $s_i(x)$, is

$$p_2(x) = \begin{cases} 1.8 + 1.1(x - 1), & 1 \leq x < 2 \\ 2.9 + 1.1(x - 2) + 0.078(x - 2)^2, & 2 \leq x < 5 \\ 6.9 + 1.567(x - 5) - 3.967(x - 5)^2, & 5 \leq x < 6 \\ 4.5 - 6.367(x - 6) + 1.548(x - 6)^2, & 6 \leq x < 10 \\ 3.8 + 6.017(x - 10) - 1.439(x - 10)^2, & 10 \leq x \leq 13 \end{cases}$$

Since $2 \le 4 < 5$, the second spline $s_2(x)$ is used to estimate $p_2(4)$.

$$p_2(4) = s_2(4) = 2.9 + 1.1(4 - 2) + 0.078(4 - 2)^2 = 5.412$$

Because $6 \le 7 < 10$, the fourth spline $s_4(x)$ is used to estimate $p_2(7)$.

$$p_2(7) = s_4(7) = 4.5 - 6.367(7 - 6) + 1.548(7 - 6)^2 = -0.319$$

Figure 3.6 depicts the quadratic splines piecewise function $p_2(x)$ of Example 3.6 with the fifth-degree DD polynomial function $f_5(x)$ of Figure 3.3. The parabolic shape of quadratic splines reintroduces approximations of extreme values outside the range of the known data, just like $f_5(x)$ but unlike the linear splines piecewise function $p_1(x)$. As shown in Figure 3.6, the range of $p_2(x)$ is $[-2, 10.1]$, roughly 1.8 times larger than the range of the known data of $[1.8, 8.9]$. Recall the range of $f_5(x)$ $[-0.3, 12.7]$ is also approximately 1.8 times larger than the range of the known data, but shifted in the positive direction.

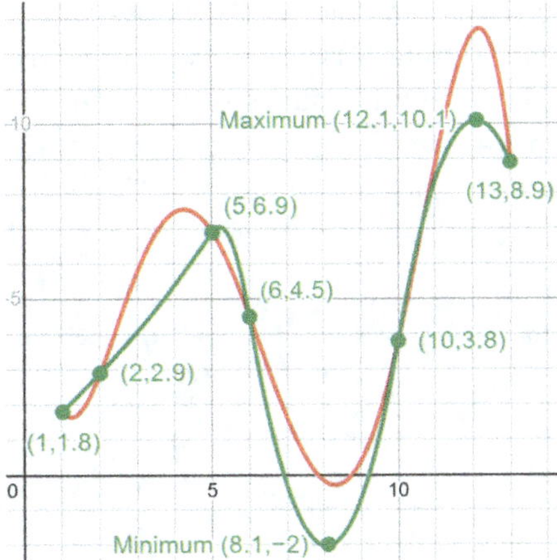

Figure 3.6: The quadratic splines piecewise function $p_2(x)$ of Example 3.6 in green with the fifth-degree DD polynomial function of Figure 3.3 in red. Both use the data of Table 3.4. $p_2(x)$ estimates the minimum and maximum values outside that data.

Figure 3.7 overlays the linear splines piecewise function $p_1(x)$ of Figure 3.5 with the quadratic splines piecewise functions $p_2(x)$ and the fifth-degree DD polynomial function $f_5(x)$ of Figure 3.6. Note the exception of making the first quadratic spline linear to generate all subsequent quadratic splines, thus the first linear and quadratic splines overlap exactly.

Extreme values are gleaned easily from a quadratic splines piecewise function because the derivatives of the splines defined by Equation (3.40) are linear and can be solved directly for x by setting $s_i'(x) = 0$. The only caveat is such x values are valid only if they fall within the interval of the restricted domain of that particular spline. For example, a spline defined by $s_j(x) = 5(x - 1) + 2(x - 1)^2$ on the interval $[1, 5)$ does *not* contain an extrema because $s_j'(x) = 5 + 4x - 4 = 0$ so that $x = -0.25$, which is not between 1 and 5.

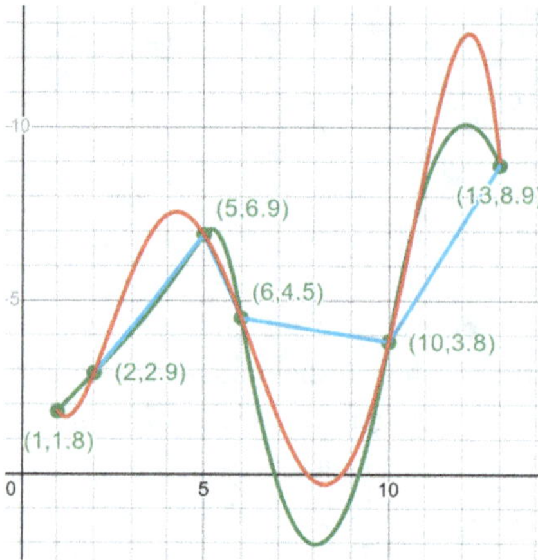

Figure 3.7: A comparison of the green quadratic splines piecewise function $p_2(x)$, the blue linear splines piecewise function $p_1(x)$ and the red fifth-degree DD polynomial function $f_5(x)$ for the data of Table 3.4.

Although not necessary, the graphs of the derivatives $s_i'(x)$ illuminate the calculus concept that extreme values of the splines $s_i(x)$ occur at x values where the derivatives $s_i'(x) = 0$ and change sign around the zero. For a maximum, this change is from positive to negative, and for linear derivatives resulting from quadratic splines, is represented by a negatively sloped line through the zero. The change from negative to positive indicates a minimum and is represented by a positively sloped line through the zero. The results of Example 3.7 in Figure 3.8 illustrate graphically identifiable extrema.

Example 3.7. Calculate the derivative $p_2'(x)$ of the quadratic splines piecewise function $p_2(x)$ of Example 3.6. Analytically use $p_2'(x)$ to identify the type of all extrema of $p_2(x)$, other than the endpoints.
Solution. Differentiate each spline of $p_2(x)$.

$$s_1'(x) = 1.1$$

$$s_2'(x) = 1.1 + 0.156(x - 2) = 0.156x + 0.788$$

$$s_3'(x) = 1.567 - 7.934(x - 5) = -7.934x + 41.237$$

$$s_4'(x) = -6.367 + 3.096(x - 6) = 3.096x - 24.943$$

$$s_5'(x) = 6.017 - 2.878(x - 10) = -2.878x + 34.797$$

These comprise the pieces of the derivative $p_2'(x)$, each corresponding to the restricted domain intervals.

$$p_2'(x) = \begin{cases} 1.1, & 1 \le x < 2 \\ 0.156x + 0.788, & 2 \le x < 5 \\ -7.934x + 41.237, & 5 \le x < 6 \\ 3.096x - 24.943, & 6 \le x < 10 \\ -2.878x + 34.797, & 10 \le x \le 13 \end{cases}$$

Potential extrema exists where $p_2'(x) = 0$, that is, where any of the spline derivatives $s_i'(x) = 0$. Because $s_1'(x) = 1.1$ is constant, $s_1'(x) = 0$ nowhere. Also, for $s_2'(x)$

$$s_2'(x) = 0.156x + 0.788 = 0$$

$$x = -5.051$$

Since $x = -5.051$ is not in $2 \leq x < 5$ there are no extrema on the first two splines. The remaining spline derivatives each have a zero in their particular intervals and are legitimate extrema.

	Zero	Interval
$s_3'(x) = -7.934x + 41.237 = 0$	$x = 5.198$	$5 \leq x < 6$
$s_4'(x) = 3.096x - 24.943 = 0$	$x = 8.057$	$6 \leq x < 10$
$s_5'(x) = -2.878x + 34.797 = 0$	$x = 12.091$	$10 \leq x \leq 13$

Because $s_3'(x)$ and $s_5'(x)$ have negative slopes -7.934 and -2.878, respectively, their function values around their zeros change from positive to negative, indicating maxima. Conversely, $s_4'(x)$ has a positive slope 3.096 and so its function values around its zero changes from negative to positive, indicating a minimum.

Figure 3.8 displays each quadratic spline $s_i(x)$ of $p_2(x)$ of Example 3.6 in a different solid color and its corresponding linear derivative $s_i'(x)$ of Example 3.7 in the same dashed color. Where applicable, if a quadratic spline $s_i(x)$ contains a relative extrema at $x = k$, it is labeled $(k, s_i(k))$ on the spline and at $(k, 0)$ on the linear derivative $s_i'(x)$, where $s_i'(k) = 0$.

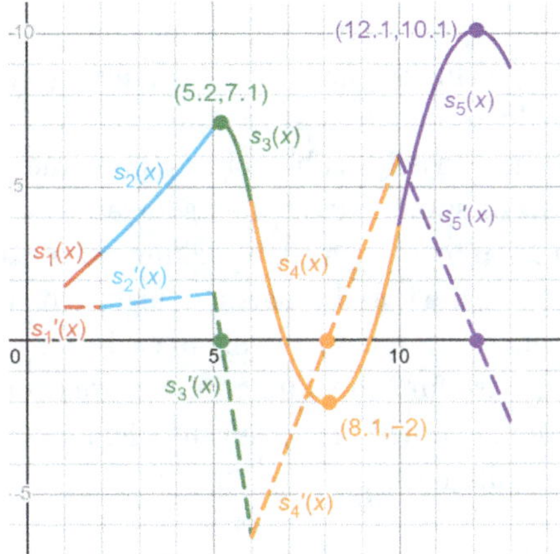

Figure 3.8: A quadratic splines piecewise function $p_2(x)$ (solid) of Example 3.6 and its derivative $p_2'(x)$ (dashed) of Example 3.7. Each color represents a particular spline. The derivatives of the splines are linear and their slopes indicate a maximum or minimum at the x-intercepts.

The sign of the slope of the linear derivative identifies the type of extremum. A negatively sloped line changes from positive to negative around the zero at $x = k$ and indicates a maximum at $x = k$. This applies to the slopes of $s_3'(x)$ in green and $s_5'(x)$ in purple, and both have relative maxima labeled on the green spline $s_3(x)$ at $(5.2, 7.1)$ and on the purple spline $s_5(x)$ at $(12.1, 10.1)$. Conversely, a positively sloped line changes from negative to positive around the zero at $x = k$ and indicates a minimum at $x = k$. The slope of $s_4'(x)$ in orange is positive and there is a relative minimum marked on the orange spline $s_4(x)$ at $(8.1, -2.0)$.

3.3.3 Cubic splines

A quadratic splines piecewise function $p_2(x)$ is smooth and differentiable everywhere on its domain by equating the first derivatives on each side of every knot during definition. At each knot $s_i'(x)$ is defined by its left-side endpoint and therefore continuous from the left side. Showing the limit from the right side is equal proves overall continuity at each knot. For instance, at $x = 6$ for $p_2'(x)$ of Example 3.7

$$\lim_{x \to 6^-} s_3'(x) = \lim_{x \to 6^-} -7.934x + 41.237 = -7.934(6) + 41.237 = -6.367$$

$$\lim_{x \to 6^+} s_4'(x) = \lim_{x \to 6^+} 3.096x - 24.943 = 3.096(6) - 24.943 = -6.367.$$

Every $p_2(x)$ is defined so that it always has a derivative that is continuous everywhere. But because the derivatives of each spline $s_i'(x)$ are linear there are likely corners on $p_2'(x)$ at the knots. From calculus, corners render the first derivative non-differentiable at the knots as shown by the dashed derivative $p_2'(x)$ in Figure 3.8. A linear splines piecewise function $p_1(x)$ is also non-differentiable at the knots, but because of jump discontinuities rather than corners.

Applications sometimes dictate a function differentiable everywhere. For instance, functions describing position of a moving object have first derivatives, representing velocity, commonly differentiable everywhere. Suppose a linear splines piecewise function represents position. The corners at the knots imply velocity is undefined, but only at the known points, probably unlikely. Similarly, using a quadratic splines piecewise function for position, finding acceleration using the second derivative is possible everywhere except at the known points. A piecewise function using cubic splines solves these problems, as well as the matter of the first quadratic spline actually being linear.

It is certainly possible to use higher-degree splines, each time making the next derivative both smooth and differentiable. However, the initial problem of using a high degree, oscillating polynomial to fit an entire data set would now apply to each spline, nullifying the usefulness of a lower-degree piecewise function. Cubic splines are commonly used in practice because a twice-differentiable function is often sufficient while the third degree is small enough to avoid large oscillations.

Each cubic spline $s_i(x)$ is defined

$$s_i(x) = a_i + b_i(x - x_i) + c_i(x - x_i)^2 + d_i(x - x_i)^3 \tag{3.45}$$

An expression using the known points for the four parameters a_i, b_i, c_i, and d_i for each cubic spline $s_i(x)$ must be determined.

1. An expression for parameter a_i

Like quadratic splines, $a_i = f(x_i)$ by Equation (3.37).

2. An expression for parameter d_i

Also, like quadratic splines, consecutive splines are equal at the point where they connect, i. e., $s_i(x_{i+1}) = s_{i+1}(x_{i+1})$. Writing each side by the expansion of Equation (3.45) extends Equation (3.38); the substitutions $f(x_{i+1}) = a_{i+1}$ and $f(x_i) = a_i$ are made in Equation (3.46) below.

$$a_i + b_i(x_{i+1} - x_i) + c_i(x_{i+1} - x_i)^2 + d_i(x_{i+1} - x_i)^3$$
$$= a_{i+1} + b_{i+1}(x_{i+1} - x_{i+1}) + c_{i+1}(x_{i+1} - x_{i+1})^2 + d_{i+1}(x_{i+1} - x_{i+1})^3$$
$$a_i + b_i(x_{i+1} - x_i) + c_i(x_{i+1} - x_i)^2 + d_i(x_{i+1} - x_i)^3 = a_{i+1} + 0 + 0 + 0$$
$$b_i(x_{i+1} - x_i) + c_i(x_{i+1} - x_i)^2 + d_i(x_{i+1} - x_i)^3 = a_{i+1} - a_i$$
$$b_i(x_{i+1} - x_i) + c_i(x_{i+1} - x_i)^2 + d_i(x_{i+1} - x_i)^3 = f(x_{i+1}) - f(x_i) \tag{3.46}$$

The first derivative of a cubic spline defined by Equation (3.45) is

$$s_i{}'(x) = b_i + 2c_i(x - x_i) + 3d_i(x - x_i)^2 \tag{3.47}$$

Like quadratic splines, consecutive first derivatives are equal at the connecting point so that $s_i{}'(x_{i+1}) = s_{i+1}{}'(x_{i+1})$. Writing each side by the expansion of Equation (3.47) extends Equation (3.41).

$$b_i + 2c_i(x_{i+1} - x_i) + 3d_i(x_{i+1} - x_i)^2$$
$$= b_{i+1} + 2c_{i+1}(x_{i+1} - x_{i+1}) + 3d_{i+1}(x_{i+1} - x_{i+1})^2$$
$$b_{i+1} = b_i + 2c_i(x_{i+1} - x_i) + 3d_i(x_{i+1} - x_i)^2 \tag{3.48}$$

The additional condition for cubic splines is the *second* derivatives of Equation (3.45) are equal at the connecting point x_{i+1} to guarantee the first derivative is smooth and differentiable at the knots. The second derivative of a cubic spline is

$$s_i{}''(x) = 2c_i + 6d_i(x - x_i) \tag{3.49}$$

For $s_i{}''(x_{i+1}) = s_{i+1}{}''(x_{i+1})$ and writing each side by the expansion of Equation (3.49)

$$2c_i + 6d_i(x_{i+1} - x_i) = 2c_{i+1} + 6d_i(x_{i+1} - x_{i+1})$$
$$c_{i+1} = c_i + 3d_i(x_{i+1} - x_i)$$

This enables finding all d_i parameters based on c_i parameters, solving for d_i.

$$d_i = \frac{c_{i+1} - c_i}{3(x_{i+1} - x_i)} \tag{3.50}$$

3. An expression for parameter b_i

Like the d_i parameters, the b_i parameters are solved in terms of the c_i parameters, with recursion based on the previous b_{i-1} parameter.

Substituting Equation (3.50) into Equation (3.48)

$$b_{i+1} = b_i + 2c_i(x_{i+1} - x_i) + 3\left(\frac{c_{i+1} - c_i}{3(x_{i+1} - x_i)}\right)(x_{i+1} - x_i)^2$$

$$b_{i+1} = b_i + 2c_i(x_{i+1} - x_i) + (c_{i+1} - c_i)(x_{i+1} - x_i)$$

$$b_{i+1} = b_i + (x_{i+1} - x_i)(2c_i + c_{i+1} - c_i)$$

$$b_{i+1} = b_i + (x_{i+1} - x_i)(c_{i+1} + c_i) \tag{3.51}$$

Except for b_1, Equation (3.51) is used to find all the b_i parameters. For the initial b_1 parameter, an alternate form of b_i is found and solved for $i = 1$. Substituting Equation (3.50) into Equation (3.46) and solving for b_i

$$b_i(x_{i+1} - x_i) + c_i(x_{i+1} - x_i)^2 + \frac{c_{i+1} - c_i}{3(x_{i+1} - x_i)}(x_{i+1} - x_i)^3 = f(x_{i+1}) - f(x_i)$$

$$b_i(x_{i+1} - x_i) + c_i(x_{i+1} - x_i)^2 + \frac{(c_{i+1} - c_i)(x_{i+1} - x_i)^2}{3} = f(x_{i+1}) - f(x_i)$$

$$b_i(x_{i+1} - x_i) + (x_{i+1} - x_i)^2\left(c_i + \frac{c_{i+1} - c_i}{3}\right) = f(x_{i+1}) - f(x_i)$$

$$b_i(x_{i+1} - x_i) + (x_{i+1} - x_i)^2\left(\frac{c_{i+1} + 2c_i}{3}\right) = f(x_{i+1}) - f(x_i)$$

$$b_i(x_{i+1} - x_i) = f(x_{i+1}) - f(x_i) - (x_{i+1} - x_i)^2\left(\frac{c_{i+1} + 2c_i}{3}\right)$$

$$b_i = \frac{f(x_{i+1}) - f(x_i)}{x_{i+1} - x_i} - \frac{(x_{i+1} - x_i)(c_{i+1} + 2c_i)}{3} \tag{3.52}$$

Equation (3.52) could be used to find all b_i parameters in addition to b_1, but is less efficient than the recursive definition of Equation (3.51), which requires less calculations. If many splines are required, the reduction in computations could be significant.

4. An expression for parameter c_i

All the c_i parameters are solved simultaneously by forming a system of equations in terms of c_i parameters only and then using Gauss–Jordan elimination. The algebra of this technique is a topic for a Linear Algebra course; graphing calculators and many online tools can perform this lengthy procedure. Briefly, the coefficient of each c_i parameter, along with the constant term from one equation, form a row of entries in a matrix. Through algebraic operations on an entire row, an augmented matrix is transformed into reduced row echelon form (RREF). A matrix in RREF conveniently contains the solutions of the c_i parameters in the last column, ordered from top to bottom. The rest of the matrix is populated by zeros, except for entries in which the row and column numbers are equal (e. g., the second row and second column or the fifth row and fifth column); these entries have value one.

First, increasing the index i by one, Equation (3.52) becomes

$$b_{i+1} = \frac{f(x_{i+2}) - f(x_{i+1})}{x_{i+2} - x_{i+1}} - \frac{(x_{i+2} - x_{i+1})(c_{i+2} + 2c_{i+1})}{3} \tag{3.53}$$

Second, substituting Equation (3.52) in Equation (3.51) produces

$$b_{i+1} = \frac{f(x_{i+1}) - f(x_i)}{x_{i+1} - x_i} - \frac{(x_{i+1} - x_i)(c_{i+1} + 2c_i)}{3} + (x_{i+1} - x_i)(c_{i+1} + c_i) \tag{3.54}$$

Equations (3.53) and (3.54) are both expressions for b_{i+1} and thus are equal.

$$\frac{f(x_{i+2}) - f(x_{i+1})}{x_{i+2} - x_{i+1}} - \frac{(x_{i+2} - x_{i+1})(c_{i+2} + 2c_{i+1})}{3}$$
$$= \frac{f(x_{i+1}) - f(x_i)}{x_{i+1} - x_i} - \frac{(x_{i+1} - x_i)(c_{i+1} + 2c_i)}{3} + (x_{i+1} - x_i)(c_{i+1} + c_i)$$
$$\frac{f(x_{i+2}) - f(x_{i+1})}{x_{i+2} - x_{i+1}} - \frac{f(x_{i+1}) - f(x_i)}{x_{i+1} - x_i}$$
$$= \frac{(x_{i+2} - x_{i+1})(c_{i+2} + 2c_{i+1})}{3} - \frac{(x_{i+1} - x_i)(c_{i+1} + 2c_i)}{3} + (x_{i+1} - x_i)(c_{i+1} + c_i)$$
$$(x_{i+2} - x_{i+1})(c_{i+2} + 2c_{i+1}) - (x_{i+1} - x_i)(c_{i+1} + 2c_i) + 3(x_{i+1} - x_i)(c_{i+1} + c_i)$$
$$= 3\left(\frac{f(x_{i+2}) - f(x_{i+1})}{x_{i+2} - x_{i+1}}\right) - 3\left(\frac{f(x_{i+1}) - f(x_i)}{x_{i+1} - x_i}\right) \tag{3.55}$$

The left side of Equation (3.55) is regrouped according to the c_i coefficients.

$$c_{i+2}(x_{i+2} - x_{i+1}) + 2c_{i+1}(x_{i+2} - x_{i+1})$$
$$- c_{i+1}(x_{i+1} - x_i) - 2c_i(x_{i+1} - x_i) + 3c_{i+1}(x_{i+1} - x_i) + 3c_i(x_{i+1} - x_i)$$
$$= c_{i+2}(x_{i+2} - x_{i+1}) + 2c_{i+1}(x_{i+2} - x_{i+1}) + 2c_{i+1}(x_{i+1} - x_i) + c_i(x_{i+1} - x_i)$$
$$= c_{i+2}(x_{i+2} - x_{i+1}) + 2c_{i+1}(x_{i+2} - x_{i+1} + x_{i+1} - x_i) + c_i(x_{i+1} - x_i)$$
$$= c_i(x_{i+1} - x_i) + 2c_{i+1}(x_{i+2} - x_i) + c_{i+2}(x_{i+2} - x_{i+1})$$

Substituting back into the left side of Equation (3.55)

$$c_i(x_{i+1} - x_i) + 2c_{i+1}(x_{i+2} - x_i) + c_{i+2}(x_{i+2} - x_{i+1})$$
$$= 3\left(\frac{f(x_{i+2}) - f(x_{i+1})}{x_{i+2} - x_{i+1}}\right) - 3\left(\frac{f(x_{i+1}) - f(x_i)}{x_{i+1} - x_i}\right) \tag{3.56}$$

Recall quadratic splines required fixing one initial parameter ($c_1 = 0$ was used) to generate the b_i parameters recursively and then the c_i parameters. Cubic splines need two such initial conditions. Again $c_1 = 0$ is used, along with $c_n = 0$. Although c_n is artificial because there are $n - 1$ splines for n points, it is needed for two reasons. First, the system of equations generated by Equation (3.56) can be solved uniquely if the number of c_i parameters equals the number of equations (a conclusion from Linear Algebra). Thus $c_n = 0$ provides that nth equation. Second, the penultimate equation in that system requires a c_n coefficient; see the below system and the equation immediately preceding $c_n = 0$. Unlike quadratic splines, letting $c_1 = 0$ and $c_n = 0$ does not result in a linear first spline because of the cubic terms associated with the d_i parameters.

Starting with $i = 1$ and using Equation (3.56) repeatedly up to $i = n - 2$, along with the initial conditions for $c_1 = 0$ and $c_n = 0$, the entire set of n equations is

$$c_1 = 0$$

$$c_1(x_2 - x_1) + 2c_2(x_3 - x_1) + c_3(x_3 - x_2) = 3\left(\frac{f(x_3) - f(x_2)}{x_3 - x_2}\right) - 3\left(\frac{f(x_2) - f(x_1)}{x_2 - x_1}\right)$$

$$c_2(x_3 - x_2) + 2c_3(x_4 - x_2) + c_4(x_4 - x_3) = 3\left(\frac{f(x_4) - f(x_3)}{x_4 - x_3}\right) - 3\left(\frac{f(x_3) - f(x_2)}{x_3 - x_2}\right)$$

$$\vdots$$

$$c_{n-2}(x_{n-1} - x_{n-2}) + 2c_{n-1}(x_n - x_{n-2}) + c_n(x_n - x_{n-1}) = 3\left(\frac{f(x_n) - f(x_{n-1})}{x_n - x_{n-1}}\right)$$
$$- 3\left(\frac{f(x_{n-1}) - f(x_{n-2})}{x_{n-1} - x_{n-2}}\right)$$

$$c_n = 0$$

This system of n equations in n unknowns is written as a matrix equation, described below.

$$
\begin{bmatrix}
1 & 0 & 0 & 0 & \cdots & 0 \\
x_2 - x_1 & 2(x_3 - x_1) & x_3 - x_2 & 0 & \cdots & 0 \\
0 & x_3 - x_2 & 2(x_4 - x_2) & x_4 - x_3 & \cdots & 0 \\
\vdots & \ddots & \ddots & \ddots & \ddots & \vdots \\
0 & \cdots & 0 & x_{n-1} - x_{n-2} & 2(x_n - x_{n-2}) & x_n - x_{n-1} \\
0 & \cdots & 0 & 0 & 0 & 1
\end{bmatrix}
$$

$$
\times
\begin{bmatrix}
c_1 \\
c_2 \\
c_3 \\
\vdots \\
c_{n-1} \\
c_n
\end{bmatrix}
=
\begin{bmatrix}
0 \\
3(\frac{f(x_3)-f(x_2)}{x_3-x_2}) - 3(\frac{f(x_2)-f(x_1)}{x_2-x_1}) \\
3(\frac{f(x_4)-f(x_3)}{x_4-x_3}) - 3(\frac{f(x_3)-f(x_2)}{x_3-x_2}) \\
\vdots \\
3(\frac{f(x_n)-f(x_{n-1})}{x_n-x_{n-1}}) - 3(\frac{f(x_{n-1})-f(x_{n-2})}{x_{n-1}-x_{n-2}}) \\
0
\end{bmatrix}
\tag{3.57}
$$

In the leftmost matrix, each row across represents the left side of one of the n equations of the system, containing the coefficients of the c_i parameters in order. In the rightmost matrix, the column down contains the right side constants of the system of equations.

The leftmost coefficient matrix augmented with the column matrix on the right side is transformed to RREF using Gauss–Jordan elimination to solve for the column matrix of all the c_i parameters. Now the d_i parameters can be solved directly using Equation (3.50), and the b_i parameters can be solved recursively using Equation (3.51), after b_1 is found using Equation (3.52). The entire process of determining parameters a_i, b_i, c_i and d_i for each cubic spline $s_i(x)$ for $i = 1, \ldots, n - 1$ is summarized in Table 3.6.

Table 3.6: Determining parameters a_i, b_i, c_i and d_i for each cubic spline $s_i(x)$.

1.	Let every $a_i = f(x_i)$	Equation (3.37)
2.	Solve every c_i with an augmented matrix and RREF	Equation (3.57)
3.	Solve for b_1 using c_1 and c_2	Equation (3.52)
4.	Solve for $b_2, \ldots, b_{n-1}$ recursively and using $c_1, \ldots, c_{n-1}$	Equation (3.51)
5.	Solve for every d_i using c_i parameters	Equation (3.50)

Writing a cubic splines piecewise function $p_3(x)$ follows the same definition as the linear and quadratic splines piecewise functions $p_1(x)$ and $p_2(x)$ of Equations (3.35) and (3.44), respectively. Using the splines $s_i(x)$ defined by Equation (3.45) and finding parameters a_i, b_i, c_i, and d_i determined by the steps in Table 3.6

$$p_3(x) = \begin{cases} s_1(x), & x_1 \le x < x_2 \\ s_2(x), & x_2 \le x < x_3 \\ \vdots & \vdots \\ s_{n-2}(x), & x_{n-2} \le x < x_{n-1} \\ s_{n-1}(x), & x_{n-1} \le x \le x_n \end{cases} \tag{3.58}$$

Example 3.8. Construct a cubic splines piecewise function $p_3(x)$ using the points in Table 3.4. Use $p_3(x)$ to estimate function values at $x = 4$ and $x = 7$.

Solution. The a_i parameters are found immediately and exactly as in Example 3.6 by using Equation (3.37), where $a_i = f(x_i)$ for $i = 1, \ldots, 5$.

$$a_1 = f(x_1) = 1.8$$
$$a_2 = f(x_2) = 2.9$$
$$a_3 = f(x_3) = 6.9$$
$$a_4 = f(x_4) = 4.5$$
$$a_5 = f(x_5) = 3.8$$

The c_i parameters must be found next because the b_i and d_i parameters are defined in terms of the c_i parameters. The matrix equation of Equation (3.57) is used in this case specifically for both $n = 6$ points and equations. The initial template is

$$
\begin{bmatrix}
1 & 0 & 0 & 0 & 0 & 0 \\
x_2 - x_1 & 2(x_3 - x_1) & x_3 - x_2 & 0 & 0 & 0 \\
0 & x_3 - x_2 & 2(x_4 - x_2) & x_4 - x_3 & 0 & 0 \\
0 & 0 & x_4 - x_3 & 2(x_5 - x_3) & x_5 - x_4 & 0 \\
0 & 0 & 0 & x_5 - x_4 & 2(x_6 - x_4) & x_6 - x_5 \\
0 & 0 & 0 & 0 & 0 & 1
\end{bmatrix}
$$

$$
\times
\begin{bmatrix} c_1 \\ c_2 \\ c_3 \\ c_4 \\ c_5 \\ c_6 \end{bmatrix}
=
\begin{bmatrix}
0 \\
3(\frac{f(x_3)-f(x_2)}{x_3-x_2}) - 3(\frac{f(x_2)-f(x_1)}{x_2-x_1}) \\
3(\frac{f(x_4)-f(x_3)}{x_4-x_3}) - 3(\frac{f(x_3)-f(x_2)}{x_3-x_2}) \\
3(\frac{f(x_5)-f(x_4)}{x_5-x_4}) - 3(\frac{f(x_4)-f(x_3)}{x_4-x_3}) \\
3(\frac{f(x_6)-f(x_5)}{x_6-x_5}) - 3(\frac{f(x_5)-f(x_4)}{x_5-x_4}) \\
0
\end{bmatrix}
$$

Substituting all the x_i and $f(x_i)$ values for $i = 1, \ldots, 6$

$$
\begin{bmatrix}
1 & 0 & 0 & 0 & 0 & 0 \\
2-1 & 2(5-1) & 5-2 & 0 & 0 & 0 \\
0 & 5-2 & 2(6-2) & 6-5 & 0 & 0 \\
0 & 0 & 6-5 & 2(10-5) & 10-6 & 0 \\
0 & 0 & 0 & 10-6 & 2(13-6) & 13-10 \\
0 & 0 & 0 & 0 & 0 & 1
\end{bmatrix}
$$

$$
\times
\begin{bmatrix} c_1 \\ c_2 \\ c_3 \\ c_4 \\ c_5 \\ c_6 \end{bmatrix}
=
\begin{bmatrix}
0 \\
3(\frac{f(5)-f(2)}{5-2}) - 3(\frac{f(2)-f(1)}{2-1}) \\
3(\frac{f(6)-f(5)}{6-5}) - 3(\frac{f(5)-f(2)}{5-2}) \\
3(\frac{f(10)-f(6)}{10-6}) - 3(\frac{f(6)-f(5)}{6-5}) \\
3(\frac{f(13)-f(10)}{13-10}) - 3(\frac{f(10)-f(6)}{10-6}) \\
0
\end{bmatrix}
$$

$$
\begin{bmatrix}
1 & 0 & 0 & 0 & 0 & 0 \\
1 & 8 & 3 & 0 & 0 & 0 \\
0 & 3 & 8 & 1 & 0 & 0 \\
0 & 0 & 1 & 10 & 4 & 0 \\
0 & 0 & 0 & 4 & 14 & 3 \\
0 & 0 & 0 & 0 & 0 & 1
\end{bmatrix}
\begin{bmatrix} c_1 \\ c_2 \\ c_3 \\ c_4 \\ c_5 \\ c_6 \end{bmatrix}
=
\begin{bmatrix}
0 \\
3(\frac{6.9-2.9}{3}) - 3(\frac{2.9-1.8}{1}) \\
3(\frac{4.5-6.9}{1}) - 3(\frac{6.9-2.9}{3}) \\
3(\frac{3.8-4.5}{4}) - 3(\frac{4.5-6.9}{1}) \\
3(\frac{8.9-3.8}{3}) - 3(\frac{3.8-4.5}{4}) \\
0
\end{bmatrix}
$$

The augmented matrix for the above matrix equation is

$$\begin{bmatrix} 1 & 0 & 0 & 0 & 0 & 0 & \vdots & 0 \\ 1 & 8 & 3 & 0 & 0 & 0 & \vdots & 0.7 \\ 0 & 3 & 8 & 1 & 0 & 0 & \vdots & -11.2 \\ 0 & 0 & 1 & 10 & 4 & 0 & \vdots & 6.675 \\ 0 & 0 & 0 & 4 & 14 & 3 & \vdots & 5.625 \\ 0 & 0 & 0 & 0 & 0 & 1 & \vdots & 0 \end{bmatrix}$$

Using Gauss–Jordan elimination to put into RREF,

$$\begin{bmatrix} 1 & 0 & 0 & 0 & 0 & 0 & \vdots & 0 \\ 0 & 1 & 0 & 0 & 0 & 0 & \vdots & 0.755 \\ 0 & 0 & 1 & 0 & 0 & 0 & \vdots & -1.78 \\ 0 & 0 & 0 & 1 & 0 & 0 & \vdots & 0.773 \\ 0 & 0 & 0 & 0 & 1 & 0 & \vdots & 0.181 \\ 0 & 0 & 0 & 0 & 0 & 1 & \vdots & 0 \end{bmatrix}$$

The rightmost column represents the values for c_i, $i = 1, \ldots, 6$, corresponding to row i. Recall the initial conditions $c_1 = 0$ and $c_6 = 0$, the latter of which is artificial and not used by any spline.

Using Equation (3.52) once for b_1,

$$b_1 = \frac{f(x_2) - f(x_1)}{x_2 - x_1} - \frac{(x_2 - x_1)(c_2 + 2c_1)}{3}$$
$$b_1 = \frac{2.9 - 1.8}{2 - 1} - \frac{(2 - 1)(0.755 + 2 \cdot 0)}{3} = 0.848$$

Using Equations (3.51) recursively for all other b_i parameters,

$$b_2 = b_1 + (x_2 - x_1)(c_2 + c_1) = 0.848 + (2 - 1)(0.755 + 0) = 1.603$$
$$b_3 = b_2 + (x_3 - x_2)(c_3 + c_2) = 1.603 + (5 - 2)(-1.78 + 0.755) = -1.472$$
$$b_4 = b_3 + (x_4 - x_3)(c_4 + c_3) = -1.472 + (6 - 5)\big(0.773 + (-1.78)\big) = -2.479$$
$$b_5 = b_4 + (x_5 - x_4)(c_5 + c_4) = -2.479 + (10 - 6)(0.181 + 0.773) = 1.337.$$

Lastly, using Equation (3.50) repeatedly for all the d_i parameters,

$$d_1 = \frac{c_2 - c_1}{3(x_2 - x_1)} = \frac{0.755 - 0}{3(2 - 1)} = 0.252$$
$$d_2 = \frac{c_3 - c_2}{3(x_3 - x_2)} = \frac{-1.78 - 0.755}{3(5 - 2)} = -0.282$$
$$d_3 = \frac{c_4 - c_3}{3(x_4 - x_3)} = \frac{0.773 - (-1.78)}{3(6 - 5)} = 0.851$$
$$d_4 = \frac{c_5 - c_4}{3(x_5 - x_4)} = \frac{0.181 - 0.773}{3(10 - 6)} = -0.049$$

$$d_5 = \frac{c_6 - c_5}{3(x_6 - x_5)} = \frac{0 - 0.181}{3(13 - 10)} = -0.02.$$

Each cubic spline $s_i(x)$ for $i = 1, \ldots, 5$ is constructed using Equation (3.45) and the corresponding a_i, b_i, c_i, and d_i parameters calculated above. Using Equation (3.58) with every such cubic spline and including the domain of each

$$p_3(x) = \begin{cases} 1.8 + 0.848(x-1) + 0.252(x-1)^3, & 1 \leq x < 2 \\ 2.9 + 1.603(x-2) + 0.755(x-2)^2 - 0.282(x-2)^3, & 2 \leq x < 5 \\ 6.9 - 1.472(x-5) - 1.78(x-5)^2 + 0.851(x-5)^3, & 5 \leq x < 6 \\ 4.5 - 2.479(x-6) + 0.773(x-6)^2 - 0.049(x-6)^3, & 6 \leq x < 10 \\ 3.8 + 1.337(x-10) + 0.181(x-10)^2 - 0.02(x-10)^3, & 10 \leq x \leq 13 \end{cases}$$

Because $2 \leq 4 < 5$, the second spline $s_2(x)$ is used to estimate $p_3(4)$.

$$p_3(4) = s_2(4) = 2.9 + 1.603(4-2) + 0.755(4-2)^2 - 0.282(4-2)^3 = 6.87$$

Since $6 \leq 7 < 10$, the fourth spline $s_4(x)$ is used to estimate $p_3(7)$.

$$p_3(7) = s_4(7) = 4.5 - 2.479(7-6) + 0.773(7-6)^2 - 0.049(7-6)^3 = 2.745$$

Figure 3.9 displays the cubic splines piecewise function $p_3(x)$ of Example 3.8 with the fifth-degree DD polynomial function $f_5(x)$ of Figure 3.3. Figure 3.10 overlays the linear $p_1(x)$, quadratic $p_2(x)$, and cubic $p_3(x)$ splines piecewise functions and the fifth-degree DD polynomial function.

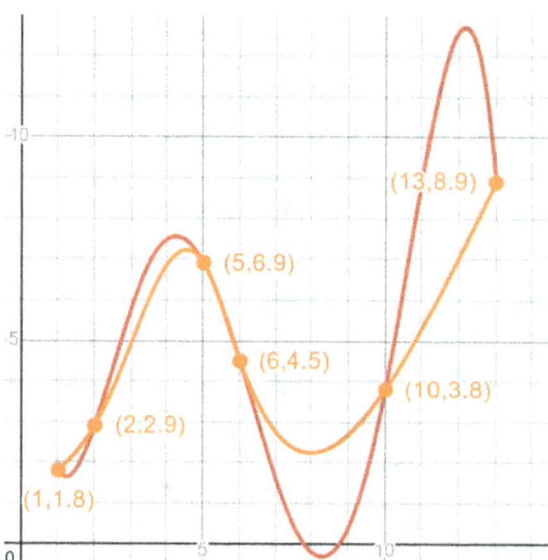

Figure 3.9: The cubic splines piecewise function $p_3(x)$ of Example 3.8 in orange with the fifth-degree DD polynomial function of Figure 3.3 in red. Both use the data of Table 3.4.

Comparing the three splines piecewise functions by intervals, on $1 \leq x < 2$ the linear, quadratic, and cubic splines are nearly identical. On both $5 \leq x < 6$ and $10 \leq x \leq 13$, the cubic and linear splines are graphically close, while the quadratic and linear splines are graphically close on $2 \leq x < 5$. The linear, quadratic, and cubic splines are most distinct on $6 \leq x < 10$.

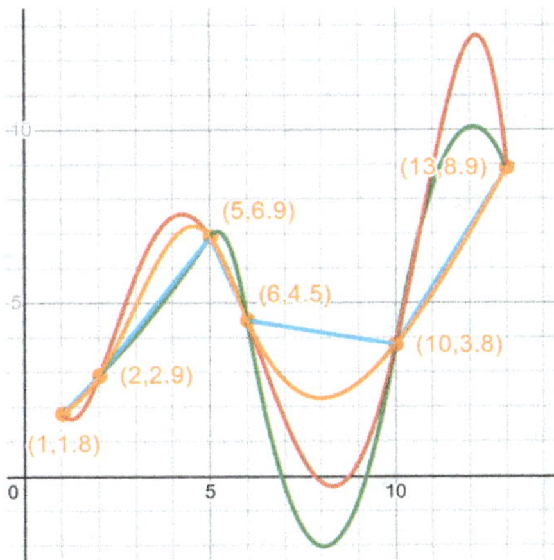

Figure 3.10: A comparison of all three splines piecewise functions, orange cubic $p_3(x)$, green quadratic $p_2(x)$, and blue linear $p_1(x)$, along with the red fifth-degree DD polynomial function $f_5(x)$. All four use the data of Table 3.4.

Examining the three splines piecewise functions in their entirety, the cubic and linear splines piecewise functions are similar with respect to their extreme values, none of which stray far from the extreme values of the known data; the extreme values of the quadratic splines piecewise function vary much more from the extreme values of the known data. However, $p_3(x)$ is similar to $p_2(x)$ because it is smooth and differentiable everywhere, unlike $p_1(x)$.

Like a quadratic splines piecewise function, extrema for a cubic splines piecewise function are found directly by setting the derivative of each spline $s_i'(x) = 0$ and solving for the x value(s). However, the derivatives $s_i'(x)$ of cubic splines are second-degree polynomials, likely requiring the quadratic formula that probably produces two distinct solutions. Only real solutions in the restricted interval on which the corresponding spline $s_i(x)$ is defined are valid; solutions outside the restricted intervals, and complex solutions, are discarded.

Determining the type of extrema algebraically is a second difference between quadratic and cubic splines piecewise functions. Recall the derivative for a quadratic spline is linear, so the sign of its slope easily determines a maximum or minimum. For cubic splines, the second derivative test from calculus is a convenient tool for making the distinction. If $x = k$ is a zero of $s_i'(x)$ and $s_i''(k) < 0$ a maximum occurs at $x = k$, and if $s_i''(k) > 0$ a minimum occurs at $x = k$. Note the second derivatives $s_i''(x)$ are always readily available and easily calculated because they are derived from the second-degree polynomials $s_i'(x)$.

Example 3.9. Calculate the derivative $p_3'(x)$ of the cubic splines piecewise function of Example 3.8. Analytically use $p_3'(x)$ to determine all extrema of $p_3(x)$, other than the endpoints. Analytically use $p_3''(x)$ to identify the type of all extrema.

Solution. Differentiating the first spline

$$s_1(x) = 1.8 + 0.848(x - 1) + 0.252(x - 1)^3$$

$$s_1{}'(x) = 0.848 + 0.756(x - 1)^2$$

$$= 0.848 + 0.756\left(x^2 - 2x + 1\right)$$

$$= 0.756x^2 - 1.512x + 1.604$$

The process for the second through fifth splines is slightly different because each contains an additional quadratic term. For the second spline,

$$s_2(x) = 2.9 + 1.603(x - 2) + 0.755(x - 2)^2 - 0.282(x - 2)^3$$

$$s_2{}'(x) = 1.603 + 1.51(x - 2) - 0.846\left(x^2 - 4x + 4\right)$$

$$= 1.603 + 1.51x - 3.02 - 0.846x^2 + 3.384x - 3.384$$

$$= -0.846x^2 + 4.894x - 4.801$$

Repeating this process for the third through fifth splines and writing the cubic splines piecewise function derivative in its entirety, including the domain of each spline

$$p_3{}'(x) = \begin{cases} 0.756x^2 - 1.512x + 1.604, & 1 \leq x < 2 \\ -0.846x^2 + 4.894x - 4.801, & 2 \leq x < 5 \\ 2.553x^2 - 29.09x + 80.153, & 5 \leq x < 6 \\ -0.147x^2 + 3.31x - 17.047, & 6 \leq x < 10 \\ -0.06x^2 + 1.562x - 8.283, & 10 \leq x \leq 13 \end{cases}$$

The quadratic formula is used for each spline to locate extrema, provided the resulting x values are real numbers and lie in the interval for which the spline is defined. For $s_1{}'(x)$,

$$x = \frac{1.512 \pm \sqrt{(-1.512)^2 - 4(0.756)(1.604)}}{2(0.756)} = \frac{1.512 \pm \sqrt{-2.564}}{1.512}$$

Since the discriminant -2.564 is negative, no real zeros exist and thus no extrema occur on the first interval.

For $s_2{}'(x)$,

$$x = \frac{-4.894 \pm \sqrt{(4.894)^2 - 4(-0.846)(-4.801)}}{2(-0.846)} = 1.252, 4.533.$$

Only the second x value lies in the interval $2 \leq x < 5$ on which $s_2(x)$ is defined, so $x = 1.252$ is discarded. Using the second derivative test for $x = 4.533$,

$$s_2{}''(x) = -1.692x + 4.894$$

$$s_2{}''(4.533) = -1.692(4.533) + 4.894 = -2.776 < 0.$$

This proves $x = 4.533$ is the location of a maximum.

For $s_3{}'(x)$,

$$x = \frac{29.09 \pm \sqrt{(-29.09)^2 - 4(2.553)(80.153)}}{2(2.553)} = 6.728, 4.666.$$

Neither of these x values lies in the interval $5 \le x < 6$ on which $s_3(x)$ is defined and therefore neither is a candidate for an extremum.

For $s_4'(x)$,

$$x = \frac{-3.31 \pm \sqrt{(3.31)^2 - 4(-0.147)(-17.047)}}{2(-0.147)} = 7.974, 14.543.$$

Only the first x value lies in the interval $6 \le x < 10$ on which $s_4(x)$ is defined, eliminating $x = 14.543$ as a potential extremum. Again, using the second derivative test, this time for $x = 7.974$,

$$s_4''(x) = -0.294x + 3.31$$

$$s_4''(7.974) = -0.294(7.974) + 3.31 = 0.966 > 0.$$

This proves a minimum occurs at $x = 7.974$.

Lastly, for $s_5'(x)$,

$$x = \frac{-1.562 \pm \sqrt{(1.562)^2 - 4(-0.06)(-8.283)}}{2(-0.06)} = 7.415, 18.619.$$

Neither of these x values lies in the interval $10 \le x \le 13$ on which $s_5(x)$ is defined and thus are not possible extrema.

The graph of $p_3'(x)$ once more reinforces the calculus concept of extrema located at x-intercepts and their type by whether, and in which direction, the sign of $s_i'(x)$ changes around the x-intercept. Furthermore, such a graph eliminates the additional work required calculating zeros of quadratic functions; those splines whose derivative do not cross the x-axis need not be considered.

Figure 3.11 shows $p_3(x)$ and its solid splines $s_i(x)$ of Example 3.8, along with its derivative $p_3'(x)$ and its dashed splines $s_i'(x)$ of Example 3.9, correspondingly colored. Only the blue second and orange fourth splines contain extrema since they are the only ones whose derivatives, $s_2'(x)$ and $s_4'(x)$, cross the x-axis. Each type is easily identified

Figure 3.11: The cubic splines piecewise function $p_3(x)$ of Example 3.8 (solid) and its derivative $p_3'(x)$ of Example 3.9 (dashed). Each color represents a particular spline. The x-intercepts of the spline derivatives $s_2'(x)$ in blue and $s_4'(x)$ in orange identify extrema, and the direction of the change in sign of their derivatives around their x-intercepts indicates a maximum and minimum, respectively. $p_3'(x)$ is smooth and differentiable everywhere.

also, a maximum at $x = 4.533$ because $s_2'(x)$ changes sign from positive to negative around $x = 4.533$, and a minimum at $x = 7.974$ because $s_4'(x)$ changes sign from negative to positive around $x = 7.974$. By the definition of its construction, the derivative $p_3'(x)$ is continuous, smooth and differentiable everywhere, including at the knots.

3.4 Inverse interpolation

Given a set of points ordered $(x, f(x))$, inverse interpolation involves approximating the x value given an $f(x)$ value, rather than estimating an $f(x)$ value given the x value. An easy solution is simply fitting a model to the inverse set of points $(f(x), x)$. However, there are mathematical restrictions and inconsistencies that make this process unsound, and likely will not produce a reliable inverse *function* to the original set of points.

Implicit is the assumption $f(x)$ represents a function not only mathematically but practically, in which $f(x)$ values are truly dependent on independent variables x. If the coordinate pairs are merely related and one is not dependent on the other definitively, switching coordinates could be a viable solution. This section details a procedure in the absence of such a solution.

3.4.1 Problems inherent using the set of inverse points

Consider the cubic splines piecewise function $p_3(x)$ of Figure 3.9. The overall function $p_3(x)$ and even some of the individual cubic splines like $s_4(x)$ do not pass the graphical horizontal line test. In the case of $s_4(x)$, the horizontal line $y = 3$ intersects $s_4(x)$ in more than one point, meaning $s_4(x)$ is not a one-to-one spline function and therefore no inverse function for this spline, $s_4^{-1}(x)$, is possible. In turn, no inverse function $p_3^{-1}(x)$ is possible either.

Second, an algorithmic issue arises using an inverse set a points in which the original function values $f(x)$ are not strictly increasing or decreasing, in which case, the points must be reordered according to the $f(x)$ values. This is precisely the case for the points of Table 3.4. When the coordinates are initially switched in Table 3.7, the third and fifth points must be interchanged so the $f(x)$ values are sequential, as in Table 3.8.

Table 3.7: The inverse set of points of Table 3.4. The third and fifth points are out of sequence now based on $f(x)$ values.

$f(x)$	1.8	2.9	6.9	4.5	3.8	8.9
x	1	2	5	6	10	13

Table 3.8: The inverse set of points of Table 3.7, reordered increasingly by $f(x)$ values.

$f(x)$	1.8	2.9	3.8	4.5	6.9	8.9
x	1	2	10	6	5	13

Nevertheless, a cubic splines piecewise function using the inverse set of points always can be generated algebraically. However, such a function is a third-degree polynomial which certainly has no guarantee it is one-to-one. If it is not one-to-one, it has no inverse, meaning it does not model the original set of points, which does represent a function.

The overlays of the graphs of two cubic splines piecewise functions, one based on the set of points $(x_i, f(x_i))$ and one on the inverse set of points $(f(x_i), x_i)$ can provide perhaps the clearest understanding of the non-inverse relationship between the two functions.

The cubic splines piecewise function $g_3(x)$ using the inverse set of points of Table 3.8 is

$$g_3(x) = \begin{cases} 1 - 2.725(x - 1.8) + 3.003(x - 1.8)^3, & 1.8 \le x < 2.9 \\ 2 + 8.176(x - 2.9) + 9.91(x - 2.9)^2 - 10.131(x - 2.9)^3, & 2.9 \le x < 3.8 \\ 10 + 1.396(x - 3.8) - 17.444(x - 3.8)^2 + 10.41(x - 3.8)^3, & 3.8 \le x < 4.5 \\ 6 - 7.724(x - 4.5) + 4.416(x - 4.5)^2 - 0.572(x - 4.5)^3, & 4.5 \le x < 6.9 \\ 5 + 3.597(x - 6.9) + 0.301(x - 6.9)^2 - 0.05(x - 6.9)^3, & 6.9 \le x \le 8.9 \end{cases} \quad (3.59)$$

True inverse functions are graphical reflections of each other in their entirety over the diagonal line $y = x$. Figure 3.12 compares the cubic splines piecewise function $p_3(x)$ of Figure 3.9, based on the original set of points of Table 3.4 in orange, and the cubic splines piecewise function $g_3(x)$ based on the inverse set of points of Table 3.8 in purple.

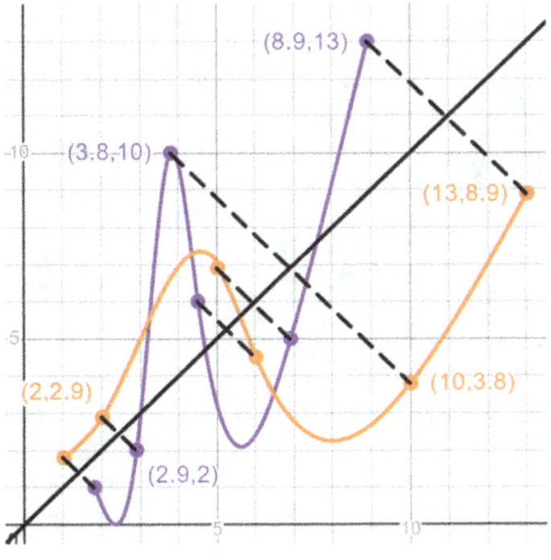

Figure 3.12: The cubic splines piecewise function using the points of Table 3.4 in orange and the one using the inverse set of points of Table 3.8 in purple are not complete reflections of each other over the black line $y = x$. They are not inverse functions of each other.

While each pair of points $(x_i, f(x_i))$ and $(f(x_i), x_i)$ for $i = 1, 2, \ldots, 6$ is reflected over the black line $y = x$, indicated by connected dashed line segments, clearly the entire orange and purple graphs are not reflections of each other over $y = x$.

3.4.2 Different yet familiar techniques

A mathematically sounder approach simply uses the original set of points and one of the techniques described in this chapter, as before, to fit the known data, i. e., a DD polynomial function $f_n(x)$, a Lagrange polynomial function $\mathcal{L}_n(x)$, or one of the three splines piecewise functions $p_1(x)$, $p_2(x)$, or $p_3(x)$. The function $f_n(x)$ or $\mathcal{L}_n(x)$, or the particular spline $s_i(x)$ of $p_n(x)$ is then set equal to the given $f(x)$ value and subtracted so that one side of the equation is zero. The problem then becomes one of finding roots, outlined in Chapter 1.

Example 3.10. Estimate the x values for the unknown function value $f(x) = 4$ on the cubic splines piecewise function $p_3(x)$ of Example 3.8 and Figure 3.9. Compare these x values to ones estimated by the cubic splines piecewise function of Equation (3.59), constructed from the inverse data of Table 3.8.

Solution. Using $p_3(x)$ of Figure 3.9 and overlaying the graph of $f(x) = 4$ produces Figure 3.13, indicating $p_3(x)$ and $f(x)$ intersect at three points, one each on the second, fourth, and fifth splines $s_2(x)$, $s_4(x)$, and $s_5(x)$, respectively.

Solving for the x value for the point on the second spline $s_2(x)$ of Example 3.8

$$s_2(x) = f(x)$$

$$2.9 + 1.603(x - 2) + 0.755(x - 2)^2 - 0.282(x - 2)^3 = 4$$

$$-1.1 + 1.603(x - 2) + 0.755(x - 2)^2 - 0.282(x - 2)^3 = 0.$$

Define the left side as

$$h_2(x) = -1.1 + 1.603(x - 2) + 0.755(x - 2)^2 - 0.282(x - 2)^3.$$

The problem now becomes one of estimating the root of $h_2(x) = 0$. The Newton–Raphson algorithm of Section 1.4 is a good choice for finding the zero, since the derivative $h_2{'}(x)$ is easily calculated.

$$h_2{'}(x) = 1.603 + 1.51(x - 2) - 0.846(x - 2)^2$$

Figure 3.13 suggests $x_1 = 2.5$ is a reasonable first estimation of the zero to start the Newton–Raphson algorithm of Equation (1.7).

$$x_2 = x_1 - \frac{h_2(x_1)}{h_2{'}(x_1)} = 2.5 - \frac{h_2(2.5)}{h_2{'}(2.5)} = 2.5 - \frac{-0.145}{2.147} = 2.568$$

$$x_3 = x_2 - \frac{h_2(x_2)}{h_2{'}(x_2)} = 2.568 - \frac{h_2(2.568)}{h_2{'}(2.568)} = 2.568 - \frac{0.001}{2.187} = 2.567$$

$$x_4 = x_3 - \frac{h_2(x_3)}{h_2{'}(x_3)} = 2.567 - \frac{h_2(2.567)}{h_2{'}(2.567)} = 2.567 - \frac{0}{2.187} = 2.567$$

Repeating the process for estimating the x value for the point on the fourth spline $s_4(x)$ of Example 3.8

$$s_4(x) = f(x)$$

$$4.5 - 2.479(x - 6) + 0.773(x - 6)^2 - 0.049(x - 6)^3 = 4$$

$$0.5 - 2.479(x - 6) + 0.773(x - 6)^2 - 0.049(x - 6)^3 = 0.$$

The left side is defined

$$h_4(x) = 0.5 - 2.479(x - 6) + 0.773(x - 6)^2 - 0.049(x - 6)^3.$$

The derivative of $h_4(x)$ is

$$h_4{}'(x) = -2.479 + 1.546(x - 6) - 0.147(x - 6)^2.$$

A reasonable first estimation for this zero using Figure 3.13 is $x_1 = 6.5$, and again using Equation (1.7),

$$x_2 = 6.5 - \frac{h_4(6.5)}{h_4{}'(6.5)} = 6.5 - \frac{-0.552}{-1.743} = 6.183$$

$$x_3 = 6.183 - \frac{h_4(6.183)}{h_4{}'(6.183)} = 6.183 - \frac{0.072}{-2.201} = 6.216$$

$$x_4 = 6.216 - \frac{h_4(6.216)}{h_4{}'(6.216)} = 6.216 - \frac{0.001}{-2.152} = 6.216.$$

Solving for the last x value for the point on the fifth spline $s_5(x)$ of Example 3.8,

$$s_5(x) = f(x)$$

$$3.8 + 1.337(x - 10) + 0.181(x - 10)^2 - 0.02(x - 10)^3 = 4$$

$$-0.2 + 1.337(x - 10) + 0.181(x - 10)^2 - 0.02(x - 10)^3 = 0.$$

Defining the left side as $h_5(x)$ and deriving $h_5{}'(x)$,

$$h_5(x) = -0.2 + 1.337(x - 10) + 0.181(x - 10)^2 - 0.02(x - 10)^3$$

$$h_5{}'(x) = 1.337 + 0.362(x - 10) - 0.06(x - 10)^2.$$

Using Figure 3.13 to obtain the reasonable estimation of $x_1 = 10.5$ and using Equation (1.7),

$$x_2 = 10.5 - \frac{h_5(10.5)}{h_5{}'(10.5)} = 10.5 - \frac{0.511}{1.503} = 10.16$$

$$x_3 = 10.16 - \frac{h_5(10.16)}{h_5{}'(10.16)} = 10.16 - \frac{0.018}{1.393} = 10.147$$

$$x_4 = 10.147 - \frac{h_5(10.147)}{h_5{}'(10.147)} = 10.147 - \frac{0}{1.389} = 10.147.$$

Summarizing, the three estimations for $f(x) = 4$ using $p_3(x)$ of Example 3.8 are $x = 2.567, 6.216$, and 10.147.

The cubic splines piecewise function of Equation (3.59) constructed from the inverse data set of Table 3.8 only produces a single estimation for $f(x) = 4$, namely, $g_3(4) = s_3(4) = 9.665$. This x estimate is not particularly close to any of the three estimations using the cubic splines piecewise function $p_3(x)$.

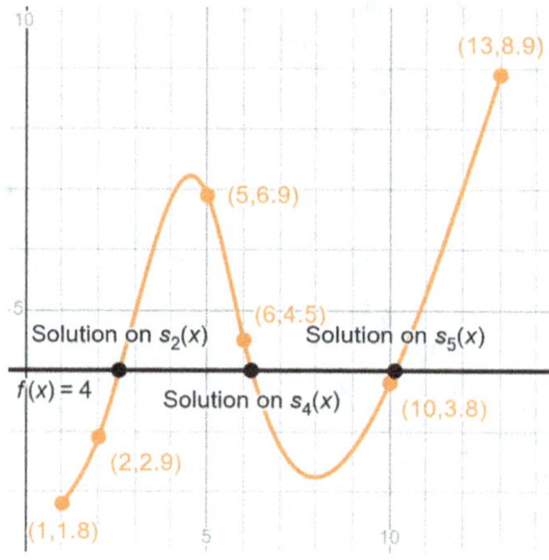

Figure 3.13: The cubic splines piecewise function $p_3(x)$ of Example 3.8 intersects the line $f(x) = 4$ at three points, one on each of the second, fourth, and fifth splines.

Interpolation concepts summary

1. Interpolation fits a function exactly through each point in a set.
2. The coefficients of an nth-degree divided difference (DD) polynomial function $f_n(x)$ are calculated recursively using DD terms up to $f[x_n, \ldots, x_0]$, inclusive.
3. $f_n(x)$ uses an additional point and the $(n + 1)$th DD term for a truncation error estimate E_t.
4. Underlying data generated by an nth-degree polynomial function produces the equivalent $f_n(x)$.
5. The Lagrange polynomial function $\mathcal{L}_n(x)$ is equivalent to $f_n(x)$.
6. Coefficients of $\mathcal{L}_n(x)$ are calculated directly from the given points, not recursively like $f_n(x)$.
7. High-degree polynomial functions fit exactly through a set of points tend to oscillate excessively and overestimate underlying extrema.
8. A splines piecewise function $p_n(x)$ avoids using a single high-degree polynomial function on a set of many points; each spline between consecutive points can be any small nth-degree polynomial function.
9. A linear splines piecewise function $p_1(x)$ never produces extrema outside the range of the given points.
10. $p_1(x)$ is not smooth or differentiable at the given points.
11. A quadratic splines piecewise function $p_2(x)$ is smooth and differentiable at the given points, but its derivative $p_2'(x)$ is neither.
12. The first spline $s_1(x)$ of $p_2(x)$ is a linear, not quadratic, function.
13. A cubic splines piecewise function $p_3(x)$ and its derivative $p_3'(x)$ are smooth and differentiable on the entire domain of the given points.
14. $p_3(x)$ is a good compromise of using spline polynomials functions which are neither too small nor too large a degree.
15. Solving the derivatives of quadratic and cubic splines $s_i'(x) = 0$ locate extrema on $p_2(x)$ and $p_3(x)$, provided solutions lie within the domain of the corresponding ith spline.
16. Inverse interpolation initially follows the same procedure of fitting a function $f(x)$ through a set of points, then recasting $f(x) = k$ as $f(x) - k = 0$ and using a zero-finding algorithm.
17. If the underlying data does not represent a one-to-one function, switching the coordinates of the given points will not produce the non-existent inverse function for inverse interpolation.
18. Switching the coordinates of the points potentially causes reordering issues for inverse interpolation.

Exercises

Section 3.1

Construct an nth-degree DD polynomial function $f_n(x)$ through the $n + 1$ given points $(x, f(x))$. Round each DD coefficient to the thousandths place. Graph $f_n(x)$ and label each given point using a computer application.

1. $(2, 3)$ $\quad$ $(4, 6)$ $\quad$ $(7, 15)$ $\quad$ $(8, 21)$
2. $(1, 2)$ $\quad$ $(2, 5)$ $\quad$ $(4, 14)$ $\quad$ $(9, 61)$
3. $(0, 4.5)$ $\quad$ $(2.5, 8)$ $\quad$ $(6.25, 14.75)$ $\quad$ $(12.5, 5)$
4. $(0, 1)$ $\quad$ $(3, 7)$ $\quad$ $(5, 15)$ $\quad$ $(10, 21)$
5. $(-3, -2)$ $\quad$ $(-1, 1.4)$ $\quad$ $(7, 15)$ $\quad$ $(13, -8.4)$
6. $(-5, 3)$ $\quad$ $(-2, 4.5)$ $\quad$ $(8, 9.5)$ $\quad$ $(13, -1.5)$
7. $(-1, 4.7)$ $\quad$ $(0, 2.6)$ $\quad$ $(2.5, -10)$ $\quad$ $(3.5, 11)$ $\quad$ $(4, 35.3)$
8. $(0, -3)$ $\quad$ $(1.5, 1.2)$ $\quad$ $(2.5, 3.3)$ $\quad$ $(4, 5.1)$ $\quad$ $(5, 7.5)$
9. $(1, 5)$ $\quad$ $(1.5, 6.7)$ $\quad$ $(3, 7.3)$ $\quad$ $(4, 8)$ $\quad$ $(6, 8.5)$
10. $(-2, -1.2)$ $\quad$ $(-1, 3.4)$ $\quad$ $(1.5, 5.8)$ $\quad$ $(2, 6.1)$ $\quad$ $(3, 7)$
11. $(0, 0.5)$ $\quad$ $(1.5, 4.4)$ $\quad$ $(2, 4.9)$ $\quad$ $(3, 4.7)$ $\quad$ $(3.5, 6.1)$
12. $(1, 2.8)$ $\quad$ $(1.4, 2)$ $\quad$ $(2.2, -2)$ $\quad$ $(3, -9.2)$ $\quad$ $(3.2, -1.6)$
13. $(2, 4.8)$ $\quad$ $(2.4, 3.8)$ $\quad$ $(2.6, 3.3)$ $\quad$ $(3, 2.3)$ $\quad$ $(3.2, 0)$ $\quad$ $(3.4, 1.3)$
14. $(0, 7.5)$ $\quad$ $(1.5, 10.8)$ $\quad$ $(3.5, 15.2)$ $\quad$ $(4, 16.3)$ $\quad$ $(5, 8)$ $\quad$ $(5.5, 6.4)$
15. $(-2, 6.2)$ $\quad$ $(-1.5, 6.2)$ $\quad$ $(0.5, 15.7)$ $\quad$ $(1, 16.7)$ $\quad$ $(1.5, 15.65)$ $\quad$ $(2, 5.5)$
16. $(-4, 8.1)$ $\quad$ $(-3, 8.1)$ $\quad$ $(-1.5, 2.7)$ $\quad$ $(1, 1.8)$ $\quad$ $(2, 12.15)$ $\quad$ $(4, 1.38)$

Construct an nth-degree DD polynomial function $f_n(x)$ through the first $n+1$ given points $(x, f(x))$. Round each DD coefficient to the thousandths place. Estimate $f(a)$. Estimate E_t and $E_{t,r}$ for $x = a$ using $(x_t, f(x_t))$. Write the interval in which the true value of $f(a)$ lies.

17. $n = 3$ $\quad$ $(1.5, 2.3)$ $\quad$ $(2, 0.7)$ $\quad$ $(3, 1.4)$ $\quad$ $(3.5, 2.56)$
    $(x_t, f(x_t)) =$ $\quad$ $(5.5, 1.26)$ $\quad$ $a = 2.5$
18. $n = 3$ $\quad$ $(1, 6)$ $\quad$ $(3, 2.5)$ $\quad$ $(8, 5.3)$ $\quad$ $(11, 3.86)$
    $(x_t, f(x_t)) =$ $\quad$ $(12, 6.352)$ $\quad$ $a = 9$
19. $n = 3$ $\quad$ $(-10, -15.9)$ $\quad$ $(-4.5, -0.83)$ $\quad$ $(2, -9.15)$ $\quad$ $(4.5, 2.95)$
    $(x_t, f(x_t)) =$ $\quad$ $(10, 2.07)$ $\quad$ $a = 3.5$
20. $n = 3$ $\quad$ $(-1, -6)$ $\quad$ $(0.5, -4.8)$ $\quad$ $(2, -6.75)$ $\quad$ $(4, 5)$
    $(x_t, f(x_t)) =$ $\quad$ $(7, 2.48)$ $\quad$ $a = 0$
21. $n = 4$ $\quad$ $(1, -0.5)$ $\quad$ $(3, -1.2)$ $\quad$ $(7, 6.4)$ $\quad$ $(8, 13.5)$
    $(11, 13.2)$ $\quad$ $(x_t, f(x_t)) =$ $\quad$ $(12, -13.26)$ $\quad$ $a = 2$
22. $n = 4$ $\quad$ $(2, 4.5)$ $\quad$ $(3, 3.2)$ $\quad$ $(4.5, 6.8)$ $\quad$ $(5, 10.2)$
    $(8, 3.3)$ $\quad$ $(x_t, f(x_t)) =$ $\quad$ $(10, 6.14)$ $\quad$ $a = 7.5$
23. $n = 5$ $\quad$ $(-3, 7)$ $\quad$ $(0, 10.6)$ $\quad$ $(2, 8)$ $\quad$ $(3, 11.5)$
    $(3.5, 19.805)$ $\quad$ $(5, 57.8)$ $\quad$ $(x_t, f(x_t)) =$ $\quad$ $(7, 19.7)$ $\quad$ $a = 4$
24. $n = 5$ $\quad$ $(0, 0.25)$ $\quad$ $(2, 1.15)$ $\quad$ $(4.5, 0.25)$ $\quad$ $(5, 6.4)$
    $(7, 2.35)$ $\quad$ $(8, 9.49)$ $\quad$ $(x_t, f(x_t)) =$ $\quad$ $(10, 36.55)$ $\quad$ $a = 6$

Construct an nth-degree DD polynomial function $f_n(x)$ through the given points. Round each DD coefficient to the thousandths place. Justify $f_n(x)$ fits all points exactly.

25. $(-6, 12.2)$ $(-2.5, 1.7)$ $(3.75, -7.3)$ $(4.25, -7.48)$ $(12.5, 1.1)$
26. $(1.5, 14.7)$ $(2, 10)$ $(3.5, 6.25)$ $(6, 34.5)$ $(7.5, 72.15)$
27. $(0, 13)$ $(4.1, 25.3)$ $(6.6, 25.87)$ $(8.6, 22.546)$ $(12, 9.184)$
28. $(-2, -8.2)$ $(1, -2.77)$ $(5.5, 3.35)$ $(8.5, 6.08)$ $(10, 7.04)$
29. $(2.5, -3.7)$ $(3.8, -7.47)$ $(5, 0)$
     $(5.5, 9.479)$ $(7, 72.674)$ $(8, 153.369)$
30. $(1, 2.2)$ $(2.25, -3.1)$ $(3, -4)$
     $(4, 1.135)$ $(6, 48.875)$ $(7, 99.88)$
31. $(0.5, 1.8)$ $(0.9, 2.62)$ $(1.4, 4.95)$ $(2, 8.076)$
     $(2.5, 6.6)$ $(2.9, -1.734)$ $(3, -5.36)$
32. $(-1, -12.1)$ $(0.5, -4.3)$ $(1.5, 0)$ $(2, 0.8)$
     $(3, -3.9)$ $(3.5, -11.38)$ $(4, -23.9)$

Section 3.2

Construct a second-degree Lagrange polynomial $\mathcal{L}_2(x)$ through the given points $(x, f(x))$. Round each decimal value to the thousandths place.

33. $(0.5, 3.6)$ $(3.5, -2.1)$ $(10.5, 5.6)$
34. $(1, 6.3)$ $(3, 2.4)$ $(5.5, 7.2)$
35. $(-2, -9.6)$ $(1.2, 28.8)$ $(6, -86.4)$
36. $(-5, -2.7)$ $(0, 10.5)$ $(17.5, 12.6)$
37. $(0, 16.5)$ $(2.5, -12.3)$ $(6.6, 0)$
38. $(-1, 63)$ $(2, 78)$ $(7, 303)$
39. $(1, 12.6)$ $(6, -1.4)$ $(10, -27)$
40. $(2.5, 1.2)$ $(5.7, 0)$ $(10, -12.9)$

Sections 3.1 and 3.2

Construct a second-degree Lagrange polynomial $\mathcal{L}_2(x)$ through the given points $(x, f(x))$. Round each decimal value to the thousandths place. Verify each $\mathcal{L}_2(x)$ is equivalent to a second-degree DD polynomial function $f_2(x)$.

41. $(2, 32)$ $(6, 3)$ $(12, 25.5)$
42. $(5, 12)$ $(15, 1)$ $(20, 33)$
43. $(0, 4.2)$ $(2.5, 13.2)$ $(10, 16.2)$
44. $(1.5, 57)$ $(3, 78)$ $(9.5, 39)$
45. $(1, -12.6)$ $(3.8, -12.6)$ $(11, 5.4)$
46. $(0.5, 11.7)$ $(9.5, -37.8)$ $(13, 0)$
47. $(-1, -31)$ $(4, 0)$ $(11.5, -96)$
48. $(2, 16.8)$ $(3.2, 16.8)$ $(6, 0)$

Section 3.3.1

Construct a linear splines piecewise function $p_1(x)$ through the given points $(x, f(x))$. Use $p_1(x)$ to estimate the two unknown x values. Round each decimal value to the thousandths place.

49.

x	$f(x)$
0	3
5	4.3
8	6.7
11	10.6
16	15.9
4	
10	

50.

x	$f(x)$
1	4.1
3	7.7
4	8.2
9	8.5
13	14.9
2	
7	

51.

x	$f(x)$
1.1	22.7
3.6	21.4
5.4	13.3
7.3	9.5
10.9	3.2
5.1	
8.2	

52.

x	$f(x)$
0	21.3
2.5	15.8
4.6	9.5
8.6	6
12.2	5.1
3.4	
9.2	

53.

x	$f(x)$
10.1	−0.95
10.2	−0.72
10.6	−0.72
10.8	0.97
11.1	1.24
10.5	
10.9	

54.

x	$f(x)$
0.2	12.7
0.5	13.3
1.1	13.3
1.3	14.8
1.7	15.4
0.6	
1.6	

55.

x	$f(x)$
5	1.64
10	−0.9
25	−1.89
35	−1.89
50	−2.1
8	
29	

56.

x	$f(x)$
1.2	153
1.8	144.9
2.8	107.6
3.2	107.6
4.2	88.7
1.5	
3.1	

57.

x	$f(x)$
0.5	4.2
1.9	6.3
2.6	2.1
3.7	−2.3
5.1	−1.6
5.9	0
2.4	
5.7	

58.

x	$f(x)$
4.1	11.8
4.4	12.7
4.6	15.1
5	9.7
5.4	6.3
5.8	4.2
4.5	
5.5	

59.

x	$f(x)$
−2	162
−1.5	154.8
0	160.5
2.5	185.7
3.5	185.7
5	152.1
8.5	157
−1.3	
4.2	

60.

x	$f(x)$
−8	−6.4
0	−7.8
6	−10.2
10	−10.2
12	−7.3
18	−5.2
28	−6.1
3.2	
13.4	

Section 3.3.2

Construct a quadratic splines piecewise function $p_2(x)$ through the given points $(x, f(x))$. Use $p_2(x)$ to estimate the two unknown x values. Round each decimal value to the thousandths place.

61.			62.			63.			64.	
x	$f(x)$		x	$f(x)$		x	$f(x)$		x	$f(x)$
0.5	3.2		5	4.2		4	29.7		0	9.7
1	5.6		6.5	8.7		8	8.4		2.4	6.1
4.5	7.7		7	9.1		24	12.8		8.6	4.488
7	10.4		9.5	12.5		32	7		14.1	5.401
2.5			6.8			12			7	
5			7.5			28.2			10.1	

65.			66.			67.			68.	
x	$f(x)$		x	$f(x)$		x	$f(x)$		x	$f(x)$
5	−3.9		2	−2.8		1	0.9		7.4	−5.1
5.2	−6		12	−5.7		9	16.3		7.9	1.8
5.6	−0.8		15	0.63		14	20.7		8.2	2.7
5.9	18.7		20	28.18		26	14.4		8.6	−0.8
5.1			4			13			8.1	
5.7			16			18.5			8.4	

69.			70.			71.			72.	
x	$f(x)$		x	$f(x)$		x	$f(x)$		x	$f(x)$
−10	13.5		1	3.4		−3	3.2		0	5.1
−6	−2.7		6	5.9		−1	6.3		3	−1.2
0	5.4		8	4.1		4	0		7	6.8
2	1.8		14	11.9		6	−2.7		9	10.4
8	−0.9		24	8.8		12	−0.6		16	12.5
1			7.3			−2.4			5.2	
3.8			20			10			9.9	

Construct a quadratic splines piecewise function $p_2(x)$ through the given points $(x, f(x))$. Round each decimal value to the thousandths place. Analytically calculate and use $p_2'(x)$ to identify the type of all extrema of $p_2(x)$, other than the endpoints.

73.			74.			75.			76.	
x	$f(x)$		x	$f(x)$		x	$f(x)$		x	$f(x)$
2	9.6		1.5	43.2		1.3	17		3.4	1.5
3.5	0.6		4	13.2		3.3	44		6.4	34.5
4.5	2.6		7	4.2		10.8	4.625		8.9	30.75
7	21.35		8	7.7		20.8	64.625		10.9	4.75
11	9.35		8.5	5.7		21.3	83.75		12.4	0.25

77.			78.			79.			80.		
x	$f(x)$		x	$f(x)$		x	$f(x)$		x	$f(x)$	
10	−12		5	−9		1	22		6	30	
16	−6		25	35		4	64		10	50	
24	50		35	82		7	61		14	54	
36	278		50	358.75		10	−2.75		18	10	
50	138		60	505.75		13	30.25		22	86	

Section 3.3.3

Construct a cubic splines piecewise function $p_3(x)$ through the given points $(x, f(x))$. Use $p_3(x)$ to estimate the two unknown x values. Round each decimal value to the thousandths place.

81.			82.			83.			84.		
x	$f(x)$		x	$f(x)$		x	$f(x)$		x	$f(x)$	
2	6		1	3		0	14.2		3.4	1.5	
7	13		4	6		2	50.7		5.8	4.4	
8	21		8	12		4.5	83.6		7.6	11.9	
12	10		10	4		6	153.1		9.2	25.3	
4			2			3.1			6.3		
9			7			5			8.6		

85.			86.			87.			88.		
x	$f(x)$		x	$f(x)$		x	$f(x)$		x	$f(x)$	
1.3	12.8		10.4	60		1	68.8		0.5	19.1	
2.1	7.8		17.6	25.7		9.4	6.4		4.7	4.4	
3.3	5.2		23.9	18		25.7	94.4		10.2	27.5	
4.8	15.4		28.3	72.1		36.5	279.2		14.6	72.6	
1.6			11.1			10.3			1.6		
4.1			24.7			26.5			5.2		

89.			90.			91.			92.		
x	$f(x)$		x	$f(x)$		x	$f(x)$		x	$f(x)$	
3	38.1		4	57.6		2.4	13.5		5	13.2	
6	14.4		8	29.1		4.1	20.3		12	32.8	
9	26.9		12	42.4		5.3	37.8		21	73.6	
12	4.5		16	18.7		5.9	31.2		29	62.4	
15	2.7		20	22.8		8.3	0.4		42	2.6	
4			7			5			12.6		
6.3			13.4			7.3			30		

93.	
x	$f(x)$
10.5	−3.9
16.3	57
22.7	28.2
29.2	39.9
35.4	21.3
12.4	
30.2	

94.	
x	$f(x)$
7.5	1.9
17.1	20.7
19.2	10.9
23.4	15.1
28.7	9.8
10.8	
24.9	

95.	
x	$f(x)$
2	21.2
7	44.4
11	30.8
18	71.4
21	52.6
28	93.9
11.8	
19	

96.	
x	$f(x)$
0	3.5
4	10.3
6	7.1
12	16.2
15	14.4
20	23.7
6.6	
13	

97.	
x	$f(x)$
10	112.8
20	31.2
30	141.4
40	74.1
50	14.3
60	124.7
20.5	
50.3	

98.	
x	$f(x)$
5	12.3
10	10.7
15	15
20	11.4
25	8.1
30	14.8
11	
26.2	

99.	
x	$f(x)$
6.4	31.8
10.7	53.3
14.6	15.6
19.1	40.5
23.9	71.7
27.9	81.5
10.8	
20.2	

100.	
x	$f(x)$
1.2	5
3.8	18
6.1	4.2
7.9	7.8
10.3	21.4
12.5	28
4.9	
9.8	

Graph $p_3(x)$ and $p_3{}'(x)$ for the data of the given exercise number above using a computer application. Use the quadratic formula to solve for x values of all relative extrema other than the endpoints, i. e., where the derivative of the spline containing the extremum equals zero, $s_i{}'(x) = 0$.

101.	#81.	102.	#82.	103.	#83.	104.	#84.
105.	#85.	106.	#86.	107.	#87.	108.	#88.
109.	#89.	110.	#90.	111.	#91.	112.	#92.
113.	#93.	114.	#94.	115.	#95.	116.	#96.
117.	#97.	118.	#98.	119.	#99.	120.	#100.

Sections 3.1, 3.3.1, and 3.3.3

Construct the following through the given points $(x, f(x))$, (a) a fifth-degree DD polynomial function $f_5(x)$, (b) a linear splines piecewise function $p_1(x)$, and (c) a cubic splines piecewise function $p_3(x)$. Graph $f_5(x)$, $p_1(x)$, and $p_3(x)$ on the same plane using a computer application with an appropriately sized window. Round all intermediate and final values to the thousandths place.

121.

x	$f(x)$
0	63.8
2	42.6
4	35.2
6	72.8
8	52.2
10	23.8

122.

x	$f(x)$
5	1
10	140.5
15	92.5
20	187
25	304
30	248.5

123.

x	$f(x)$
1	130
7	385
12	955
16	835
25	588.4
30	295.3

124.

x	$f(x)$
0.5	191.4
3.5	88.2
6	134.2
10	62.2
12.5	43.2
13	10.3

125.

x	$f(x)$
10	520
18	505.6
30	474.4
35	407
43	513.4
45	79

126.

x	$f(x)$
2	743.5
8	1046.5
11	1117
15	854
22	975.1
27	353

127.

x	$f(x)$
1.2	110
1.8	125
2.6	96
3.6	143
4	180
4.6	195

128.

x	$f(x)$
4.6	39
5.2	36
5.6	28.6
6.6	19.2
7.4	25
8	22

Section 3.4

Graph a quadratic $p_2(x)$ or cubic $p_3(x)$ splines piecewise function for the data of the given exercise number above using a computer application. Use the graph to determine which spline(s) intersect the given $p_n(x)$ value and estimate x. Use the quadratic formula for the splines $s_i(x)$ of $p_2(x)$, rounded to the thousandths place, and the Newton–Raphson algorithm (Section 1.4) for the splines $s_i(x)$ of $p_3(x)$, starting at the initial x_1 estimate(s) and accurate to the thousandths place.

129. $p_2(x)$ for #73. $p_2(x) = 6$

130. $p_2(x)$ for #74. $p_2(x) = 5$

131. $p_2(x)$ for #79. $p_2(x) = 12$

132. $p_2(x)$ for #80. $p_2(x) = 18$

133. $p_3(x)$ for #95. $p_3(x) = 48$ $x_1 = 12$

134. $p_3(x)$ for #96. $p_3(x) = 13$ $x_1 = 7.5$

135. $p_3(x)$ for #97. $p_3(x) = 132$ $x_1 = 22$ and $x_1 = 36$

136. $p_3(x)$ for #98. $p_3(x) = 9.5$ $x_1 = 23$ and $x_1 = 27$

4 Least squares regression

The goal of interpolation methods in the previous chapter is fitting a particular function through a given set of points exactly. While this function $f(x)$ may be a single nth-degree polynomial or a piecewise function of multiple nth-degree polynomials, if (x_i, y_i) is a known point, then $y_i = f(x_i)$ must be satisfied. The goal of regression methods in this chapter also entails fitting a function $r(x)$ through a given set of points, but without the stipulation that $y_i = r(x_i)$. Rather, a function is found which minimizes the accumulated error between the known points and their approximated value based on the function $r(x)$.

Finding the optimal regression function is not as well-defined as finding interpolation functions. An interpolation function satisfying all points accomplishes its goal. But because regression functions need not satisfy every point exactly, quantifying the most appropriate fit is nebulous unless specified under particular assumptions. While assumptions are crucial for the validity of the model, such topics are covered fully in a statistics text. Nevertheless, statistical justifications are cited in this chapter, such as the use of specific accumulation errors.

Least squares is one particular metric for the accumulation error, initially defined, discussed, and used for each type of regression function in the following sections. The first such type explained in detail is the two-parameter linear function. The linear function is then used as motivation for developing the logarithmic regression function, also defined by two parameters. The quadratic regression function defined by three parameters provides a third possibility. The chapter closes considering when to apply each of these three regression functions.

4.1 Least squares accumulation error

Gathering and recording data is inherently inexact to some degree; at minimum, slight variations in the dependent values are assumed routinely. Such perturbations may obscure a true underlying function, if one even exists. Suppose an experiment produces a given set of points, displayed in Figure 4.1. A single linear or logarithmic function will not fit exactly through all the points, but the data suggests a possible corresponding *trend*. Given slight variations in the imprecise dependent values, a true underlying linear or logarithmic model may actually exist, in which case the regression function approximates that model. But even if no underlying model exists explicitly, the regression function still can describe the trend concisely with a single expression. In either case, the regression function does not need to pass through each point exactly, as Figures 4.2 and 4.3 illustrate.

Naturally, there are an infinite number of linear, logarithmic, and other functions which might reasonably and visually describe the overall trend of the data. Least squares (LS) is one measurement that quantifies the best-fit function of a particular type.

https://doi.org/10.1515/9783112221051-004

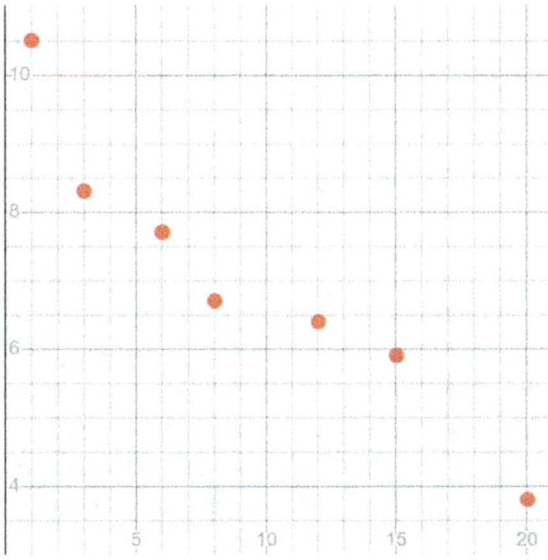

Figure 4.1: Data observed from an experiment in which the recording process allows for some inexactness in the dependent values.

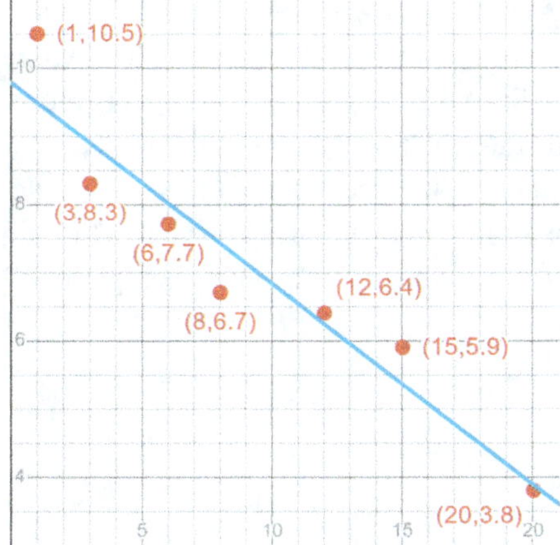

Figure 4.2: The blue linear function does not pass through all points exactly but describes the overall trend of the data of Figure 4.1.

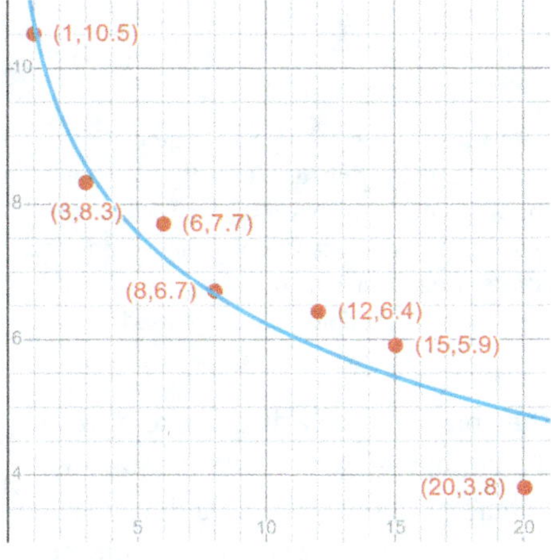

Figure 4.3: Like Figure 4.2, the blue logarithmic function describes the overall trend of the data of Figure 4.1 without passing through all points exactly.

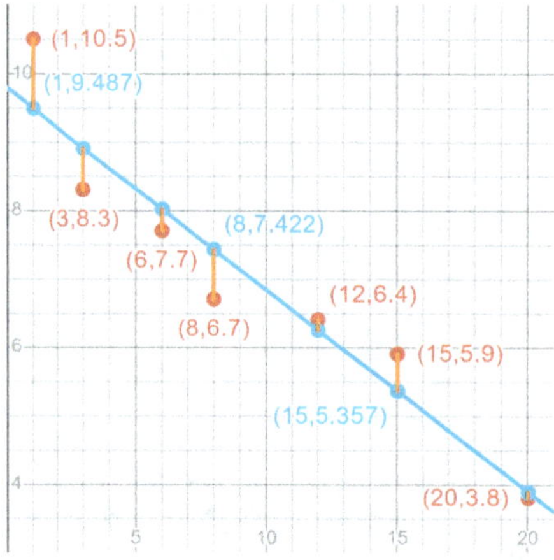

Figure 4.4: For each *x* value, the red points contain the observed *y* values and the blue points use the approximated *y* values from the blue linear regression function. The length of the orange vertical segments represents the error of each approximated dependent value.

Consider the blue linear regression function $r_1(x)$ in Figure 4.4. For $i = 1,\ldots,7$, each x_i corresponds to both a known y_i, indicated by a red point, and the approximated value $r_1(x_i)$ of y_i on the line, indicated by a blue point. Each difference $d_i = y_i - r_1(x_i)$ is represented by an orange vertical segment and quantifies the error between the known and approximated value at x_i. Changing the slope m and y-intercept b of $r_1(x) = mx + b$ decreases some d_i values and increases others, as shown in Figure 4.5. The goal is finding the m and b that minimizes some composite accumulation error E_a which comprises all differences d_i.

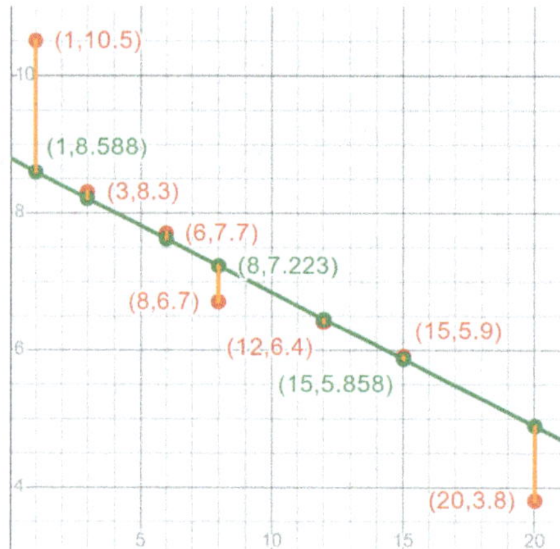

Figure 4.5: For the same points of Figure 4.4, a green linear regression function with a different slope and y-intercept is constructed. Each of the five orange vertical distances of the interior points has decreased, while both of the orange vertical distances of the endpoints have increased.

Simply adding all the d_i values is not quite the correct approach because d_i is positive for $y_i > r_1(x_i)$ and negative for $y_i < r_1(x_i)$. Potentially, large positive and negative d_i values could produce a small E_a, even though some or many individual d_i values may be quite large. Adding the absolute values of each d_i is one way of making all differences positive, and the minimization of the resulting positive E_a meaningful. However, the LS method employs the square of each difference d_i for the E_a.

Like the absolute value of each difference $|d_i|$, each squared difference d_i^2 produces all positive values for the E_a. However, as the difference d_i increases, the corresponding absolute value error increases linearly, whereas the squared value error increases quadratically. Thus, the difference penalty is increasing at an increasing rate.

For example, if the difference d_i increases by a factor of 1.6, using the absolute value of the difference $|d_i|$ increases the error by the same linear factor, since $|1.6d_i| = 1.6|d_i|$. But the same increase of 1.6 using the squared difference d_i^2 increases the error by a factor of 2.56 because $(1.6d_i)^2 = (1.6)^2 d_i^2 = 2.56 d_i^2$. The error of the absolute value of the difference d_i is said to have order $O(d_i)$ while the error of the squared difference is referred to as order $O(d_i^2)$. The use of this standard order notation recurs in Chapter 6 to compare differentiation approximations.

Each of the remaining sections of this chapter define a particular regression function $r(x)$. In each instance, the accumulation error E_a for LS is defined consistently.

$$E_a = \sum_{i=1}^{n} d_i = \sum_{i=1}^{n} (y_i - r(x_i))^2. \tag{4.1}$$

Defining E_a by squared terms rather than absolute value terms is advantageous because the former is differentiable everywhere, unlike the latter, and E_a is minimized in each section of this chapter using the standard calculus technique of solving $E_a{}' = 0$.

While the convenience of differentiability and the imposition of greater penalties on larger deviations seemingly warrant defining E_a using LS, a stronger mathematical justification exists. By applying maximum likelihood estimators (MLE) from Statistics that include specific assumptions about the data, particular equivalent regression functions can be proved as optimal fits.

In practice it is not unusual differences d_i are presumed drawn from a normal distribution with specific characteristics surrounding the distribution function, in which case using the LS E_a with a linear regression function is the best approximation model. An alternate, albeit less common occurrence, are differences d_i that approximate a Laplace distribution, which is fit better by a model with the E_a defined by absolute value differences rather than LS.

The next three sections are structured identically. First, the specific regression function used, such as the linear $r_1(x_i)$ of Section 4.2, replaces $r(x_i)$ in Equation (4.1). Second, the value E_a is minimized. Section 4.5 then compares the appropriateness of each regression function according to the characteristics or shape of the data. Note in applications in which the truly best-fit regression function is critical, the choice of that regression function rests on multiple factors, but should always include meaningful justification for the assumptions about the model and the statistical properties of the data.

Like the interpolation functions of Chapter 3, once a particular regression function $r(x)$ is constructed, it is used for approximating unknown dependent values $y_k = r(x_k)$ for any x_k where (x_k, y_k) is unknown.

4.2 Linear regression functions

The linear regression function $r_1(x)$ is a line defined by two parameters, the slope m and the y-intercept b.

$$r_1(x) = mx + b \tag{4.2}$$

Substituting this specific function for the general function $r(x_i)$ for the E_a of Equation (4.1) and expanding

$$E_a = \sum_{i=1}^{n} (y_i - r_1(x_i))^2 = \sum_{i=1}^{n} (y_i - (mx_i + b))^2$$

$$= \sum_{i=1}^{n} (y_i^2 - 2mx_iy_i - 2by_i + m^2x_i^2 + 2mbx_i + b^2). \tag{4.3}$$

Now, this accumulation error E_a is minimized. From calculus, minimizing a function requires its first derivative equal to zero. But note Equation (4.3) contains two variables, the parameters m and b; all x_i and y_i values are given and known. Therefore, a partial derivative with respect to (w. r. t.) m and a partial derivative w. r. t. b must be taken and simultaneously equal zero.

When taking a partial derivative w. r. t. a certain variable, all other variables are held constant. So for the following, when the partial derivative w. r. t. m is taken, b is treated as a constant and when the partial derivative w.r.t b is taken, m is treated as a constant; the given x_i and y_i values are all constants.

The partial derivative of E_a w. r. t. m is

$$\frac{\partial(E_a)}{\partial m} = \sum_{i=1}^{n} (-2x_iy_i + 2mx_i^2 + 2bx_i)$$

Equating to zero, dividing by two, and keeping only the m and b terms on the left side

$$\sum_{i=1}^{n} (-2x_iy_i + 2mx_i^2 + 2bx_i) = 0$$

$$\sum_{i=1}^{n} (mx_i^2 + bx_i) = \sum_{i=1}^{n} x_iy_i$$

$$m \sum_{i=1}^{n} x_i^2 + b \sum_{i=1}^{n} x_i = \sum_{i=1}^{n} x_iy_i. \tag{4.4}$$

The partial derivative of E_a w. r. t. b is

$$\frac{\partial(E_a)}{\partial b} = \sum_{i=1}^{n}(-2y_i + 2mx_i + 2b).$$

Again, equating to zero, dividing by two, and keeping only the m and b terms on the left side

$$\sum_{i=1}^{n}(-2y_i + 2mx_i + 2b) = 0$$

$$\sum_{i=1}^{n}(mx_i + b) = \sum_{i=1}^{n} y_i$$

$$m\sum_{i=1}^{n} x_i + b\sum_{i=1}^{n} 1 = \sum_{i=1}^{n} y_i$$

$$m\sum_{i=1}^{n} x_i + bn = \sum_{i=1}^{n} y_i. \tag{4.5}$$

Note that $\sum_{i=1}^{n} 1 = n$, so that the middle term in the third equation simplifies to just bn in the fourth equation.

Equations (4.4) and (4.5) form a system in the variables m and b. The b terms are eliminated easily, providing the solution for m. Multiplying Equation (4.4) by $-n$ on both sides,

$$-nm\sum_{i=1}^{n} x_i^2 - nb\sum_{i=1}^{n} x_i = -n\sum_{i=1}^{n} x_i y_i. \tag{4.6}$$

Equation (4.5) is multiplied by $\sum_{i=1}^{n} x_i$ on both sides.

$$m\sum_{i=1}^{n} x_i \sum_{i=1}^{n} x_i + bn\sum_{i=1}^{n} x_i = \sum_{i=1}^{n} y_i \sum_{i=1}^{n} x_i$$

$$m\left(\sum_{i=1}^{n} x_i\right)^2 + nb\sum_{i=1}^{n} x_i = \sum_{i=1}^{n} x_i \sum_{i=1}^{n} y_i. \tag{4.7}$$

Adding Equations (4.6) and (4.7) eliminates the middle terms,

$$-nm\sum_{i=1}^{n} x_i^2 + m\left(\sum_{i=1}^{n} x_i\right)^2 = -n\sum_{i=1}^{n} x_i y_i + \sum_{i=1}^{n} x_i \sum_{i=1}^{n} y_i.$$

Lastly, solving for m,

$$m\left[-n\sum_{i=1}^{n} x_i^2 + \left(\sum_{i=1}^{n} x_i\right)^2\right] = -n\sum_{i=1}^{n} x_i y_i + \sum_{i=1}^{n} x_i \sum_{i=1}^{n} y_i$$

$$m = \frac{n \sum_{i=1}^{n} x_i y_i - \sum_{i=1}^{n} x_i \sum_{i=1}^{n} y_i}{n \sum_{i=1}^{n} x_i^2 - \left(\sum_{i=1}^{n} x_i\right)^2}. \tag{4.8}$$

Once m is found it is substituted in Equation (4.5) and b is solved.

$$bn = \sum_{i=1}^{n} y_i - m \sum_{i=1}^{n} x_i$$

$$b = \frac{\sum_{i=1}^{n} y_i - m \sum_{i=1}^{n} x_i}{n}. \tag{4.9}$$

Equation (4.9) is used to calculate b in the examples, but a common equivalent computation in terms of the mean value of x and y, by further rewriting, is

$$b = \frac{\sum_{i=1}^{n} y_i}{n} - m \frac{\sum_{i=1}^{n} x_i}{n} = \bar{y}_i - m\bar{x}_i.$$

Summarizing, Equations (4.8) and (4.9) define the parameters m and b for the linear regression function $r_1(x)$ of Equation (4.2). $r_1(x)$ minimizes the accumulation error E_a comprised of the sum of the squared differences between each known dependent value and its approximated dependent value, one difference for each independent value. Table 4.1 is a template for organizing all values needed for Equations (4.8) and (4.9) and the linear regression function $r_1(x)$.

Table 4.1: The linear regression function of Equation (4.2) requires two parameters, m and b, which are found using the value of n and the four sums according to Equations (4.8) and (4.9).

i	x_i	y_i	x_i^2	$x_i y_i$
1	x_1	y_1	x_1^2	$x_1 y_1$
2	x_2	y_2	x_2^2	$x_2 y_2$
⋮	⋮	⋮	⋮	⋮
n	x_n	y_n	x_n^2	$x_n y_n$
	$\sum_{i=1}^{n} x_i$	$\sum_{i=1}^{n} y_i$	$\sum_{i=1}^{n} x_i^2$	$\sum_{i=1}^{n} x_i y_i$

Example 4.1. Construct the linear regression function $r_1(x)$ for the points of Figure 4.2. Use $r_1(x)$ to approximate function values at $x = 2$ and $x = 13.8$.
Solution. The known points are $(1, 10.5)$, $(3, 8.3)$, $(6, 7.7)$, $(8, 6.7)$, $(12, 6.4)$, $(15, 5.9)$, and $(20, 3.8)$. The Table 4.1 template organizes all necessary sums.

i	x_i	y_i	x_i^2	x_iy_i
1	1	10.5	1	10.5
2	3	8.3	9	24.9
3	6	7.7	36	46.2
4	8	6.7	64	53.6
5	12	6.4	144	76.8
6	15	5.9	225	88.5
7	20	3.8	400	76.0
	65	49.3	879	376.5

Parameter m is found first by Equation (4.8) using the sums from the table.

$$m = \frac{n \sum_{i=1}^n x_iy_i - \sum_{i=1}^n x_i \sum_{i=1}^n y_i}{n \sum_{i=1}^n x_i^2 - (\sum_{i=1}^n x_i)^2} = \frac{7(376.5) - (65)(49.3)}{7(879) - (65)^2} = -0.295$$

Using this m, the sums from the table, and Equation (4.9) for parameter b

$$b = \frac{\sum_{i=1}^n y_i - m \sum_{i=1}^n x_i}{n} = \frac{49.3 - (-0.295)(65)}{7} = 9.782$$

The linear regression function $r_1(x)$ according to Equation (4.2) is therefore

$$r_1(x) = -0.295x + 9.782.$$

This is precisely the blue line depicted in Figure 4.2.

For $x = 2$ and $x = 13.8$, the approximate dependent values are

$$r_1(2) = -0.295(2) + 9.782 = 9.192$$
$$r_1(13.8) = -0.295(13.8) + 9.782 = 5.711$$

Both points $(2, 9.192)$ and $(13.8, 5.711)$ lie on the blue line of Figure 4.2.

4.3 Logarithmic regression functions

A logarithmic regression function $r_{L_\beta}(x)$ for $x > 0$ is defined most generally by five parameters, c_1, c_2, c_3, c_4, and β.

$$r_{L_\beta}(x) = c_1 + c_2 \log_\beta(c_3x + c_4) \tag{4.10}$$

The coefficient of x, parameter c_3, is restricted to positive values so that $c_3 > 0$. This concession reduces the cases in which the logarithmic regression function should be employed, but there are graphical and algebraic justifications, described in Section 4.3.2. Furthermore, the positive restriction on c_3 allows a reduction in the number of parameters, not only simplifying the logarithmic regression function but also drawing a satisfying parallel to the linear regression function.

Parameter c_4 shifts the graph of $y = r_{L_\beta}(x)$ horizontally, but since $c_3 > 0$ any shifting can be handled vertically by the parameter c_1, making c_4 superfluous. So simplifying by letting $c_4 = 0$ and expanding Equation (4.10) by the property of a logarithm of a product,

$$r_{L_\beta}(x) = c_1 + c_2 \log_\beta(c_3 x)$$

$$r_{L_\beta}(x) = c_1 + c_2 \log_\beta c_3 + c_2 \log_\beta x. \tag{4.11}$$

Because $c_3 > 0$, $\log_\beta c_3$ is always defined. Also, the two terms $c_1 + c_2 \log_\beta c_3$ are constants and can be represented by a single constant parameter b.

$$b = c_1 + c_2 \log_\beta c_3 \tag{4.12}$$

Another property of logarithms is any logarithm can be written in terms of any base. Simplifying the last term on the right side of Equation (4.11) using the natural logarithm ln of base e (more on this choice in Subsection 4.3.1)

$$c_2 \log_\beta x = c_2 \frac{\ln x}{\ln \beta} = \frac{c_2}{\ln \beta} \ln x. \tag{4.13}$$

The term $\frac{c_2}{\ln \beta}$ also can be represented by a single constant parameter m, so that the left side of Equation (4.13) can be written

$$c_2 \log_\beta x = \frac{c_2}{\ln \beta} \ln x = m \ln x \tag{4.14}$$

By substituting Equations (4.12) and (4.14) in Equation (4.11), the five-parameter logarithmic regression function is simplified to a two-parameter logarithmic regression function. Because the natural logarithm is used, ln replaces L_β in the notation for clarity.

$$r_{L_\beta}(x) = b + m \ln x$$

$$r_{\ln}(x) = m \ln x + b \tag{4.15}$$

The significance of simplifying Equation (4.11) to Equation (4.15) is the similarity between the latter and the linear regression function of Equation (4.2). The only difference is $\ln x$ replaces x. This substitution is now shown simply carrying through in every step until the final forms of m and b, exactly as in Equations (4.8) and (4.9) of the linear regression function $r_1(x)$.

The accumulation error E_a for $r_{\ln}(x)$ using Equation (4.1) is

$$E_a = \sum_{i=1}^{n} (y_i - r_{\ln}(x_i))^2 = \sum_{i=1}^{n} (y_i - (m \ln x_i + b))^2$$

This expands to the same E_a of Equation (4.3), with $\ln x_i$ in lieu of x_i.

$$E_a = \sum_{i=1}^{n} \left(y_i^2 - 2m(\ln x_i)y_i - 2by_i + m^2(\ln x_i)^2 + 2mb\ln x_i + b^2 \right)$$

Recall E_a is minimized by taking two partial derivatives w.r.t. m and b. Because the x_i values are given and thus constant, the $\ln x_i$ values are constant. Therefore, the two partial derivatives w.r.t. m and b of this E_a are identical to those of the E_a of $r_1(x)$, with $\ln x_i$ replacing x_i.

$$\frac{\partial(E_a)}{\partial m} = \sum_{i=1}^{n} \left(-2(\ln x_i)y_i + 2m(\ln x_i)^2 + 2b\ln x_i \right)$$

$$\frac{\partial(E_a)}{\partial b} = \sum_{i=1}^{n} \left(-2y_i + 2m\ln x_i + 2b \right)$$

Equating each partial derivative to zero and solving in the same manner as Equations (4.4) and (4.5) produces a system of equations.

$$m\sum_{i=1}^{n}(\ln x_i)^2 + b\sum_{i=1}^{n}\ln x_i = \sum_{i=1}^{n}(\ln x_i)y_i$$

$$m\sum_{i=1}^{n}\ln x_i + bn = \sum_{i=1}^{n}y_i. \tag{4.16}$$

This system is solved by eliminating the middle terms and following the analogous steps that produced Equations (4.6) and (4.7), with the constant x_i values substituted by the constant $\ln x_i$ values. Multiplying the first equation on both sides by $-n$, the second equation by $\sum_{i=1}^{n}\ln x_i$ on both sides, and adding the resulting equations

$$-nm\sum_{i=1}^{n}(\ln x_i)^2 - nb\sum_{i=1}^{n}\ln x_i = -n\sum_{i=1}^{n}(\ln x_i)y_i$$

$$m\sum_{i=1}^{n}\ln x_i \sum_{i=1}^{n}\ln x_i + bn\sum_{i=1}^{n}\ln x_i = \sum_{i=1}^{n}y_i \sum_{i=1}^{n}\ln x_i$$

$$-nm\sum_{i=1}^{n}(\ln x_i)^2 + m\left(\sum_{i=1}^{n}\ln x_i\right)^2 = -n\sum_{i=1}^{n}(\ln x_i)y_i + \sum_{i=1}^{n}\ln x_i \sum_{i=1}^{n}y_i.$$

Solving for m produces the same result as Equation (4.8), with $\ln x_i$ replacing x_i.

$$m = \frac{n\sum_{i=1}^{n}(\ln x_i)y_i - \sum_{i=1}^{n}\ln x_i \sum_{i=1}^{n}y_i}{n\sum_{i=1}^{n}(\ln x_i)^2 - \left(\sum_{i=1}^{n}\ln x_i\right)^2} \tag{4.17}$$

Lastly, substituting this m in Equation (4.16) and solving for b,

$$b = \frac{\sum_{i=1}^{n}y_i - m\sum_{i=1}^{n}\ln x_i}{n}. \tag{4.18}$$

Note that Equation (4.18) matches Equation (4.9), with $\ln x_i$ in place of x_i.

In summary, Equations (4.17) and (4.18) define the parameters m and b for the logarithmic regression function $r_{\ln}(x)$ of Equation (4.15). The Table 4.2 template organizes all values needed for Equations (4.17) and (4.18) and a logarithmic regression function $r_{\ln}(x)$.

Table 4.2: The logarithmic regression function of Equation (4.15) needs two parameters m and b, found using the value of n and the four sums of Equations (4.17) and (4.18).

i	x_i	y_i	$\ln x_i$	$(\ln x_i)^2$	$(\ln x_i)y_i$
1	x_1	y_1	$\ln x_1$	$(\ln x_1)^2$	$(\ln x_1)y_1$
2	x_2	y_2	$\ln x_2$	$(\ln x_2)^2$	$(\ln x_2)y_2$
$\vdots$	$\vdots$	$\vdots$	$\vdots$	$\vdots$	$\vdots$
n	x_n	y_n	$\ln x_n$	$(\ln x_n)^2$	$(\ln x_n)y_n$
		$\sum_{i=1}^{n} y_i$	$\sum_{i=1}^{n} \ln x_i$	$\sum_{i=1}^{n}(\ln x_i)^2$	$\sum_{i=1}^{n}(\ln x_i)y_i$

Example 4.2. Construct the logarithmic regression function $r_{\ln}(x)$ for the points of Figure 4.3. Use $r_{\ln}(x)$ to approximate function values at $x = 3.2$ and $x = 17$.

Solution. The given points are $(1, 10.5)$, $(3, 8.3)$, $(6, 7.7)$, $(8, 6.7)$, $(12, 6.4)$, $(15, 5.9)$, and $(20, 3.8)$. The Table 4.2 template is used for all needed sums.

i	x_i	y_i	$\ln x_i$	$(\ln x_i)^2$	$(\ln x_i)y_i$
1	1	10.5	0	0	0
2	3	8.3	1.099	1.208	9.122
3	6	7.7	1.792	3.211	13.798
4	8	6.7	2.079	4.322	13.929
5	12	6.4	2.485	6.175	15.904
6	15	5.9	2.708	7.333	15.977
7	20	3.8	2.996	8.976	11.385
		49.3	13.159	31.225	80.115

First, parameter m is found by Equation (4.17) with the sums from the table.

$$m = \frac{n \sum_{i=1}^{n}(\ln x_i)y_i - \sum_{i=1}^{n} \ln x_i \sum_{i=1}^{n} y_i}{n \sum_{i=1}^{n}(\ln x_i)^2 - \left(\sum_{i=1}^{n} \ln x_i\right)^2} = \frac{7(80.115) - (13.159)(49.3)}{7(31.225) - (13.159)^2} = -1.936$$

Using this m, the sums from the table, and Equation (4.18) for parameter b

$$b = \frac{\sum_{i=1}^{n} y_i - m \sum_{i=1}^{n} \ln x_i}{n} = \frac{49.3 - (-1.936)(13.159)}{7} = 10.682.$$

The logarithmic regression function $r_{\ln}(x)$ according to Equation (4.15) is thus

$$r_{\ln}(x) = -1.936 \ln x + 10.682$$

This is the same blue curve of Figure 4.3.

For $x = 3.2$ and $x = 17$, the dependent value approximations are

$$r_{\ln}(3.2) = -1.936 \ln 3.2 + 10.682 = -1.936(1.163) + 10.682 = 8.430$$

$$r_{\ln}(17) = -1.936 \ln 17 + 10.682 = -1.936(2.833) + 10.682 = 5.197$$

The blue curve of Figure 4.3 contains these points, $(3.2, 8.430)$ and $(17, 5.197)$.

4.3.1 The general logarithmic regression function

In the derivation $r_{\ln}(x)$ of Equation (4.15), the natural logarithm of base e, $\ln x$, was substituted in place of the general $\log_\beta x$ in Equation (4.13) to remove the base β parameter. However, *any* logarithm of a known base can be substituted in the same manner, not just $\ln x$. For example, the common logarithm of base 10, $\log x$, could replace $\ln x$ in Equation (4.13) so that

$$c_2 \log_\beta x = c_2 \frac{\log x}{\log \beta} = \frac{c_2}{\log \beta} \log x \tag{4.19}$$

Representing $\frac{c_2}{\log \beta}$ by a single constant parameter m as before, Equation (4.19) becomes

$$c_2 \log_\beta x = \frac{c_2}{\log \beta} \log x = m \log x \tag{4.20}$$

Substituting Equations (4.12) and now (4.20) in Equation (4.11) produces a logarithmic regression function using base 10, notated $\log x$. The change is reflected in the subscript for base 10 specifically, in place of the general L_β or ln of Equation (4.15) as

$$r_{L_\beta}(x) = b + m \log x$$

$$r_{\log}(x) = m(\log x) + b. \tag{4.21}$$

Certainly, parameters m and b depend on the base of the logarithm substituted for base β at the Equation (4.13) step. While logarithms of base e and 10 are typical, the general form of the logarithmic regression function using any known base β, instead of $\ln x$ of Equation (4.15), or $\log x$ of Equation (4.21), is defined

$$r_{L_\beta}(x) = m_\beta(\log_\beta x) + b_\beta \tag{4.22}$$

Equation (4.22) uses the analogous the general forms of Equation (4.17) and (4.18), respectively, for parameters m_β and b_β using any known base β.

$$m_\beta = \frac{n \sum_{i=1}^{n}(\log_\beta x_i)y_i - \sum_{i=1}^{n} \log_\beta x_i \sum_{i=1}^{n} y_i}{n \sum_{i=1}^{n}(\log_\beta x_i)^2 - (\sum_{i=1}^{n} \log_\beta x_i)^2} \tag{4.23}$$

$$b_\beta = \frac{\sum_{i=1}^{n} y_i - m \sum_{i=1}^{n} \log_\beta x_i}{n} \tag{4.24}$$

Lastly, the Table 4.2 template using $\ln x$ is restated for the general case using any known base β as Table 4.3.

Table 4.3: The general logarithmic regression function for any known base β of Equation (4.22) employs two parameters m_β and b_β, using the value n and the four sums of Equations (4.23) and (4.24).

i	x_i	y_i	$\log_\beta x_i$	$(\log_\beta x_i)^2$	$(\log_\beta x_i)y_i$
1	x_1	y_1	$\log_\beta x_1$	$(\log_\beta x_1)^2$	$(\log_\beta x_1)y_1$
2	x_2	y_2	$\log_\beta x_2$	$(\log_\beta x_2)^2$	$(\log_\beta x_2)y_2$
$\vdots$	$\vdots$	$\vdots$	$\vdots$	$\vdots$	$\vdots$
n	x_n	y_n	$\log_\beta x_n$	$(\log_\beta x_n)^2$	$(\log_\beta x_n)y_n$
		$\sum_{i=1}^{n} y_i$	$\sum_{i=1}^{n} \log_\beta x_i$	$\sum_{i=1}^{n} (\log_\beta x_i)^2$	$\sum_{i=1}^{n} (\log_\beta x_i)y_i$

The next two examples demonstrate graphically the identical nature of the logarithmic regression functions using different bases, according to Equations (4.21) and (4.22).

Example 4.3. Construct the logarithmic regression function $r_{\log}(x)$ for the points of Figure 4.3. Use $r_{\log}(x)$ to approximate function values at $x = 3.2$ and $x = 17$. Graph $r_{\log}(x)$.
Solution. By the Table 4.3 template with $\beta = 10$ and $\log x$ the required sums are found.

i	x_i	y_i	$\log x_i$	$(\log x_i)^2$	$(\log x_i)y_i$
1	1	10.5	0	0	0
2	3	8.3	0,477	0.228	3.959
3	6	7.7	0.778	0.605	5.991
4	8	6.7	0.903	0.815	6.050
5	12	6.4	1.079	1.164	6.906
6	15	5.9	1.176	1.383	6.938
7	20	3.8	1.301	1.693	4.944
		49.3	5.714	5.888	34.788

Using Equation (4.23) with $\beta = 10$, $\log x$, and the sums from the table for parameter m_β

$$m_\beta = \frac{n \sum_{i=1}^{n} (\log x_i)y_i - \sum_{i=1}^{n} \log x_i \sum_{i=1}^{n} y_i}{n \sum_{i=1}^{n} (\log x_i)^2 - (\sum_{i=1}^{n} \log x_i)^2}$$

$$m_{10} = \frac{7(34.788) - (5.174)(49.3)}{7(5.888) - (5.174)^2} = -4.458.$$

Using this m_β, $\beta = 10$, $\log x$, the sums from the table, and Equation (4.24) for parameter b_β

$$b_\beta = \frac{\sum_{i=1}^{n} y_i - m_\beta \sum_{i=1}^{n} \log x_i}{n}$$

$$b_{10} = \frac{49.3 - (-4.458)(5.174)}{7} = 10.682.$$

The logarithmic regression function $r_{L_\beta}(x)$ for $\beta = 10$ using Equation (4.22) is

$$r_{L_\beta}(x) = r_{L_{10}}(x) = -4.458(\log_{10} x) + 10.682.$$

Since the base $\beta = 10$ this is equivalent to Equation (4.21).

$$r_{L_{10}}(x) = r_{\log}(x) = -4.458 \log x + 10.682.$$

The orange dotted curve of Figure 4.6 displays this function.

The approximations $r_{\log}(3.2)$ and $r_{\log}(17)$ are

$$r_{\log}(3.2) = -4.458 \log 3.2 + 10.682 = -4.458(0.505) + 10.682 = 8.431$$
$$r_{\log}(17) = -4.458 \log 17 + 10.682 = -4.458(1.230) + 10.682 = 5.199$$

Figure 4.6 graphically supports the identical nature of logarithmic regression functions regardless of the base. The orange dotted curve $r_{\log}(x)$ of Example 4.3 lies directly on the blue curve $r_{\ln}(x)$ of Example 4.2. Furthermore, comparing the algebraic results of Examples 4.3 and 4.2, the absolute errors of the dependent values for $x = 3.2$ and $x = 17$ are

$$\left| r_{\log}(3.2) - r_{\ln}(3.2) \right| = |8.431 - 8.430| = 0.001$$
$$\left| r_{\log}(17) - r_{\ln}(17) \right| = |5.199 - 5.197| = 0.002.$$

These discrepancies occur due to rounding intermediate values to the thousandths place. Retaining more decimal places throughout the entire process produces 8.430 and 5.197, respectively, for $x = 3.2$ and $x = 17$ using either $r_{\log}(x)$ or $r_{\ln}(x)$.

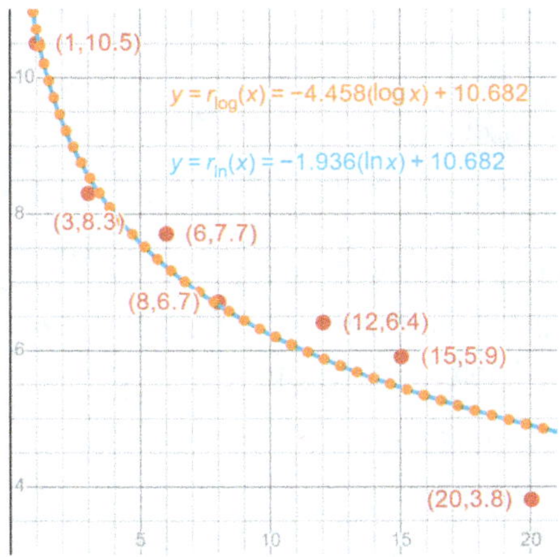

Figure 4.6: The logarithmic regression functions $r_{\log}(x)$ and $r_{\ln}(x)$ lie on top of each other and produce identical results.

Example 4.4. Construct the logarithmic regression function $r_{L_\beta}(x)$ for base $\beta = 2$ and the points of Figure 4.3. Use $r_{L_2}(x)$ to approximate function values at $x = 3.2$ and $x = 17$. Graph $r_{L_2}(x)$.
Solution. The Table 4.3 template is used with $\beta = 2$ for the necessary sums.

i	x_i	y_i	$\log_2 x_i$	$(\log_2 x_i)^2$	$(\log_2 x_i)y_i$
1	1	10.5	0	0	0
2	3	8.3	1.585	2.512	13.156
3	6	7.7	2.585	6.682	19.905
4	8	6.7	3.000	9.000	20.100
5	12	6.4	3.585	12.852	22.944
6	15	5.9	3.907	15.265	23.051
7	20	3.8	4.322	18.680	16.424
		49.3	18.984	64.991	115.580

According to Equation (4.23), parameter m_β for $\beta = 2$ using the sums from the table is

$$m_\beta = \frac{n \sum_{i=1}^n (\log_2 x_i)y_i - \sum_{i=1}^n \log_2 x_i \sum_{i=1}^n y_i}{n \sum_{i=1}^n (\log_2 x_i)^2 - (\sum_{i=1}^n \log_2 x_i)^2}$$

$$m_2 = \frac{7(115.580) - (18.984)(49.3)}{7(64.991) - (18.984)^2} = -1.342$$

Using this $m_\beta, \beta = 2$, the sums from the table, and Equation (4.24) for parameter b_β

$$b_\beta = \frac{\sum_{i=1}^n y_i - m_\beta \sum_{i=1}^n \log_2 x_i}{n}$$

$$b_2 = \frac{49.3 - (-1.342)(18.984)}{7} = 10.682.$$

By Equation (4.22), the logarithmic regression function $r_{L_2}(x)$ is

$$r_{L_\beta}(x) = r_{L_2}(x) = -1.342 \log_2 x + 10.682.$$

Figure 4.7 displays this function with a black dotted curve.
The approximations $r_{L_2}(3.2)$ and $r_{L_2}(17)$ are

$$r_{L_2}(3.2) = -1.342 \log_2 3.2 + 10.682 = -1.342(1.678) + 10.682 = 8.430$$
$$r_{L_2}(17) = -1.342 \log_2 17 + 10.682 = -1.342(4.087) + 10.682 = 5.197.$$

The black dotted curve $r_{L_2}(x)$ in Figure 4.7 of Example 4.4, just like the orange dotted curve $r_{\log}(x)$ in Figure 4.6 of Example 4.3, lies directly on the blue curve $r_{\ln}(x)$ of Example 4.2. Also, the algebraic results of Examples 4.4 and 4.2 are identical.

$$r_{L_2}(3.2) = r_{\ln}(3.2) = 8.430$$
$$r_{L_2}(17) = r_{\ln}(17) = 5.197$$

Because base $\beta = 2$ is closer than base $\beta = 10$ is to base $e \approx 2.718$ of the natural logarithm ln, the algebraic approximations using $r_{L_2}(x)$ and $r_{\ln}(x)$ round to the same thousandths place, even with rounding the intermediate values in the process.

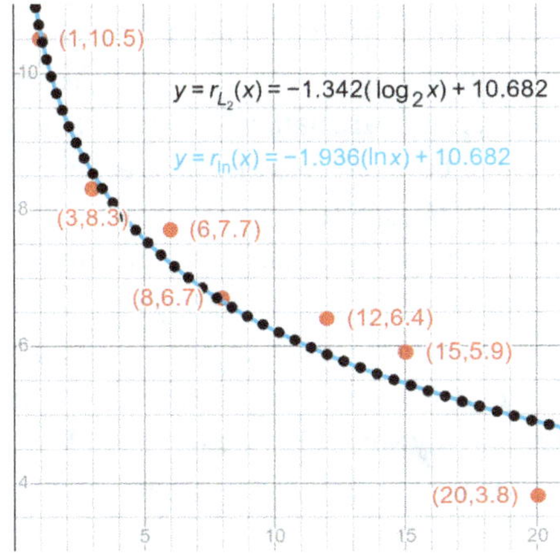

Figure 4.7: The logarithmic regression functions $r_{L_2}(x)$ and $r_{\ln}(x)$ lie on top of each other, just like $r_{\log}(x)$ and $r_{\ln}(x)$ of Figure 4.6.

4.3.2 Justification for $c_3 > 0$ on $r_{L_\beta}(x)$

In developing the logarithmic regression function at the outset of Section 4.3, the restriction $c_3 > 0$ was placed on Equation (4.10). The graphical and algebraic implications of permitting $c_3 < 0$ and why these negative values were removed from the logarithmic regression function definition are now considered. Recall the domain of the independent variable is positive, $x > 0$.

In general, the graphs of $f(x)$ and $g(x) = f(-x)$ are vertical reflections of each other over the vertical line $x = 0$. Parameter c_3 of Equation (4.10) is the coefficient of x so that if $c_3 < 0$, the logarithmic regression function reflects vertically. Consider the reflected graph of $g(x) = \ln(-2x)$ in Figure 4.8. Since $c_3 = -2$, the graph is a vertical reflection of $f(x) = \ln(2x)$ over the vertical line $x = 0$, a vertical asymptote of both functions.

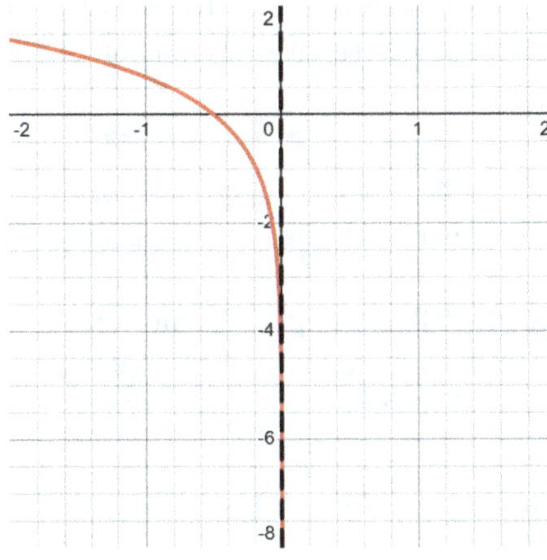

Figure 4.8: The red graph $g(x) = \ln(-2x)$ is a reflection of $f(x) = \ln(2x)$ over the vertical asymptote $x = 0$, lying to the right of $g(x)$.

The significance of allowing $c_3 < 0$ is the vertical asymptote will lie to the right of the logarithmic regression function, creating an arbitrary upper bound on x. Although in contrast $c_3 > 0$ forces a lower bound on x, many scenarios are naturally conducive to a lower bounded, and conversely, an unrestricted upper bounded function. Two examples are x quantifying a measurement such as length or time elapsed.

Furthermore, as logarithmic regression functions approach an upper bound on x at the vertical asymptote, small changes in the x values generate increasingly very large changes relatively in the dependent values $r_{L_\beta}(x)$, which tend towards positive or negative infinity. These two factors justify restricting $c_3 > 0$ and keeping the bound on the logarithmic regression function to its left. Consequently, logarithmic regression functions are considered for data that is generally either increasing and concave down or decreasing and concave up.

4.4 Quadratic regression functions

The quadratic regression function $r_2(x)$ is a second-degree polynomial defined by three parameters a, b, and c.

$$r_2(x) = ax^2 + bx + c. \tag{4.25}$$

Substituting this particular function for the general function $r(x_i)$ for the accumulation error E_a of Equation (4.1) and expanding

$$E_a = \sum_{i=1}^{n}(y_i - r_2(x_i))^2 = \sum_{i=1}^{n}(y_i - (ax_i^2 + bx_i + c))^2$$

$$E_a = \sum_{i=1}^{n}(y_i^2 - 2y_i(ax_i^2 + bx_i + c) + (ax_i^2 + bx_i + c)^2).$$

E_a is now minimized by taking partial derivatives w. r. t. each of the three parameters, equating each to zero, and grouping the variable terms a, b, and c on the left side. This mimics the procedure for minimizing the E_a of the linear regression function of Section 4.2, but which required one less parameter.

The partial derivative of E_a w. r. t. a is

$$\frac{\partial(E_a)}{\partial a} = \sum_{i=1}^{n}(-2x_i^2 y_i + 2(ax_i^2 + bx_i + c)x_i^2)$$

Equating to zero, dividing every term by two, and distributing the x_i^2 factor

$$\sum_{i=1}^{n}(-2x_i^2 y_i + 2(ax_i^2 + bx_i + c)x_i^2) = 0$$

$$\sum_{i=1}^{n}(-x_i^2 y_i + ax_i^4 + bx_i^3 + cx_i^2) = 0.$$

Distributing the summation and moving the non-variable term to the right side

$$a \sum_{i=1}^{n} x_i^4 + b \sum_{i=1}^{n} x_i^3 + c \sum_{i=1}^{n} x_i^2 = \sum_{i=1}^{n} x_i^2 y_i. \tag{4.26}$$

Repeating these steps, the partial derivative of E_a w. r. t. b is

$$\frac{\partial(E_a)}{\partial b} = \sum_{i=1}^{n} (-2x_i y_i + 2(ax_i^2 + bx_i + c)x_i).$$

Equating to zero, dividing every term by two, and distributing the x_i factor,

$$\sum_{i=1}^{n} (-2x_i y_i + 2(ax_i^2 + bx_i + c)x_i) = 0$$

$$\sum_{i=1}^{n} (-x_i y_i + ax_i^3 + bx_i^2 + cx_i) = 0.$$

Distributing the summation and moving the non-variable term to the right side,

$$a \sum_{i=1}^{n} x_i^3 + b \sum_{i=1}^{n} x_i^2 + c \sum_{i=1}^{n} x_i = \sum_{i=1}^{n} x_i y_i. \tag{4.27}$$

Lastly, for the partial derivative of E_a w. r. t. c,

$$\frac{\partial(E_a)}{\partial c} = \sum_{i=1}^{n} (-2y_i + 2(ax_i^2 + bx_i + c)(1)).$$

Equating to zero and dividing every term by two,

$$\sum_{i=1}^{n} (-2y_i + 2(ax_i^2 + bx_i + c)) = 0$$

$$\sum_{i=1}^{n} (-y_i + ax_i^2 + bx_i + c) = 0.$$

Distributing the summation, moving the non-variable term to the right side, and making the substitution $\sum_{i=1}^{n} 1 = n$ in the third term

$$a \sum_{i=1}^{n} x_i^2 + b \sum_{i=1}^{n} x_i + c \sum_{i=1}^{n} 1 = \sum_{i=1}^{n} y_i$$

$$a \sum_{i=1}^{n} x_i^2 + b \sum_{i=1}^{n} x_i + cn = \sum_{i=1}^{n} y_i. \tag{4.28}$$

Equations (4.26), (4.27), and (4.28) form a system of equations in three variables, parameters a, b, and c. As a matrix equation this system is written

$$\begin{bmatrix} \sum_{i=1}^{n} x_i^4 & \sum_{i=1}^{n} x_i^3 & \sum_{i=1}^{n} x_i^2 \\ \sum_{i=1}^{n} x_i^3 & \sum_{i=1}^{n} x_i^2 & \sum_{i=1}^{n} x_i \\ \sum_{i=1}^{n} x_i^2 & \sum_{i=1}^{n} x_i & n \end{bmatrix} \times \begin{bmatrix} a \\ b \\ c \end{bmatrix} = \begin{bmatrix} \sum_{i=1}^{n} x_i^2 y_i \\ \sum_{i=1}^{n} x_i y_i \\ \sum_{i=1}^{n} y_i \end{bmatrix} \tag{4.29}$$

Appending the rightmost column matrix to the leftmost coefficient matrix, both of Equation (4.29), forms the augmented matrix.

$$\begin{bmatrix} \sum_{i=1}^{n} x_i^4 & \sum_{i=1}^{n} x_i^3 & \sum_{i=1}^{n} x_i^2 & \vdots & \sum_{i=1}^{n} x_i^2 y_i \\ \sum_{i=1}^{n} x_i^3 & \sum_{i=1}^{n} x_i^2 & \sum_{i=1}^{n} x_i & \vdots & \sum_{i=1}^{n} x_i y_i \\ \sum_{i=1}^{n} x_i^2 & \sum_{i=1}^{n} x_i & n & \vdots & \sum_{i=1}^{n} y_i \end{bmatrix} \tag{4.30}$$

Transforming Equation (4.30) into RREF by Gauss–Jordan elimination solves for the parameters a, b, and c of Equation (4.25), the quadratic regression function that minimizes E_a. RREF was described briefly in Step 4 of Section 3.3.3. Table 4.4 summarizes all summations needed to populate Equation (4.30).

Table 4.4: The quadratic regression function of Equation (4.25) needs three parameters a, b, and c, found using the value n and the seven sums in the matrix of Equation (4.30).

x_i	y_i	x_i^2	x_i^3	x_i^4	$x_i^2 y_i$	$x_i y_i$
x_1	y_1	x_1^2	x_1^3	x_1^4	$x_1^2 y_1$	$x_1 y_1$
x_2	y_2	x_2^2	x_2^3	x_2^4	$x_2^2 y_2$	$x_2 y_2$
$\vdots$	$\vdots$	$\vdots$	$\vdots$	$\vdots$	$\vdots$	$\vdots$
x_n	y_n	x_n^2	x_n^3	x_n^4	$x_n^2 y_n$	$x_n y_n$
$\sum_{i=1}^{n} x_i$	$\sum_{i=1}^{n} y_i$	$\sum_{i=1}^{n} x_i^2$	$\sum_{i=1}^{n} x_i^3$	$\sum_{i=1}^{n} x_i^4$	$\sum_{i=1}^{n} x_i^2 y_i$	$\sum_{i=1}^{n} x_i y_i$

Example 4.5. Construct the quadratic regression function $r_2(x)$ for the points of Figure 4.1. Use $r_2(x)$ to approximate function values at $x = 2$ and $x = 13.8$.
Solution. The known points are $(1, 10.5)$, $(3, 8.3)$, $(6, 7.7)$, $(8, 6.7)$, $(12, 6.4)$, $(15, 5.9)$, and $(20, 3.8)$. The Table 4.4 template organizes all needed sums.

x_i	y_i	x_i^2	x_i^3	x_i^4	$x_i^2 y_i$	$x_i y_i$
1	10.5	1	1	1	10.5	10.5
3	8.3	9	27	81	74.7	24.9
6	7.7	36	216	1296	277.2	46.2
8	6.7	64	512	4096	428.8	53.6
12	6.4	144	1728	20736	921.6	76.8
15	5.9	225	3375	50625	1327.5	88.5
20	3.8	400	8000	160000	1520.0	76.0
65	49.3	879	13859	236835	4560.3	376.5

Substituting the value of each sum from the table and $n = 7$ in the augmented matrix of Equation (4.30),

$$\begin{bmatrix} 236835 & 13859 & 879 & \vdots & 4560.3 \\ 13859 & 879 & 65 & \vdots & 376.5 \\ 879 & 65 & 7 & \vdots & 49.3 \end{bmatrix}$$

Transforming to RREF by Gauss–Jordan elimination,

$$\begin{bmatrix} 1 & 0 & 0 & \vdots & 0.006 \\ 0 & 1 & 0 & \vdots & -0.417 \\ 0 & 0 & 1 & \vdots & 10.176 \end{bmatrix}$$

Referencing Equation (4.29), the parameters a, b, and c are thus in order, $a = 0.006$, $b = -0.417$, and $c = 10.176$. According to Equation (4.25), the quadratic regression function is

$$r_2(x) = 0.006x^2 - 0.417x + 10.176.$$

The blue curve of Figure 4.9 illustrates $r_2(x)$.

For $x = 2$ and $x = 13.8$, the approximate dependent values are

$$r_2(2) = 0.006(2)^2 - 0.417(2) + 10.176 = 9.366$$
$$r_2(13.8) = 0.006(13.8)^2 - 0.417(13.8) + 10.176 = 5.564.$$

Both points $(2, 9.366)$ and $(13.8, 5.564)$ lie on the blue curve of Figure 4.9.

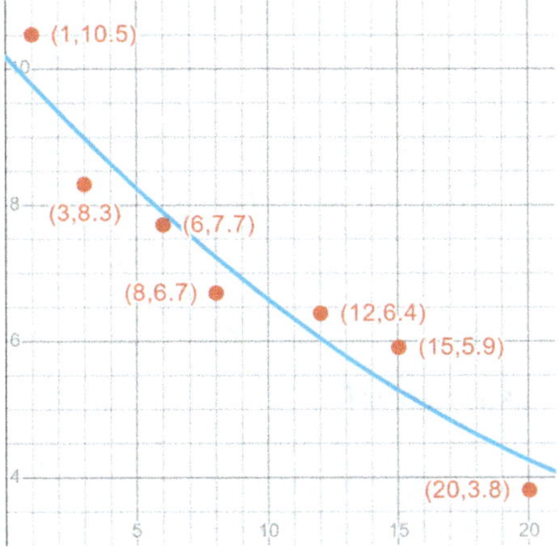

Figure 4.9: The quadratic regression function $r_2(x)$ for the data of Figure 4.1.

4.5 Comparison of regression functions

The relatively small coefficient 0.006 of the x^2 term for the quadratic regression function $r_2(x)$ of Example 4.5 and Figure 4.9 gives it a linear appearance similar to the linear regression function $r_1(x)$ of Figure 4.2. Both are noticeably different from the curved logarithmic regression functions $r_{\ln}(x)$, $r_{\log}(x)$, and $r_{L_\beta}(x)$ of Figures 4.3, 4.6, and 4.7, respectively. While all three function types are reasonable fits to the data visually, a comparison of the accumulation error E_a for each function using all known points is one algebraic measure for determining definitively which type fits best.

The linear, logarithmic, and quadratic regression functions of Examples 4.1, 4.2, and 4.5, fit to the data of Figure 4.1, are

$$r_1(x) = -0.295x + 9.782$$
$$r_{\ln}(x) = -1.936 \ln x + 10.682$$
$$r_2(x) = 0.006x^2 - 0.417x + 10.176.$$

Recall from Section 4.3.1 that any base can be used in the logarithmic regression function, so comparing to $r_{\log}(x)$ or $r_{L_\beta}(x)$ is redundant, although either could be used in lieu of $r_{\ln}(x)$.

Table 4.5 contains the seven known points, the approximated values of the dependent values according to the linear regression function $r_1(x)$, and the squared differences between the known and approximated values.

Table 4.5: Values used for the E_a for $r_1(x)$ and the data of Figure 4.1.

i	x_i	y_i	$r_1(x_i)$	$(y_i - r_1(x_i))^2$
1	1	10.5	9.487	1.026
2	3	8.3	8.897	0.356
3	6	7.7	8.012	0.097
4	8	6.7	7.422	0.521
5	12	6.4	6.242	0.025
6	15	5.9	5.357	0.295
7	20	3.8	3.882	0.007

Adding all values of the last column, the accumulation error for $r_1(x)$ is

$$E_a = \sum_{i=1}^{n}(y_i - r_1(x_i))^2 = 2.327$$

Table 4.6 contains the known points, the approximated values of the dependent values according to the logarithmic regression function $r_{\ln}(x)$, and the squared differences between the known and approximated values.

Table 4.6: Values used for the E_a for $r_{\ln}(x)$ and the data of Figure 4.1.

i	x_i	y_i	$r_{\ln}(x_i)$	$(y_i - r_{\ln}(x_i))^2$
1	1	10.5	10.682	0.033
2	3	8.3	8.555	0.065
3	6	7.7	7.213	0.237
4	8	6.7	6.656	0.002
5	12	6.4	5.871	0.280
6	15	5.9	5.439	0.213
7	20	3.8	4.882	1.171

Summing all values of the last column, the accumulation error for $r_{\ln}(x)$ is

$$E_a = \sum_{i=1}^{n}(y_i - r_{\ln}(x_i))^2 = 2.001.$$

Lastly, Table 4.7 contains the known points, the approximated values of the dependent values according to the quadratic regression function $r_2(x)$, and the squared differences between the known and approximated values.

Table 4.7: Values used for the E_a for $r_2(x)$ and the data of Figure 4.1.

i	x_i	y_i	$r_2(x_i)$	$(y_i - r_2(x_i))^2$
1	1	10.5	9.765	0.540
2	3	8.3	8.979	0.461
3	6	7.7	7.890	0.036
4	8	6.7	7.224	0.275
5	12	6.4	6.036	0.132
6	15	5.9	5.271	0.396
7	20	3.8	4.236	0.190

The accumulation error for $r_2(x)$, adding all values in the last column is

$$E_a = \sum_{i=1}^{n}(y_i - r_2(x_i))^2 = 2.030.$$

For this particular data, the quadratic function is slightly better than the linear function, while the logarithmic function is better than both.

Table 4.9 compares the approximated dependent value of each regression function to each other and the known dependent value, or y_i value, for every x_i value, highlighting the comparable fit of each function type to this data.

The value of E_a is relative to both the number and magnitude of the independent and dependent values. A set of many points consisting of large values will have a much

Table 4.8: A comparison of the E_a per function for the Figure 4.1 data.

Function	E_a
$r_{\ln}(x)$	2.001
$r_2(x)$	2.030
$r_1(x)$	2.327

Table 4.9: The linear, logarithmic, and quadratic regression functions produce similar dependent value approximations to each other and the known y_i value for each x_i.

i	x_i	y_i	$r_1(x_i)$	$r_{\ln}(x_i)$	$r_2(x_i)$
1	1	10.5	9.487	10.682	9.765
2	3	8.3	8.897	8.555	8.979
3	6	7.7	8.012	7.213	7.890
4	8	6.7	7.422	6.656	7.224
5	12	6.4	6.242	5.871	6.036
6	15	5.9	5.357	5.439	5.271
7	20	3.8	3.882	4.882	4.236

greater E_a than a set with few points and small values. Therefore, using the E_a as a comparison metric among regression functions works in context on the same data, but should not be used across different sets of data. For example, $E_a = 10$ on data consisting of 50 points with values in the hundreds may constitute an appropriate fit, whereas $E_a = 10$ on data of ten points with values in the single digits is likely a poor fit. However, if three regression functions have E_a values of 10, 10.1, and 10.3 on the same data, the first can be justified as the best fit of the three within the context of that particular data.

Each particular set of points determines the regression function that best fits the data. However, certain data may be ill-suited for certain types of regression functions based on the inherent characteristics of the function type. Eliminating one or more options removes the need to find the E_a for every type of function. The following three subsections note some pros and cons of each type of regression function.

4.5.1 Linear regression function considerations

The simplest regression function type is the linear $r_1(x)$. The extended advantage of this simplicity is all required sums used for parameters m and b involve either a sum of just the x_i or y_i values, or a sum of a single product of the known x_i and y_i values, i. e., x_i^2 and $x_i y_i$. As such, all sums are less likely to grow much larger or smaller relative to the given data versus the sums used by other regression functions. Thus constructing $r_1(x)$ can use data as given, without needing to scale it or round decimal values at intermediate steps.

However, any linear function increases or decreases at a constant rate over its domain. Therefore the linear regression function $r_1(x)$ is a poor choice for data that does not appear monotonic overall. While an $r_1(x)$ can always be constructed, it does not preclude better fitting options such as the logarithmic and quadratic regression functions, which by their curved nature do not increase or decrease at a constant rate.

Consider the data of Figure 4.10. The dependent values increase and decrease around a maximum at $(35, 9.4)$. Nevertheless, the next example finds the best linear regression function $r_1(x)$ and suggests better options exist.

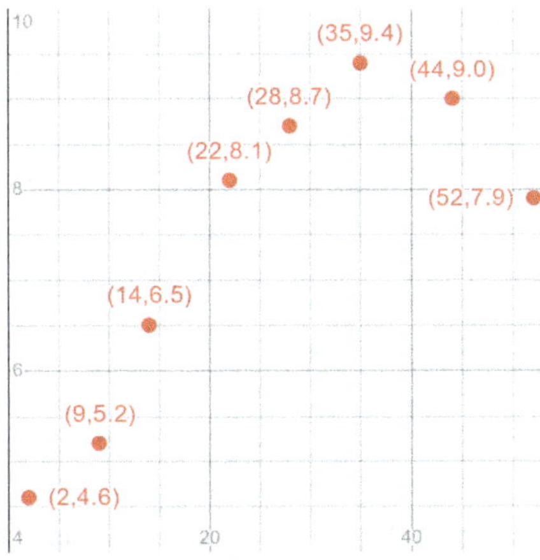

Figure 4.10: This data is not monotonic overall, increasing to a maximum at $(35, 9.4)$ and then decreasing.

Example 4.6. Construct $r_1(x)$ for the data of Figure 4.10 and calculate E_a.

Solution. The Table 4.1 template is used to find all needed sums.

i	x_i	y_i	x_i^2	$x_i y_i$
1	2	4.6	4	9.2
2	9	5.2	81	46.8
3	14	6.5	196	91.0
4	22	8.1	484	178.2
5	28	8.7	784	243.6
6	35	9.4	1225	329.0
7	44	9.0	1936	396.0
8	52	7.9	2704	410.8
	206	59.4	7414	1704.6

Parameter m is found first using Equation (4.8) and the sums from the table.

$$m = \frac{n\sum_{i=1}^{n} x_i y_i - \sum_{i=1}^{n} x_i \sum_{i=1}^{n} y_i}{n\sum_{i=1}^{n} x_i^2 - (\sum_{i=1}^{n} x_i)^2} = \frac{8(1704.6) - (206)(59.4)}{8(7414) - (206)^2} = 0.083$$

Using this m, the sums from the table, and Equation (4.9) for parameter b

$$b = \frac{\sum_{i=1}^{n} y_i - m \sum_{i=1}^{n} x_i}{n} = \frac{59.3 - (-0.083)(206)}{8} = 5.288$$

The linear regression function $r_1(x)$, according to Equation (4.2) and represented by the blue line in Figure 4.11 is

$$r_1(x) = 0.083x + 5.288.$$

Finding E_a algebraically confirms the poor fit.

$$E_a = \sum_{i=1}^{n} \left(y_i - r_1(x_i)\right)^2 = 7.949$$

i	x_i	y_i	$r_1(x_i)$	$(y_i - r_1(x_i))^2$
1	2	4.6	5.454	0.729
2	9	5.2	6.035	0.697
3	14	6.5	6.450	0.002
4	22	8.1	7.114	0.972
5	28	8.7	7.612	1.184
6	35	9.4	8.193	1.457
7	44	9.0	8.940	0.004
8	52	7.9	9.604	2.904

Figure 4.11 illustrates the drawback of a linear regression function on data which is not monotonic. All points to the left and leading to the maximum at $(35, 9.4)$ have increasing y_i values, warranting an $r_1(x)$ with a positive slope parameter m. But after the maximum, the y_i values decrease and force a compromise between the maximum and minimum at the endpoint. In the previous table the yellow cells highlight the two biggest contributing terms to the E_a, precisely at the maximum $(35, 9.4)$ and the right endpoint $(52, 7.9)$.

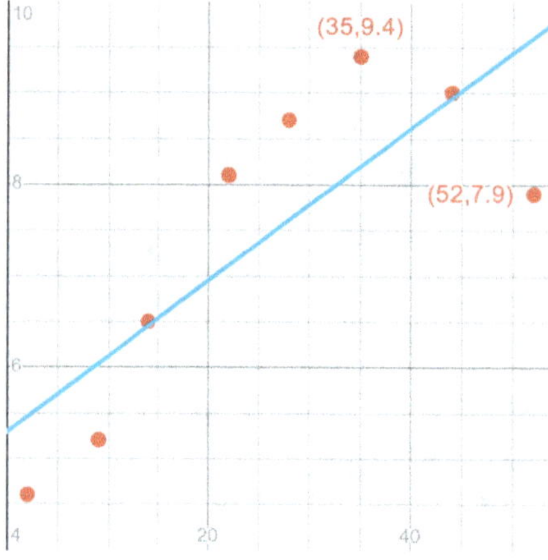

Figure 4.11: The best linear regression function $r_1(x)$ through the data of Figure 4.10, which is not very good.

Conversely, the $r_1(x)$ for the monotonic data of Figure 4.1, with an $E_a = 2.327$ per Table 4.8, is a much better fit than the $r_1(x)$ for the Figure 4.10 data.

Even if the dependent values are not strictly monotonic but the overall trend of the data is, an $r_1(x)$ may be adequate. For instance, consider the data of Figure 4.12. There are two instances of a dependent values decreasing, the points in green, but most dependent values are increasing, particularly at the beginning and ending points. The linear regression function through this data is the blue line in Figure 4.13, $r_1(x) = 0.073x + 8.061$ with an $E_a = 0.548$.

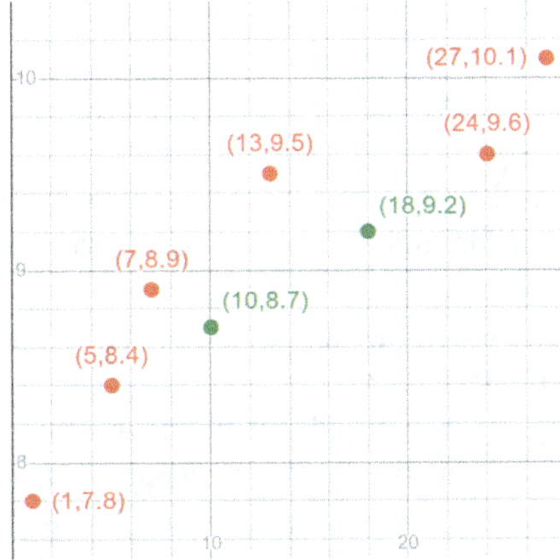

Figure 4.12: This data is not strictly monotonic because of the green points.

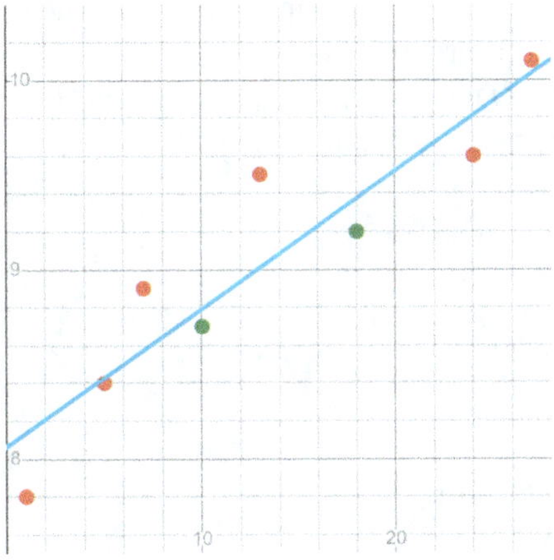

Figure 4.13: The blue linear regression function $r_1(x)$ may be suitable for data which is mostly, but not strictly, monotonic.

4.5.2 Logarithmic regression function considerations

Section 4.3.1 demonstrated any base β for the general logarithmic regression function $r_{L_\beta}(x)$ is viable; the natural logarithmic regression function $r_{\ln}(x)$ is used throughout this section for simplicity.

Recall the similarity between the linear and logarithmic regression functions of Equations (4.2) and (4.15).

$$r_1(x) = mx + b$$
$$r_{\ln}(x) = m \ln x + b$$

Parameters m and b are found analogously, with $\ln x$ replacing x throughout the calculations of Equations (4.17) and (4.18), and (4.8) and (4.9), respectively. More than just maintaining a simple function form, using $\ln x$ rather than x eliminates the constant rate of increase or decrease of the regression function and gives $r_{\ln}(x)$ curvature that $r_1(x)$ lacks. The curved $r_{\ln}(x)$ can lead to an improved fit.

Additionally, because the forms for m and b are retained, the magnitude of all sums remains relatively similar. In fact, logarithmic functions have a slow rate of change, so using $\ln x$ keeps the magnitude of values in the sums from rapidly changing. Lastly, in extreme cases for very small magnitudes of m, $r_{\ln}(x)$ can appear linear if applicable.

But a prominent drawback of $r_{\ln}(x)$ is also linked to its relation to $r_1(x)$. Although curved, the term $m \ln x$ is strictly increasing for $m > 0$ and strictly decreasing for $m < 0$. Therefore, data which is not monotonic might pose a problematic fit for $r_{\ln}(x)$. A second drawback, as previously discussed in Section 4.3.2, is $r_{\ln}(x)$ is limited to data which is generally either increasing and concave down or decreasing and concave up.

The following example again uses the data of Figure 4.10. An $r_{\ln}(x)$ is fit and shows both an improvement over $r_1(x)$ and the lingering issue of nonmonotonic data.

Example 4.7. Construct $r_{\ln}(x)$ for the data of Figure 4.10 and calculate E_a.
Solution. The Table 4.2 template organizes all necessary sums.

i	x_i	y_i	$\ln x_i$	$(\ln x_i)^2$	$(\ln x_i)y_i$
1	2	4.6	0.693	0.480	3.188
2	9	5.2	2.197	4.828	11.426
3	14	6.5	2.639	6.965	17.154
4	22	8.1	3.091	9.555	25.037
5	28	8.7	3.332	11.104	28.990
6	35	9.4	3.555	12.640	33.420
7	44	9.0	3.784	14.320	34.058
8	52	7.9	3.951	15.612	31.215
		59.4	23.243	75.504	184.488

Parameter m is found first using Equation (4.17) and the sums from the table.

$$m = \frac{n \sum_{i=1}^{n}(\ln x_i)y_i - \sum_{i=1}^{n}\ln x_i \sum_{i=1}^{n}y_i}{n \sum_{i=1}^{n}(\ln x_i)^2 - (\sum_{i=1}^{n}\ln x_i)^2} = \frac{8(184.488) - (23.243)(59.4)}{8(75.504) - (23.243)^2} = 1.494$$

Using this m, the sums from the table, and Equation (4.18) for parameter b

$$b = \frac{\sum_{i=1}^{n}y_i - m \sum_{i=1}^{n}\ln x_i}{n} = \frac{59.4 - (1.494)(23.243)}{8} = 3.085$$

By Equation (4.15), the logarithmic regression function $r_{\ln}(x)$, represented by the blue curve in Figure 4.14, is

$$r_{\ln}(x) = 1.494 \ln x + 3.085$$

The corresponding E_a algebraically verifies a better fit than the linear $r_1(x)$.

$$E_a = \sum_{i=1}^{n}\big(y_i - r_{\ln}(x_i)\big)^2 = 4.694$$

i	x_i	y_i	$r_{\ln}(x_i)$	$(y_i - r_{\ln}(x_i))^2$
1	2	4.6	4.121	0.230
2	9	5.2	6.368	1.363
3	14	6.5	7.028	0.279
4	22	8.1	7.703	0.158
5	28	8.7	8.063	0.405
6	35	9.4	8.397	1.007
7	44	9.0	8.739	0.068
8	52	7.9	8.988	1.184

Figure 4.14 displays a better navigation of $r_{\ln}(x)$ between the maximum at $(35, 9.4)$ and the decreased right endpoint $(52, 7.9)$ than the $r_1(x)$ of Figure 4.11, although there remains ample room for improvement using another function type. The green cells in

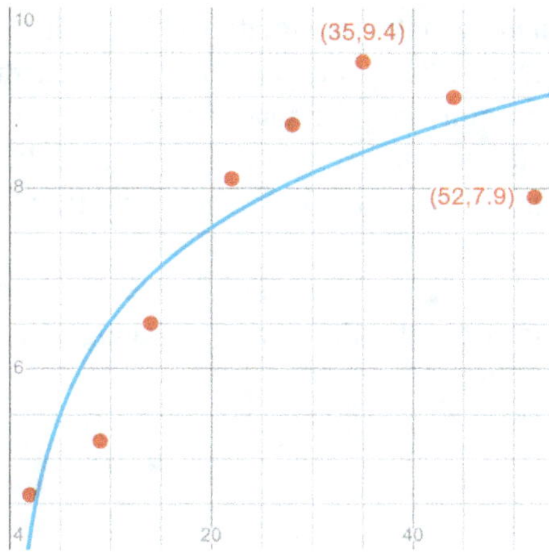

Figure 4.14: The best logarithmic regression function $r_{\ln}(x)$ through the data of Figure 4.10, an overall improvement to the linear regression function $r_1(x)$ of Figure 4.11.

the table show a decrease in the value of the LS terms at those same two points versus the yellow cells in the table of Example 4.6; the E_a has improved from 7.949 to 4.694.

However, note the better fit of the $r_{\ln}(x)$ of Figure 4.1 containing monotonic data, with an $E_a = 2.001$ per Table 4.8, to the $r_{\ln}(x)$ for the Figure 4.10 data. But revisiting the data of Figure 4.12, which is not strictly monotonic, using $r_{\ln}(x)$ is reasonable, as depicted by the blue curve $r_{\ln}(x) = 0.635 \ln x + 7.613$ of Figure 4.15 with an $E_a = 0.518$.

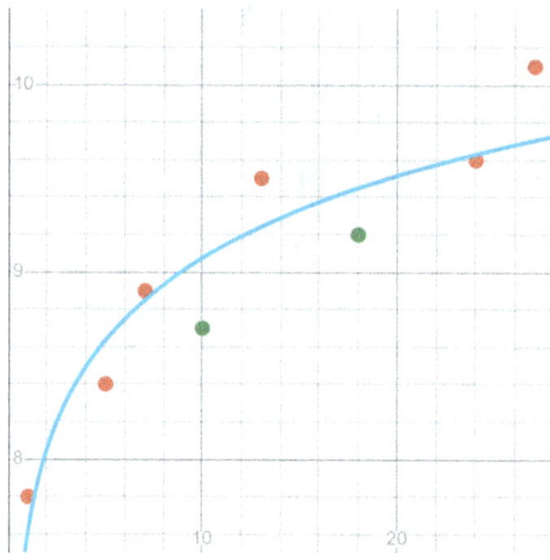

Figure 4.15: The blue logarithmic regression function $r_{\ln}(x)$ may be suitable for data which is mostly, but not strictly, monotonic.

4.5.3 Quadratic regression function considerations

One considerable limitation of the linear and logarithmic regression functions is modeling data which is not generally monotonic. The quadratic regression function $r_2(x)$ potentially offers a better solution since it contains a extremum which can fit around the change in the data. And as Example 4.5 and Figure 4.9 demonstrated, a small coefficient for the x^2 term allows $r_2(x)$ to behave linearly if appropriate.

A possible problem exists solving the augmented matrix of Equation (4.30) with the added sums used for calculating the parameters a, b, and c of $r_2(x)$. Third and fourth powers of the independent values could become quite large or small and lead to very large or small sum entries, especially relative to the other sum entries and the value of n. Reducing a matrix to RREF form with entries of large discrepancies could lead to rounding and accuracy errors.

Returning to the data of Figure 4.10, $\sum_{i=1}^{8} x_i{}^4 > 13$ million, while $\sum_{i=1}^{8} y_i = 59.4$ and $n = 8$. Thus in the following example, the x_i values are scaled first to close this magnitude gap.

Example 4.8. Construct $r_2(x)$ for the data of Figure 4.10 and calculate E_a.

Solution. As justified in the previous paragraph, all x_i values initially are divided by ten. The Table 4.4 template specifies the required sums.

x_i	y_i	x_i^2	x_i^3	x_i^4	$x_i^2 y_i$	$x_i y_i$
0.2	4.6	0.04	0.008	0.002	0.184	0.92
0.9	5.2	0.81	0.729	0.656	4.212	4.68
1.4	6.5	1.96	2.744	3.842	12.740	9.10
2.2	8.1	4.84	10.648	23.426	39.204	17.82
2.8	8.7	7.84	21.952	61.466	68.208	24.36
3.5	9.4	12.25	42.875	150.063	115.150	32.90
4.4	9.0	19.36	85.184	374.810	174.240	39.60
5.2	7.9	27.04	140.608	731.162	213.616	41.08
20.6	59.4	74.14	304.748	1345.427	627.554	170.46

Using all sums from the table and $n = 8$ in the Equation (4.30) augmented matrix

$$\begin{bmatrix} 1345.427 & 304.748 & 74.14 & \vdots & 627.554 \\ 304.748 & 74.14 & 20.6 & \vdots & 170.46 \\ 74.14 & 20.6 & 8 & \vdots & 59.4 \end{bmatrix}$$

Reducing this matrix to RREF with Gauss–Jordan elimination solves the parameters a, b, and c simultaneously.

$$\begin{bmatrix} 1 & 0 & 0 & \vdots & -0.395 \\ 0 & 1 & 0 & \vdots & 2.963 \\ 0 & 0 & 1 & \vdots & 3.459 \end{bmatrix}$$

In order, $a = -0.395$, $b = 2.963$, and $c = 3.459$. According to Equation (4.25) the quadratic regression function $r_2(x)$ graphed in blue in Figure 4.16 is

$$r_2(x) = -0.395x^2 + 2.963x + 3.459$$

The associated E_a algebraically confirms the best fit yet.

$$E_a = \sum_{i=1}^{n}\left(y_i - r_2(x_i)\right)^2 = 1.071$$

i	x_i	y_i	$r_2(x_i)$	$(y_i - r_2(x_i))^2$
1	0.2	4.6	4.036	0.318
2	0.9	5.2	5.806	0.367
3	1.4	6.5	6.833	0.111
4	2.2	8.1	8.066	0.001
5	2.8	8.7	8.659	0.002
6	3.5	9.4	8.991	0.167
7	4.4	9.0	8.849	0.023
8	5.2	7.9	8.186	0.082

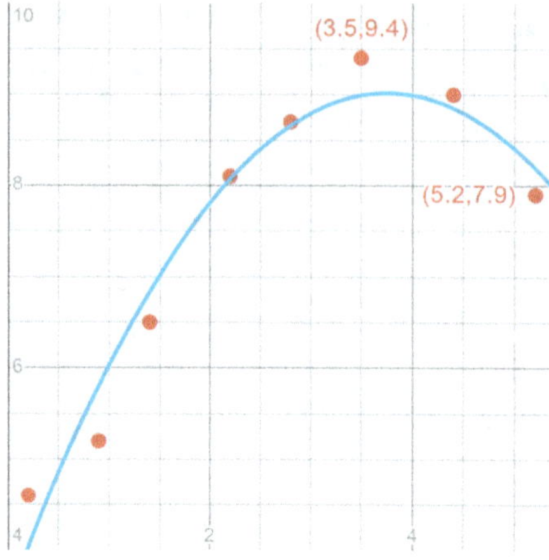

Figure 4.16: With a built-in extremum, the quadratic regression function $r_2(x)$ through the non-monotonic data of Figure 4.10 is a better fit than the linear regression function $r_1(x)$ of Figure 4.11 and the logarithmic regression function $r_{\ln}(x)$ of Figure 4.14.

Both the linear and logarithmic regression functions are better fits for the monotonic data of Figure 4.1 than for the nonmonotonic data of Figure 4.10. The reverse is true in this case for quadratic regression function, which has an $E_a = 1.071$ for the Figure 4.10 data and an $E_a = 2.028$ for the Figure 4.1 data. However, no such correlation between the regression function type and data type is implied in general.

Least squares regression concepts summary

1. Unlike interpolation, the goal of a regression function is minimization of the differences d_i between given and approximated points; the regression function does not need to fit any point exactly.
2. The least squares (LS) accumulation error E_a squares each d_i, thus removing signs and penalizing greater values.
3. An E_a defined by $|d_i|$ values has order $O(d_i)$; the LS E_a has order $O(d_i^2)$.
4. Linear $r_1(x)$ and logarithmic $r_{L_\beta}(x)$ regression functions can be defined similarly by two parameters while the quadratic regression function $r_2(x)$ requires three parameters.
5. Any base β of $r_{L_\beta}(x)$ can be used without affecting the fit of a logarithmic regression function because of the change of base property.
6. $r_1(x)$ functions are best for data which generally exhibits a constant rate of change.
7. $r_{L_\beta}(x)$ functions are best for data which is generally either monotonically increasing and concave down, or monotonically decreasing and concave up.
8. $r_2(x)$ functions are best for data which is generally nonmonotonic with a single extremum.
9. Initially scaling large independent values may benefit an $r_2(x)$ function fit, removing magnitude discrepancies between the third and fourth powers of the independent values and their sums, and the other values in the matrix equation.
10. E_a is relative to the number and size of the data; it can be used to compare regression functions of different types on the same data but not across different data.

Exercises

Section 4.2

Construct the linear regression function $r_1(x)$ for the given points $(x, f(x))$. Use $r_1(x)$ to approximate function values at the two unknown x values.

1.

x	$f(x)$
0	12
5	27
10	48
15	64
20	79
6	
18	

2.

x	$f(x)$
2	91
3	84
4	77
5	73
6	62
7	55
2.7	
5.1	

3.

x	$f(x)$
1.4	24.6
3.2	21.2
3.9	19.1
4.5	16.0
5.7	12.9
7.1	10.7
8.6	4.4
10.3	2.8
3.4	
8.2	

4.

x	$f(x)$
12.5	0.9
21.7	2.3
33.8	3.1
50.1	4.7
67.2	6.2
71.3	6.5
78.6	8.4
60.0	
70.5	

5.

x	$f(x)$
3	2.4
6	1.9
9	1.6
12	1.2
15	0.7
18	0.5
21	0.3
4.3	
14.1	

6.

x	$f(x)$
10	15.4
20	30.2
30	45.3
40	61.7
50	75.1
22.6	
37.8	

7.

x	$f(x)$
20.0	2.5
20.5	7.2
21.0	11.4
21.5	16.1
22.0	23.6
22.5	28.9
23.0	32.3
23.5	36.8
24.0	39.7
21.4	
23.6	

8.

x	$f(x)$
8.2	66.1
9.6	56.9
11.0	49.3
12.4	41.4
13.8	34.5
15.2	23.4
16.6	15.7
18.0	7.1
8.8	
13.0	

9.		10.		11.		12.	
x	$f(x)$	x	$f(x)$	x	$f(x)$	x	$f(x)$
0.8	65	1.5	2.6	2	−50	5.2	38.4
2.1	112	4.0	5.4	4	−63	5.3	30.9
2.9	97	6.5	17.3	6	−82	5.4	31.3
4.3	126	9.0	12.9	8	−76	5.5	24.5
5.4	163	11.5	25.2	10	−94	5.6	25.7
6.7	152	14.0	37.8	12	−105	5.7	11.6
9.5	190	4.7		14	−111	5.8	14.0
3.9		13.1		16	−128	5.9	8.2
7.5				18	−119	6.0	5.8
				20	−133	5.45	
				9.9		5.77	
				12.7			

Section 4.3

Construct the logarithmic regression function $r_{\ln}(x)$ for the given points $(x, f(x))$. Use $r_{\ln}(x)$ to approximate function values at the two unknown x values.

13.		14.		15.		16.	
x	$f(x)$	x	$f(x)$	x	$f(x)$	x	$f(x)$
1	14.3	4	124.0	5	4.5	1	10.8
2	19.0	8	133.5	15	2.45	4	8.4
3	27.5	12	136.5	25	2.1	7	6.5
4	28.7	16	141.0	35	1.75	10	4.3
5	30.2	20	143.5	45	1.6	13	3.6
6	31.6	6.8		55	1.45	16	2.9
1.6		18.5		65	1.2	19	2.0
4.9				20		22	1.9
				57		5.9	
						14.7	

17.

x	$f(x)$
8.00	7.2
9.25	7.5
10.50	7.8
11.75	8.0
13.00	8.2
14.25	8.5
15.50	8.8
16.75	9.0
11.9	
14.75	

18.

x	$f(x)$
3.00	47.8
3.75	32.3
4.50	24.6
5.25	19.7
6.00	15.9
6.75	13.4
7.50	11.2
5.75	
6.4	

19.

x	$f(x)$
4.5	19.25
14.9	15.50
27.8	13.00
40.3	11.50
52.4	10.25
65.7	9.50
79.6	8.50
92.0	7.50
104.1	7.00
113.3	6.50
30.5	
84.9	

20.

x	$f(x)$
11.0	−10.8
17.1	−11.1
20.8	−11.3
25.2	−11.7
29.6	−12.0
34.0	−12.2
38.4	−12.6
42.7	−12.9
47.3	−13.1
21.2	
40.3	

21.

x	$f(x)$
4	35.2
12	34.6
20	34.9
28	34.0
36	33.2
44	33.5
52	32.7
60	33.3
68	32.4
38.3	
66.1	

22.

x	$f(x)$
2	3.6
8	4.4
14	4.3
20	4.1
26	4.5
32	5.2
38	5.8
44	6.1
15.3	
33.4	

23.

x	$f(x)$
0.5	8
8.0	17
15.5	24
23.0	22
30.5	28
38.0	33
45.5	31
53.0	30
60.5	35
68.0	39
75.5	42
83	43
26.6	
70.8	

24.

x	$f(x)$
3	19.7
15	15.4
27	15.8
39	13.6
51	12.3
63	12.9
4.2	
59.7	

Section 4.4

Construct the quadratic regression function $r_2(x)$ for the given points $(x, f(x))$. Use $r_2(x)$ to approximate function values at the two unknown x values.

25.

x	$f(x)$
0	8.2
2	42.9
4	65.1
6	78.4
8	90.3
10	91.0
12	79.5
14	54.7
16	27.6
3.5	
12.3	

26.

x	$f(x)$
7.5	51
9.0	93
10.5	115
12.0	124
13.5	126
15.0	112
16.5	97
18.0	66
8.0	
14.9	

27.

x	$f(x)$
4.4	15.6
5.0	16.4
5.6	17.8
6.2	23.1
6.8	34.5
7.4	40.9
8.0	61.7
5.1	
6.6	

28.

x	$f(x)$
1	96.4
4	90.9
7	88.3
10	81.5
13	77.2
16	62.1
19	57.8
22	37.6
25	28.7
28	10.3
10.3	
25.8	

29.

x	$f(x)$
10	81.2
15	72.8
20	66.3
25	60.7
30	51.4
35	39.0
13.8	
32.0	

30.

x	$f(x)$
0	144.7
8	156.3
16	167.5
24	180.9
32	199.1
11.7	
20.0	

31.

x	$f(x)$
110	70.8
120	48.2
130	35.7
140	29.4
150	32.9
160	44.0
170	59.1
180	87.3
190	114.5
200	145.6
149	
195	

32.

x	$f(x)$
25	315.4
75	243.9
125	207.3
175	202.8
225	192.5
275	201.7
325	235.6
375	287.0
425	330.2
270	
352.5	

Hint: For #35 and #36, scale by a prime number that evenly rounds all x values to the tenth decimal place.

33.

x	$f(x)$
1.5	44.6
2.0	18.8
2.5	20.5
3.0	12.7
3.5	29.3
4.0	45.1
4.5	42.4
5.0	68.9
2.3	
4.8	

34.

x	$f(x)$
19.5	15.1
21.0	8.3
22.0	0.9
23.5	1.2
24.0	0.5
25.5	5.4
26.5	12.6
22.8	
23.9	

35.

x	$f(x)$
27.6	2.8
41.4	22.7
52.9	21.0
62.1	34.1
73.6	33.6
80.5	42.5
89.7	45.3
101.2	41.2
110.4	32.8
117.3	33.4
126.5	20.6
133.4	11.3
78.2	
131.1	

36.

x	$f(x)$
85	465.3
119	329.4
153	218.6
187	118.5
221	125.1
255	136.9
289	110.7
323	312.8
357	279.0
391	413.2
136	
278.8	

Section 4.5

Graph each set of points. Construct the one or two most appropriate linear, logarithmic, or quadratic regression function(s). Calculate the accumulation error E_a for each regression function found. In the case of two possible functions, use E_a to determine the better fit.

37.

x	$f(x)$
3	17.6
4	23.2
5	29.8
6	37.6
7	42.7
8	50.1
9	55.8

38.

x	$f(x)$
1	168.8
3	160.9
5	152.3
7	146.1
9	137.6
11	129.7
13	121.8
15	115.5
17	106.4

39.

x	$f(x)$
4	9.8
8	26.5
12	53.4
16	75.2
20	91.9
24	97.0
28	94.3
32	88.6
36	83.1
40	76.7

40.

x	$f(x)$
6	65.3
9	50.2
12	45.7
15	43.4
18	40.8
21	41.9
24	44.6
27	47.1

41.

x	$f(x)$
2	10.8
7	16.4
15	19.5
27	21.6
40	24.7
48	25.1
57	26.2
64	26.9
74	26.3

42.

x	$f(x)$
0.5	80.2
8.0	54.3
15.5	46.3
23.0	42.5
30.5	40.9
38.0	38.4
45.5	36.9

43.

x	$f(x)$
15	22.1
20	18.9
25	17.0
30	14.8
35	14.2
40	12.7
45	12.0
50	10.6
55	10.2
60	9.4

44.

x	$f(x)$
2	8.7
5	10.8
9	12.6
11	13.3
12	13.5
18	14.2
23	15.0
25	15.1

45.

x	$f(x)$
7.5	254
10.0	230
12.5	199
15.0	168
17.5	132
20.0	105
22.5	71
25.0	35

46.

x	$f(x)$
4.2	33.8
7.8	80.4
11.4	140.1
15.0	191.0
18.6	255.7
22.2	304.0

47.

x	$f(x)$
10.0	12.4
16.4	8.3
22.8	6.7
29.2	5.0
35.6	4.7
42.0	3.4
48.4	3.2

48.

x	$f(x)$
8.0	52.3
8.5	63.8
9.0	77.1
9.5	84.2
10.0	88.6
10.5	91.9
11.0	92.4
11.5	89.6
12.0	86.5

5 Integration

Differentiation of closed-form functions of a single variable $f(x)$ is always possible. Applying elementary derivative rules, and possibly more advanced rules such as those involving products and composites (chains) of functions, is straightforward for combinations of polynomial, trigonometric, logarithmic, and exponential functions. Integration is not as straightforward and often impossible because many functions $f(x)$ do not have an analytic closed-form antiderivative $F(x)$ consisting of elementary functions in which $F(x) = \int f(x)\, dx$. An example often cited is the relatively simple $\int e^{x^2}\, dx$.

A multitude of practical models and problems stem from definite integrals which necessitate methods for evaluating them, whether or not an antiderivative exists. This chapter first explores three options along with expected errors that arise by applying one or more applications of the algorithms. All these methods begin with the premise that a particular nth-degree polynomial function $f_n(x)$ satisfactorily estimates a function $f(x)$. Whereas $f(x)$ cannot be antidifferentiated, $f_n(x)$ always can by using the simple power rule for all x^n terms, or perhaps more efficiently. The chapter concludes with a completely different approach that employs a random simulation.

5.1 Trapezoidal rule

Recall the definite integral derives from using Riemann sums and rectangles that estimate the area between a function and the x-axis. As the width of the rectangles approaches zero, coinciding with the number of rectangles approaching infinity, the Riemann sum equals the true value of the area enclosed, provided the function satisfies certain conditions. Other than a horizontal line, the simplest polynomial used as an approximation for a function $f(x)$ is linear and has degree one. Graphically this changes the approximating shape from a rectangle to a trapezoid. Figures 5.1 and 5.2 depict a single trapezoidal approximation of $f(x)$, while Figure 5.3 through 5.5 preview using multiple trapezoids.

5.1.1 Single application

Equation (3.6) defined a linear polynomial function through two points $(x_0, f(x_0))$ and $(x_1, f(x_1))$ with the 1st DD of Equation (3.4). Using that DD polynomial function $f_1(x)$ as an approximation of $f(x)$, the definite integral of $f(x)$ over $[x_0, x_1]$ becomes

$$\int_{x_0}^{x_1} f(x)\, dx \approx \int_{x_0}^{x_1} f_1(x)\, dx = \int_{x_0}^{x_1} \left(f(x_0) + \frac{f(x_1) - f(x_0)}{x_1 - x_0}(x - x_0) \right) dx$$

https://doi.org/10.1515/9783112221051-005

Using a common denominator for the first term and distributing the fraction of the second term expands as

$$= \int_{x_0}^{x_1} \left(\frac{x_1 f(x_0) - x_0 f(x_0)}{x_1 - x_0} + \frac{f(x_1) - f(x_0)}{x_1 - x_0} x + \frac{-x_0 f(x_1) + x_0 f(x_0)}{x_1 - x_0} \right) dx.$$

The first and third terms are constants which are combined and integrated by multiplying by x; the second term is a constant multiplied by x and is integrated with the power rule.

$$= \left(\frac{x_1 f(x_0) - x_0 f(x_1)}{x_1 - x_0} x + \frac{f(x_1) - f(x_0)}{2(x_1 - x_0)} x^2 \right) \Big|_{x_0}^{x_1}$$

Substituting the limits of integration and subtracting,

$$= \frac{x_1 f(x_0) - x_0 f(x_1)}{x_1 - x_0} (x_1 - x_0) + \frac{f(x_1) - f(x_0)}{2(x_1 - x_0)} (x_1^2 - x_0^2).$$

The denominator factor $x_1 - x_0$ can be reduced in each term (the second term has a difference of squares in the numerator), leaving a common denominator of just two.

$$= \frac{2x_1 f(x_0) - 2x_0 f(x_1)}{2} + \frac{(f(x_1) - f(x_0))(x_1 + x_0)}{2}$$

$$= \frac{x_1 f(x_0) - x_0 f(x_1) + x_1 f(x_1) - x_0 f(x_0)}{2}$$

$$= \frac{f(x_0)(x_1 - x_0) + f(x_1)(-x_0 + x_1)}{2}$$

One final factoring produces the integral estimation equivalent to the area of a trapezoid, or one application of trapezoidal rule A_T, where height $h = x_1 - x_0$ and the two bases are $b_1 = f(x_0)$ and $b_2 = f(x_1)$.

$$\int_{x_0}^{x_1} f(x)\, dx \approx \int_{x_0}^{x_1} f_1(x)\, dx = A_T = \frac{(x_1 - x_0)(f(x_0) + f(x_1))}{2} \tag{5.1}$$

Note the graph will *not* produce a trapezoid if the bases differ in sign, but rather two triangles, one above and one below the x-axis. But the integral estimation is still valid because it was derived algebraically without any restrictions on the signs of $f(x_0)$ and $f(x_1)$. That it sometimes corresponds to the area of a trapezoid is a visually helpful particular case of the general integration of the line $f_1(x)$.

Example 5.1. Use the trapezoidal rule to estimate the area between $f(x) = 2e^{-\sqrt{x-1}}$ and the x-axis over the interval $[1.25, 3]$.

Solution. Using Equation (5.1), the area estimation is

$$\int_{1.25}^{3} 2e^{-\sqrt{x-1}}\,dx \approx A_T = \frac{(3-1.25)(f(1.25)+f(3))}{2} = \frac{(1.75)(1.213+0.486)}{2} = 1.487.$$

Figure 5.1 shows the true area between $f(x)$ and the x-axis in red. The discrepancy of the estimation using one trapezoid is the green area, representing the truncation error $E_{t,T}$ defined in Section 5.1.2. If reasonably estimated, $E_{t,T}$ can be used in conjunction with the trapezoidal rule for a better estimation, as demonstrated also in Section 5.1.2. Because of the concavity of $f(x)$, the green area included in the trapezoidal estimation indicates an overestimation.

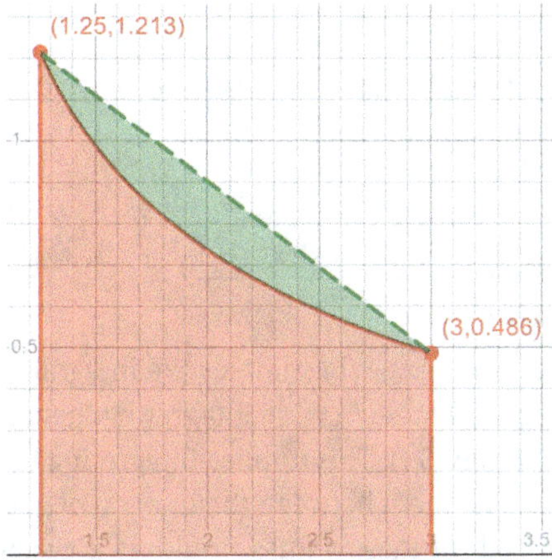

Figure 5.1: The area between $f(x) = 2e^{-\sqrt{x-1}}$ and the x-axis over the interval $[1.25, 3]$ is red. One trapezoid overestimates this area, according to the green area.

5.1.2 Single application, truncation error adjusted

A function $f(x)$ represented by a Taylor series centered at x_0 can be written

$$f(x) = f(x_0) + f'(x_0)(x - x_0) + \frac{f''(c)(x - x_0)^2}{2}$$

The first two terms are the first-degree Taylor polynomial approximation $T_1(x)$ and the third term is the Lagrange form of the remainder $R_1(x)$. Approximating $f(x)$ involves approximating both $T_1(x)$ and $R_1(x)$, not centered around one point $(x_0, f(x_0))$ by itself but in conjunction with a second point $(x_1, f(x_1))$.

First, the tangent line slope $f'(x_0)$ at one point $(x_0, f(x_0))$ is approximated by the secant line slope through that same point and another point $(x_1, f(x_1))$ by

$$f'(x_0) \approx \frac{f(x_1) - f(x_0)}{x_1 - x_0}$$

In turn the Taylor polynomial $T_1(x)$ is approximated according to

$$T_1(x) \approx f(x_0) + \frac{f(x_1) - f(x_0)}{x_1 - x_0}(x - x_0).$$

Second, an approximation for $R_1(x)$ also incorporates this second point. It follows the approximation for $R_1(x)$, which by definition is centered around just one point, is approximated using the two points thusly

$$(x - x_0)^2 \approx (x - x_0)(x - x_1)$$

As such the remainder term $R_1(x)$ is approximated by

$$R_1(x) \approx \frac{f''(c)(x - x_0)(x - x_1)}{2}.$$

Now, expressing $f(x)$ using the approximations for both $T_1(x)$ and $R_1(x)$ above

$$f(x) = f(x_0) + f'(x_0)(x - x_0) + \frac{f''(c)(x - x_0)^2}{2}$$

$$\approx f(x_0) + \frac{f(x_1) - f(x_0)}{x_1 - x_0}(x_1 - x_0) + \frac{f''(c)(x - x_0)(x - x_1)}{2}. \tag{5.2}$$

Integrating the entire right side of Equation (5.2) over $[x_0, x_1]$ estimates the integral of $f(x)$ over that interval. The first two terms are equivalent to $f_1(x)$, so integrating them is precisely the trapezoidal rule estimation of Equation (5.1). This implies integrating the remainder term produces the truncation error adjustment between the integral of $f(x)$ and the trapezoidal rule. It also implies using both of these values produces a closer estimation to the true integral value. The calculation of $f''(c)$ is explained afterward, but note it is a constant value, so that defining and integrating the remainder term

$$E_{t,T} = \int_{x_0}^{x_1} \frac{f''(c)(x - x_0)(x - x_1)}{2}\, dx$$

$$= \frac{f''(c)}{2} \int_{x_0}^{x_1} (x^2 - x_1 x - x_0 x + x_1 x_0)\, dx$$

$$\frac{f''(c)}{2}\left(\frac{x^3}{3} - \frac{x_1 x^2}{2} - \frac{x_0 x^2}{2} + x_1 x_0 x \right)\Big|_{x_0}^{x_1}$$

$$= \frac{f''(c)}{2}\left(\left(\frac{x_1^3}{3} - \frac{x_1^3}{2} - \frac{x_0 x_1^2}{2} + x_1^2 x_0 \right) - \left(\frac{x_0^3}{3} - \frac{x_1 x_0^2}{2} - \frac{x_0^3}{2} + x_1 x_0^2 \right) \right)$$

$$= \frac{f''(c)}{2}\left(\left(-\frac{x_1^3}{6} + \frac{3x_1^2 x_0}{6} \right) - \left(\frac{3x_1 x_0^2}{6} - \frac{x_0^3}{6} \right) \right)$$

$$= -\frac{f''(c)}{12}(x_1^3 - 3x_1^2 x_0 + 3x_1 x_0^2 - x_0^3)$$

The four terms in parentheses are the cubic expansion of the binomial $x_1 - x_0$, which condenses the truncation error adjustment to

$$E_{t,T} = -\frac{f''(c)(x_1 - x_0)^3}{12}.$$ (5.3)

If $f(x)$ is linear, $f''(x) = 0$ and the trapezoidal rule will fit $f(x)$ exactly with $E_{t,T} = 0$. Otherwise, the suitably small interval $[x_0, x_1]$ used to estimate $f(x)$ leaves little room for changes in concavity so that $f''(x)$ can be treated as a constant. By extension, $f''(c)$ is then constant because $x_0 \le c \le x_1$. To estimate this constant, the mean of continuous $f''(x)$ values on $[x_0, x_1]$ is used.

$$f''(c) \approx \overline{f''(x)} = \frac{\int_{x_0}^{x_1} f''(x)\,dx}{x_1 - x_0}.$$

Integrating the second derivative is equivalent to the first derivative so that

$$f''(c) \approx \frac{\int_{x_0}^{x_1} f''(x)\,dx}{x_1 - x_0} = \frac{f'(x)|_{x_0}^{x_1}}{x_1 - x_0} = \frac{f'(x_1) - f'(x_0)}{x_1 - x_0}.$$ (5.4)

Substituting the right side of Equation (5.4) for $f''(c)$ in Equation (5.3) defines the final form of the truncation error adjustment for a single application of the trapezoidal rule $E_{t,T}$.

$$E_{t,T} = -\frac{f''(c)(x_1 - x_0)^3}{12} \approx -\frac{f'(x_1) - f'(x_0)}{x_1 - x_0}\frac{(x_1 - x_0)^3}{12}$$

$$E_{t,T} = -\frac{(f'(x_1) - f'(x_0))(x_1 - x_0)^2}{12}$$ (5.5)

Summarizing, Equation (5.2) is a complete approximation of $f(x)$. Integrating Equation (5.2) produces the trapezoidal rule area estimation of Equation (5.1) plus the truncation error adjustment of Equation (5.5). Together they represent an improved trapezoidal rule $A_{T,t}$ of $f(x)$ than just the trapezoidal rule of Equation (5.1).

$$\int_{x_0}^{x_1} f(x)\,dx \approx A_{T,t} = A_T + E_{t,T}$$ (5.6)

Note the truncation error adjustment $E_{t,T}$ can be positive or negative and indicates whether the trapezoidal rule underestimated or overestimated the true integral value. If $E_{t,T} > 0$, truncating omitted a positive value and adding $E_{t,T}$ to the trapezoidal rule estimation means by itself the trapezoidal rule underestimated the true integral value. Conversely if $E_{t,T} < 0$, truncating omitted a negative value and adding, or equivalently subtracting, $E_{t,T}$ to the trapezoidal rule estimation means by itself the trapezoidal rule overestimated the true integral value.

Example 5.2. Calculate $E_{t,T}$ for the definite integral of Example 5.1. Apply $E_{t,T}$ for the improved trapezoidal rule estimation $A_{T,t}$. Use $E_{t,T}$ to determine whether the trapezoidal rule alone A_T overestimates or underestimates the true integral value.

Solution. Using $f'(x) = \dfrac{-e^{-\sqrt{x-1}}}{\sqrt{x-1}}$ and Equation (5.5)

$$E_{t,T} = -\frac{(f'(3) - f'(1.25))(3 - 1.25)^2}{12} = -\frac{(-0.172 + 1.213)(1.75)^2}{12} = -0.266$$

The trapezoidal rule estimation $A_T = 1.487$ of Example 5.1 is improved by adding this negative adjustment.

$$A_{T,t} = A_T + E_{t,T} = 1.487 + (-0.266) = 1.221$$

Because $E_{t,T} < 0$, the trapezoidal rule alone overestimated the true integral value and was modified by decreasing A_T. Figure 5.1 supports this conclusion.

Example 5.3. Use A_T and $A_{T,t}$ to estimate $\int_{0.5}^{3.5} \ln(6x - x^2)\,dx$. Use $E_{t,T}$ to determine if A_T overestimates or underestimates the true integral value.

Solution. Using the trapezoidal rule of Equation (5.1),

$$A_T = \frac{(3.5 - 0.5)(f(0.5) + f(3.5))}{2} = \frac{(3)(2.169 + 1.012)}{2} = 4.772.$$

Using $f'(x) = \dfrac{6-2x}{6x-x^2}$ and the truncation error adjustment of Equation (5.5),

$$E_{t,T} = -\frac{(f'(3.5) - f'(0.5))(3.5 - 0.5)^2}{12} = -\frac{(-0.114 - 1.818)(3)^2}{12} = 1.449.$$

The adjusted estimation $A_{T,t}$ adds the truncation error adjustment to the trapezoidal rule estimation.

$$A_{T,t} = A_T + E_{t,T} = 4.772 + 1.449 = 6.221.$$

$E_{t,T} > 0$ indicates the trapezoidal rule alone A_T underestimates the true integral value. Figure 5.2 verifies the algebra.

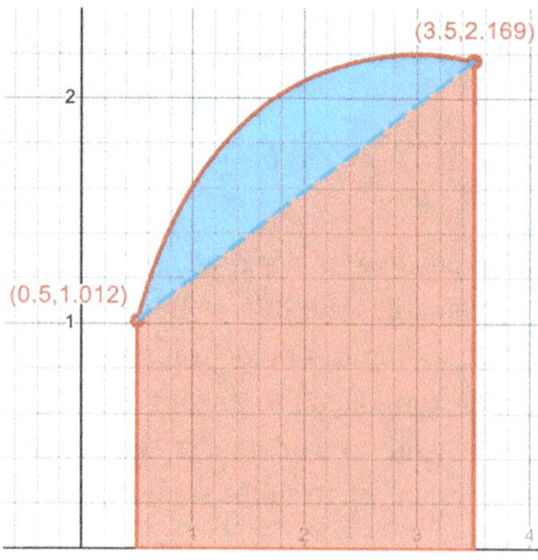

Figure 5.2: The trapezoidal rule underestimates the true integral value by not including the blue area above the trapezoid and below the graph of $f(x) = \ln(6x - x^2)$ over the interval $[0.5, 3.5]$. The true integral value is the red area plus the blue area.

5.1.3 Multiple applications

A single application of the trapezoidal rule is improved easily by dividing the entire interval $[x_0, x_1]$ into n smaller, equal subintervals and applying the trapezoidal rule to each. Figure 5.3 through 5.5 divide an interval into $n = 4$, $n = 8$, and $n = 16$ subintervals and trapezoids.

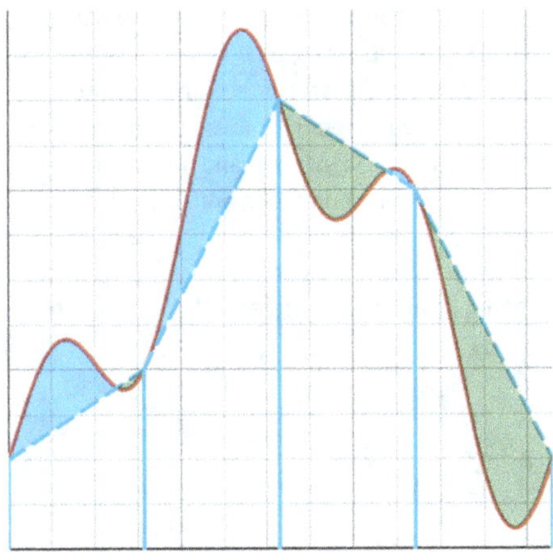

Figure 5.3: Four trapezoids estimate the area between the red $f(x)$ and the x-axis. If the blue dashed line of the trapezoid lies below $f(x)$, the difference of blue area indicates an underestimation. Conversely, where the blue dashed line of the trapezoid lies above $f(x)$, the difference of green area produces an overestimation.

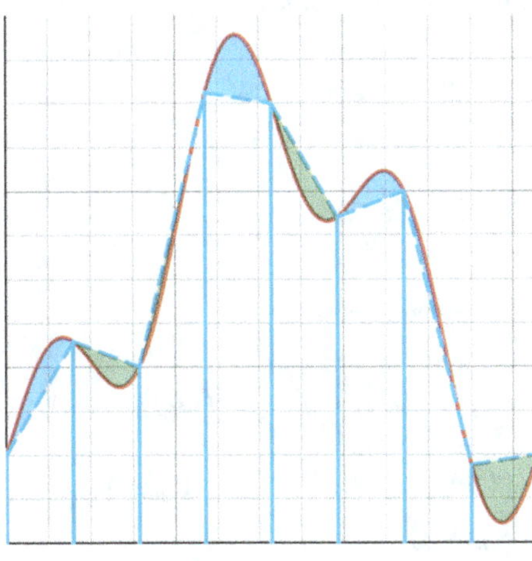

Figure 5.4: Eight trapezoids estimate the area between the red $f(x)$ and the x-axis. Blue area underestimates the true area, while green area overestimates the true area. As the number of trapezoids has doubled from four in Figure 5.3, the total amount of both blue and green area has decreased.

The single trapezoid height $h = x_1 - x_0$ was defined over the entire interval in Equation (5.1). Each of the multiple trapezoid heights h is now defined as the difference between consecutive, equally spaced x values for $i = 1, \ldots, n$.

$$h = x_i - x_{i-1}$$

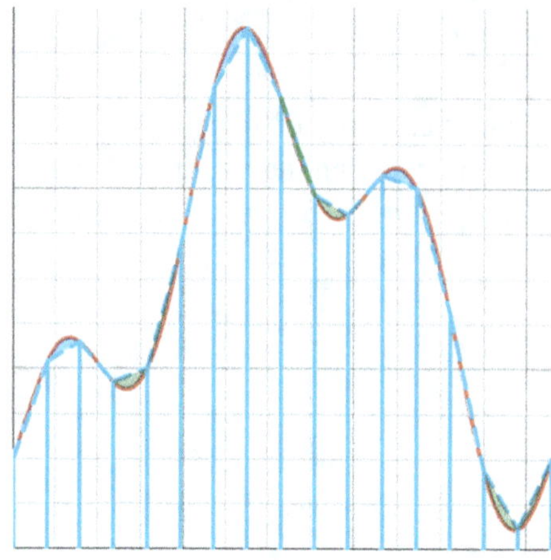

Figure 5.5: 16 trapezoids estimate the area between the red $f(x)$ and the x-axis; blue and green area underestimates and overestimates the true area, respectively. The number of trapezoids has doubled again, from eight in Figure 5.4, and now the total amount of both blue and green area is minimal.

The sum of all h values for $i = 1,\ldots,n$ is the length of the entire interval from x_0 to x_n. Since there are n equal h values

$$x_n - x_0 = \sum_{i=1}^{n}(x_i - x_{i-1}) = \sum_{i=1}^{n} h = nh$$

$$h = \frac{x_n - x_0}{n}$$

Using Equation (5.1) the integral of each of the multiple trapezoids is

$$\int_{x_{i-1}}^{x_i} f(x)\,dx \approx \frac{(x_i - x_{i-1})(f(x_{i-1}) + f(x_i))}{2}$$

$$= \frac{h(f(x_{i-1}) + f(x_i))}{2} = \frac{(x_n - x_0)(f(x_{i-1}) + f(x_i))}{2n}$$

Summing all n trapezoids over the entire interval $[x_0, x_n]$,

$$\sum_{i=1}^{n} \frac{(x_n - x_0)(f(x_{i-1}) + f(x_i))}{2n} = \frac{x_n - x_0}{2n}\big((f(x_0) + f(x_1)) + (f(x_1) + f(x_2))$$

$$+ (f(x_2) + f(x_3)) + \cdots + (f(x_{n-2}) - f(x_{n-1}))$$

$$+ (f(x_{n-1}) - f(x_n))\big)$$

The function values at the endpoints, $f(x_0)$ and $f(x_n)$, occur once in the expansion on the right, while each interior function value $f(x_i)$ for $i = 1,\ldots,n-1$ occurs twice. The right side can be rewritten

$$\sum_{i=1}^{n} \frac{(x_n - x_0)(f(x_{i-1}) + f(x_i))}{2n} = \frac{x_n - x_0}{2n}\left(f(x_0) + f(x_n) + \sum_{i=1}^{n-1} 2f(x_i)\right).$$

Lastly, the factor $\frac{1}{2}$ on the right side is distributed to minimize the number of operations. Rather than multiplying each interior term by two and then dividing by two, the first and last terms are simply divided by two. The final form of the trapezoidal rule using n multiple applications $A_T(n)$ is defined

$$\int_{x_0}^{x_n} f(x)\,dx \approx A_T(n) = \frac{x_n - x_0}{n}\left(\frac{f(x_0) + f(x_n)}{2} + \sum_{i=1}^{n-1} f(x_i)\right). \tag{5.7}$$

If the integrand $f(x)$ is continuous on the interval $[x_0, x_n]$, the trapezoidal rule on the right side of Equation (5.7) will converge exactly to $\int_{x_0}^{x_n} f(x)\,dx$ as $n \to \infty$. Figures 5.3 through 5.5 imply such convergence, verified by the following algebraic proof that invokes the limit of Riemann sums.

Theorem 5.1. *For $f(x)$ continuous on the interval $[x_0, x_n]$,*

$$\lim_{n\to\infty} \frac{x_n - x_0}{n}\left(\frac{f(x_0) + f(x_n)}{2} + \sum_{i=1}^{n-1} f(x_i)\right) = \int_{x_0}^{x_n} f(x)\,dx.$$

Proof. Divide the interval $[x_0, x_n]$ into subintervals $[x_{i-1}, x_i]$ for $i = 1, 2, \ldots, n$. Define a left endpoint Riemann sum.

$$L = \frac{x_n - x_0}{n} \sum_{i=1}^{n} f(x_{i-1})$$

Define a right endpoint Riemann sum.

$$R = \frac{x_n - x_0}{n} \sum_{i=1}^{n} f(x_i).$$

In calculus, it is proved that

$$\lim_{n\to\infty} L = \lim_{n\to\infty} R = \int_{x_0}^{x_n} f(x)\,dx.$$

Also, from calculus, since each limit exists, they can be algebraically combined such that

$$\lim_{n\to\infty} \frac{L + R}{2} = \frac{\int_{x_0}^{x_n} f(x)\,dx + \int_{x_0}^{x_n} f(x)\,dx}{2} = \frac{2\int_{x_0}^{x_n} f(x)}{2}\,dx = \int_{x_0}^{x_n} f(x)\,dx.$$

To prove multiple applications of the trapezoidal rule $A_T(n)$ of Equation (5.7) also converges to the exact integral value as $n \to \infty$, Equation (5.7) is now shown to be equivalent to the above expression $\frac{L+R}{2}$.

Adding the definitions of the left endpoint and right endpoint Riemann sums

$$L + R = \frac{x_n - x_0}{n} \sum_{i=1}^{n} f(x_{i-1}) + \frac{x_n - x_0}{n} \sum_{i=1}^{n} f(x_i).$$

Factoring out the leading fraction and expanding each summation

$$L + R = \frac{x_n - x_0}{n} \left((f(x_0) + f(x_1) + f(x_2) + \cdots + f(x_{n-1})) \right.$$
$$\left. + (f(x_1) + f(x_2) + \cdots + f(x_{n-1}) + f(x_n)) \right)$$

Across both summation expansions, the values $f(x_0)$ and $f(x_n)$ occur once each, while each interior function value $f(x_i)$ for $i = 1, \ldots, n-1$ occurs twice. Rewriting accordingly,

$$L + R = \frac{x_n - x_0}{n} \left(f(x_0) + f(x_n) + \sum_{i=1}^{n-1} 2f(x_i) \right).$$

Dividing each side by two,

$$\frac{L + R}{2} = \frac{x_n - x_0}{n} \left(\frac{f(x_0) + f(x_n)}{2} + \sum_{i=1}^{n-1} f(x_i) \right).$$

Since the limit of the left side was shown to converge to the exact integral value and the right side is exactly Equation (5.7), i. e., multiple applications of the trapezoidal rule, then as $n \to \infty$, multiple applications of the trapezoidal rule $A_T(n)$ converge to the exact integral value. $\qquad\square$

The implications of this proof will be revisited in Section 5.3.

Example 5.4. Use $n = 4$ applications of the trapezoidal rule to estimate $\int_{0.5}^{3.5} \ln(6x - x^2)\, dx$.
Solution. Dividing the entire interval $[0.5, 3.5]$ into four subintervals means

$$h = \frac{3.5 - 0.5}{4} = 0.75.$$

Starting with $x_0 = 0.5$ and repeatedly adding $h = 0.75$ marks the bounds of each subinterval.

$$x_0 = 0.5, \ x_1 = 1.25, \ x_2 = 2, \ x_3 = 2.75, \text{ and } x_4 = 3.5$$

Using Equation (5.7),

$$\int_{0.5}^{3.5} \ln(6x - x^2)\, dx \approx A_T(4) = \frac{x_4 - x_0}{4} \left(\frac{f(x_0) + f(x_4)}{2} + \sum_{i=1}^{3} f(x_i) \right)$$

$$= \frac{3.5 - 0.5}{4} \left(\frac{f(0.5) + f(3.5)}{2} + f(1.25) + f(2) + f(2.75) \right)$$

$$= 0.75 \left(\frac{1.012 + 2.169}{2} + 1.781 + 2.079 + 2.190 \right) = 5.730$$

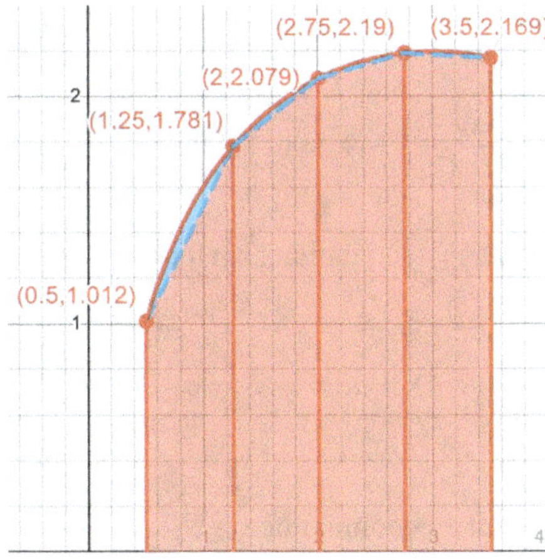

Figure 5.6: The red area of the four trapezoids estimates $\int_{0.5}^{3.5} \ln(6x - x^2)\,dx$. The four blue areas comprise the truncation error and indicate the trapezoids underestimate the true integral value.

Figure 5.6 depicts the four trapezoids used in the estimation of Example 5.4. An estimation of the same integral using one trapezoid in Figure 5.2 of Example 5.3 illustrates using even a small number of trapezoids can improve an estimation, perhaps significantly. Figure 5.6 displays much less blue area, which equals the truncation error, than the blue area of Figure 5.2.

Comparing the results using one trapezoid in Example 5.3 for A_T to using four trapezoids in Example 5.4 for $A_T(4)$, the relative error is

$$\left|\frac{A_T - A_T(4)}{A_T(4)}\right| = \left|\frac{4.772 - 5.730}{5.730}\right| \times 100\,\% = 16.72\,\%$$

Because the four-trapezoids estimation is closer than the single-trapezoid estimation to the true integral value by Theorem 5.1, $A_T(4)$ is used as the relative divisor.

As with a single application of the trapezoidal rule, a truncation error adjustment with multiple applications in Section 5.1.4 offers additional improvement of the estimation.

5.1.4 Multiple applications, truncation error adjusted

Using multiple applications of the trapezoidal rule, the truncation error adjustment of Equation (5.3) is applied to each of the n trapezoids and defined.

$$E_{t,T}(n) = \sum_{i=1}^{n} -\frac{f''(c_i)(x_i - x_{i-1})^3}{12}$$

All x_i values are equally spaced over the entire interval $[x_0, x_n]$, so that

$$x_i - x_{i-1} = \frac{x_n - x_0}{n}.$$

Substituting into $E_{t,T}(n)$ yields

$$E_{t,T}(n) = \sum_{i=1}^{n} -\frac{f''(c_i)\left(\frac{x_n-x_0}{n}\right)^3}{12} = -\frac{(x_n-x_0)^3}{12n^3}\sum_{i=1}^{n} f''(c_i). \tag{5.8}$$

The mean value is used again for each $f''(c_i)$ according to Equation (5.4) for the summation on the right side of Equation (5.8).

$$\sum_{i=1}^{n} f''(c_i) \approx \sum_{i=1}^{n} \frac{\int_{x_{i-1}}^{x_i} f''(x)\,dx}{x_i - x_{i-1}}. \tag{5.9}$$

Expanding the right side and again substituting in the denominator $x_i - x_{i-1} = \frac{x_n-x_0}{n}$ for the equally spaced x values

$$\frac{\int_{x_0}^{x_1} f''(x)\,dx}{\frac{x_n-x_0}{n}} + \frac{\int_{x_1}^{x_2} f''(x)\,dx}{\frac{x_n-x_0}{n}} + \cdots + \frac{\int_{x_{n-2}}^{x_{n-1}} f''(x)\,dx}{\frac{x_n-x_0}{n}} + \frac{\int_{x_{n-1}}^{x_n} f''(x)\,dx}{\frac{x_n-x_0}{n}}. \tag{5.10}$$

The common denominator can be factored out and written as its reciprocal. Also, repeatedly using the summation property of integrals from calculus, which states $\int_a^b f(x)\,dx + \int_b^c f(x)\,dx = \int_a^c f(x)\,dx$, the sum of all the integrals of Equation (5.10) condenses to a single integral.

$$\int_{x_0}^{x_1} f''(x)\,dx + \int_{x_1}^{x_2} f''(x)\,dx + \cdots + \int_{x_{n-2}}^{x_{n-1}} f''(x)\,dx + \int_{x_{n-1}}^{x_n} f''(x)\,dx = \int_{x_0}^{x_n} f''(x)\,dx$$

Equation (5.10) can thus be rewritten

$$\frac{n}{x_n - x_0}\int_{x_0}^{x_n} f''(x)\,dx. \tag{5.11}$$

Since Equation (5.11) is equivalent to Equation (5.10), which is the expansion of the summation of Equation (5.9), Equation (5.11) can be substituted in Equation (5.9).

$$\sum_{i=1}^{n} f''(c_i) \approx \frac{n\int_{x_0}^{x_n} f''(x)\,dx}{x_n - x_0} \tag{5.12}$$

Integrating the second derivative is equivalent to the first derivative, so that

$$\sum_{i=1}^{n} f''(c_i) \approx \frac{n\int_{x_0}^{x_n} f''(x)\,dx}{x_n - x_0} = \frac{nf'(x)\big|_{x_0}^{x_n}}{x_n - x_0} = \frac{n(f'(x_n) - f'(x_0))}{x_n - x_0}. \tag{5.13}$$

Substituting the right side of Equation (5.13) in Equation (5.8) and reducing produces the final form of the truncation error adjustment for n multiple applications of the trapezoidal rule $E_{t,T}(n)$.

$$E_{t,T}(n) = -\frac{(x_n - x_0)^3}{12n^3} \sum_{i=1}^{n} f''(c_i) \approx -\frac{(x_n - x_0)^3}{12n^3} \cdot \frac{n(f'(x_n) - f'(x_0))}{x_n - x_0}$$

$$= -\frac{(x_n - x_0)^2 (f'(x_n) - f'(x_0))}{12n^2} \tag{5.14}$$

The truncation error adjustment for a single application of the trapezoidal rule of Equation (5.5) now can be viewed as simply a special case of Equation (5.14) for $n = 1$.

In summary, the adjusted estimation $A_{T,t}(n)$ using n multiple applications of the trapezoidal rule of Equation (5.7) with the truncation error adjustment of Equation (5.14) is defined

$$A_{T,t}(n) = A_T(n) + E_{t,T}(n). \tag{5.15}$$

Example 5.5. Calculate $E_{t,T}(n)$ for the integral of Example 5.4. Apply $E_{t,T}(n)$ for the improved trapezoidal rule estimation $A_{T,t}(n)$. Use $E_{t,T}(n)$ to verify the graphical evidence of Figure 5.6 that multiple applications of the trapezoidal rule alone $A_T(n)$ in this example underestimates the true integral value.

Solution. Using $f'(x) = \frac{6-2x}{6x-x^2}$ and Equation (5.14)

$$E_{t,T}(4) = -\frac{(x_4 - x_0)^2 (f'(x_4) - f'(x_0))}{12(4)^2} = -\frac{(3.5 - 0.5)^2 \left(\frac{6-2(3.5)}{6(3.5)-(3.5)^2} - \frac{6-2(0.5)}{6(0.5)-(0.5)^2} \right)}{12(16)} = 0.091$$

$E_{t,T}(4) > 0$ indicates four applications of the trapezoidal rule underestimates the true integral value. Adding this truncation error adjustment to the previous integral estimation $A_T(4)$ of Example 5.4, the improved integral estimation is

$$A_{T,t}(4) = A_T(4) + E_{t,T}(4) = 5.730 + 0.091 = 5.821$$

Multiple applications of the trapezoidal rule with the truncation error adjustment $E_{t,T}(n)$ can produce very reasonable results depending on the function, even for a small number of applications like the four used in Example 5.5. Software routines that retain a large number of significant digits calculate a very close estimation for the integral of Example 5.5 as 5.817, relatively close to the above value $A_{T,t}(4) = 5.821$.

5.1.5 Trapezoidal rule variations, evaluated by ε_T

Estimations produced by the four trapezoidal rules of Sections 5.1.1 through 5.1.4 are now compared to the exact definite integral value I, solved analytically. The exact relative percentage error for trapezoidal rules ε_T is defined

$$\varepsilon_T = \left| \frac{I - A_{\text{Trap}}}{I} \right| \times 100\,\% \tag{5.16}$$

A_{Trap} is an estimation type using either single or multiple applications of the trapezoidal rule of Section 5.1.1 or 5.1.3, or a truncation error-adjusted estimation of Section 5.1.2 or 5.1.4. Table 5.1 summarizes these options.

Table 5.1: Estimations for the integral A_{Trap} used for ε_T of Equation (5.16) can be defined by any of the four rows, each a variation on the trapezoidal rules of Sections 5.1.1 through 5.1.4.

Type	Applications	Error Adjusted?	Equation	Uses Equations
A_T	single	no	(5.1)	
$A_{T,t}$	single	yes, by $E_{t,T}$	(5.6)	(5.1) and (5.5)
$A_T(n)$	multiple	no	(5.7)	
$A_{T,t}(n)$	multiple	yes, by $E_{t,T}(n)$	(5.15)	(5.7) and (5.14)

Example 5.6. Analytically solve $I = \int_1^4 (x^2 - 6.3x + 7.74\sqrt{x})\,dx$. Solve each of the four trapezoidal rule variations of Table 5.1 to estimate I. Calculate the exact relative percentage error ε_T for each estimation.

Solution. Analytically, using an antiderivative for $f(x) = x^2 - 6.3x + 7.74\sqrt{x}$

$$I = \int_1^4 f(x)\,dx = \left(\frac{x^3}{3} - 3.15x^2 + 5.16x^{1.5} \right)\Bigg|_1^4$$

$$= \frac{64}{3} - 3.15(16) + 5.16(8) - \left(\frac{1}{3} - 3.15 + 5.16 \right) = 9.87$$

Case 1. Applying the trapezoidal rule once with no adjustment requires the two following values.

$$f(1) = (1)^2 - 6.3(1) + 7.74\sqrt{1} = 1 - 6.3 + 7.74 = 2.44$$

$$f(4) = (4)^2 - 6.3(4) + 7.74\sqrt{4} = 16 - 25.2 + 15.48 = 6.28$$

By Equation (5.1),

$$\int_1^4 f(x)\,dx \approx A_T = \frac{(4-1)(f(1)+f(4))}{2} = \frac{(3)(2.44 + 6.28)}{2} = 13.08.$$

The exact relative percentage error of Equation (5.16) is

$$\varepsilon_T = \left| \frac{I - A_T}{I} \right| \times 100\,\% = \left| \frac{9.87 - 13.08}{9.87} \right| \times 100\,\% = 32.52\,\%.$$

Case 2. The truncation error adjustment for one application of the trapezoidal rule needs the first derivative of $f(x)$, evaluated at the limits of integration.

$$f'(x) = 2x - 6.3 + \frac{3.87}{\sqrt{x}}$$

$$f'(1) = 2(1) - 6.3 + \frac{3.87}{\sqrt{1}} = 2 - 6.3 + 3.87 = -0.43$$

$$f'(4) = 2(4) - 6.3 + \frac{3.87}{\sqrt{4}} = 8 - 6.3 + 1.935 = 3.635$$

Using Equation (5.5) for the truncation error adjustment,

$$E_{t,T} = -\frac{(f'(4) - f'(1))(4 - 1)^2}{12} = -\frac{(3.635 - (-0.43))(9)}{12} = -3.049.$$

The adjusted estimation $A_{T,t}$ of Equation (5.6) is thus

$$A_{T,t} = A_T + E_{t,T} = 13.08 + (-3.049) = 10.031.$$

The exact relative percentage error by Equation (5.16) is

$$\varepsilon_T = \left| \frac{I - A_{T,t}}{I} \right| \times 100\,\% = \left| \frac{9.87 - 10.031}{9.87} \right| \times 100\,\% = 1.63\,\%.$$

Case 3. The trapezoidal rule is now applied multiple times for a selected $n = 5$, so that $x_0 = 1$ and $x_5 = 4$. Equally spacing the points means $h = \frac{4-1}{5} = 0.6$ and

$$x_1 = 1.6, \ x_2 = 2.2, \ x_3 = 2.8, \ \text{and} \ x_4 = 3.4$$

Evaluating $f(x)$ at all the x values,

$$f(x_0) = f(1) = (1)^2 - 6.3(1) + 7.74\sqrt{1} = 2.440$$
$$f(x_1) = f(1.6) = (1.6)^2 - 6.3(1.6) + 7.74\sqrt{1.6} = 2.270$$
$$f(x_2) = f(2.2) = (2.2)^2 - 6.3(2.2) + 7.74\sqrt{2.2} = 2.460$$
$$f(x_3) = f(2.8) = (2.8)^2 - 6.3(2.8) + 7.74\sqrt{2.8} = 3.151$$
$$f(x_4) = f(3.4) = (3.4)^2 - 6.3(3.4) + 7.74\sqrt{3.4} = 4.412$$
$$f(x_5) = f(4) = 4^2 - 6.3(4) + 7.74\sqrt{4} = 6.280.$$

According to Equation (5.7),

$$\int_1^4 f(x)\,dx \approx A_T(5) = \frac{x_5 - x_0}{5} \left(\frac{f(x_0) + f(x_5)}{2} + \sum_{i=1}^4 f(x_i) \right)$$

$$= \frac{4-1}{5} \left(\frac{2.44 + 6.28}{2} + f(x_1) + f(x_2) + f(x_3) + f(x_4) \right)$$

$$= 0.6(4.36 + 2.27 + 2.46 + 3.151 + 4.412) = 9.992.$$

The exact relative percentage error of Equation (5.16) is

$$\varepsilon_T = \left| \frac{I - A_T(5)}{I} \right| \times 100\,\% = \left| \frac{9.87 - 9.992}{9.87} \right| \times 100\,\% = 1.24\,\%.$$

Case 4. The truncation error adjustment for multiple applications of the trapezoidal rule uses the same values found in Case 2, the first derivative $f'(x)$ at the limits of integration. While the value at the upper limit is the same, it is relabeled from x_1 in the single application case to x_n in the multiple application case. Since $n = 5$, here $x_n = x_5 = x_1 = 4$, and it follows $f'(x_5) = f'(x_1) = 3.635$. By Equation (5.14), the truncation error adjustment is

$$E_{t,T}(5) = -\frac{(x_5 - x_0)^2 (f'(x_5) - f'(x_0))}{12n^2} = -\frac{(4-1)^2 (3.635 + 0.43)}{12(5)^2} = -0.122$$

Using Equation (5.15) for the adjusted estimation $A_{T,t}(5)$,

$$A_{T,t}(5) = A_T(5) + E_{t,T}(5) = 9.992 + (-0.122) = 9.870$$

This is the true integral value I, accurate to the thousandths place, so the exact relative percentage error is simply 0 %.

Note the consistent decrease in the exact relative percentage error ε_T from Case 1 through Case 4 of Example 5.6.

Table 5.2 supports the order of increased efficacy of the four variations of the trapezoidal rule. Improvements of Case 2 over Case 1 and Case 4 over Case 3 are reasonable since each applies a truncation error adjustment in both the single and multiple application variations, respectively. Furthermore, Theorem 5.1 established convergence to the exact integral value as the number of applications of the trapezoidal rule increases, hence Cases 3 and 4 are superior to Cases 1 and 2. Perhaps more subtly, this implies the sounder approach to improving an integral estimation with trapezoidal rule variations is using multiple applications, even without a truncation error adjustment, rather than using a single application with a truncation error adjustment.

Table 5.2: The decrease in the exact relative percentage error ε_T relates to the improvement of each trapezoidal rule variation over the one in the previous row.

Case	Type	ε_T, %
1	A_T	32.52
2	$A_{T,t}$	1.63
3	$A_T(5)$	1.24
4	$A_{T,t}(5)$	0

5.2 Simpson's rule

Trapezoidal rules like Equation (5.1) use a linear first-degree polynomial to approximate the integrand $f(x)$. Simpson's rule instead uses a curved second-degree polynomial function $f_2(x) = ax^2 + bx + c$, presumably a closer approximation of a nonlinear $f(x)$.

Just as the trapezoidal rule was extended, employing multiple $f_2(x)$ functions improves an estimation. Figure 5.7 depicts four distinct second-degree polynomials with blue dashed curves, each approximating the red curve on four subintervals divided by blue solid horizontal lines. Blue and green areas represent underestimation and overestimation, respectively, of area between the red $f(x)$ and the x-axis. Compared to the same $f(x)$ and four subintervals of Figure 5.3, which used first-degree polynomials for approximation and the trapezoidal rules, Figure 5.7 contains less total blue and green area.

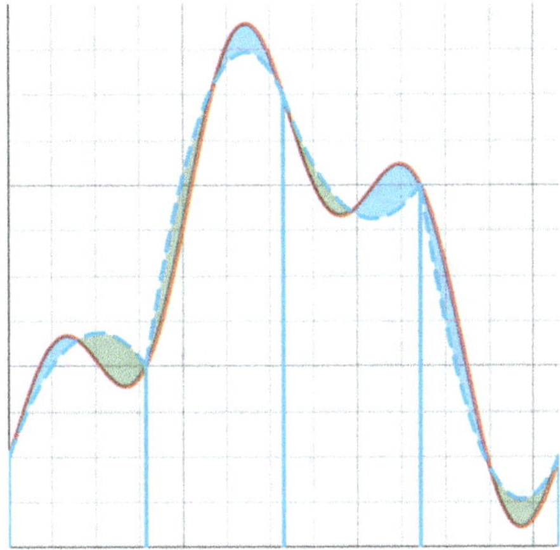

Figure 5.7: Each of the four subintervals uses a different second-degree polynomial, indicated by a blue dashed curve, to estimate the red $f(x)$ and the area between $f(x)$ and the x-axis. If the blue dashed curve lies below $f(x)$, the difference of blue area indicates an underestimation. Conversely, where the blue dashed curve lies above $f(x)$, the difference of green area produces an overestimation.

5.2.1 Single application

Three points at x_0, x_1, and x_2 are necessary to fit a unique $f_2(x)$. Equal spacing is not required for a unique fit, but applied to derive Simpson's rule. Furthermore, shifting the x values will correspondingly shift the $f_2(x)$ values so that the integral value will not change, and graphically, the areas are the same. It is convenient therefore to let $x_1 = 0$ so that $x_0 = -x_2$ when the points are equally spaced.

The first step in the derivation of Simpson's rule is integrating the second-degree polynomial function $f_2(x)$ over a shifted interval as described above.

$$\int_{x_0}^{x_2} f(x)\, dx \approx \int_{x_0}^{x_2} f_2(x)\, dx = \int_{-x_2}^{x_2} \left(ax^2 + bx + c\right) dx$$

$$= \left(\frac{ax^3}{3} + \frac{bx^2}{2} + cx\right)\Bigg|_{-x_2}^{x_2} = \frac{ax_2^3}{3} + \frac{bx_2^2}{2} + cx_2 - \left(\frac{-ax_2^3}{3} + \frac{bx_2^2}{2} - cx_2\right)$$

$$= \frac{2ax_2^3}{3} + 2cx_2 = \frac{1}{3}\left(2ax_2^3 + 6cx_2\right) \tag{5.17}$$

The second step in finding Simpson's rule involves eliminating the parameters a and c on the right side of Equation (5.17) by writing them in terms of the known points; parameter b was already eliminated in the previous definite integral steps. The three shifted points satisfy $f_2(x)$ such that

$$f_2(x_0) = f(-x_2) = a(-x_2)^2 + b(-x_2) + c = ax_2^2 - bx_2 + c$$
$$f_2(x_1) = a(0)^2 + b(0) + c = c$$
$$f_2(x_2) = ax_2^2 + bx_2 + c$$

Adding the first and third of these equations yields

$$f(x_0) + f(x_2) = 2ax_2^2 + 2c.$$

Multiplying all terms by x_2 and solving for $2ax_2^3$,

$$x_2(f(x_0) + f(x_2)) = 2ax_2^3 + 2cx_2$$

$$2ax_2^3 = x_2(f(x_0) + f(x_2)) - 2cx_2$$

Substituting this expression for $2ax_2^3$ on the right side of Equation (5.17) eliminates parameter a.

$$\int_{x_0}^{x_2} f_2(x)\, dx = \frac{1}{3}(2ax_2^3 + 6cx_2)$$

$$= \frac{1}{3}(x_2(f(x_0) + f(x_2)) - 2cx_2 + 6cx_2)$$

$$= \frac{1}{3}(x_2(f(x_0) + f(x_2)) + 4cx_2)$$

$$= \frac{x_2}{3}(f(x_0) + 4c + f(x_2)) \tag{5.18}$$

Parameter c is eliminated by substituting $f_2(x_1) = c$ as established above. A final substitution is made in anticipation of extending Simpson's rule from one to multiple applications. Since the width of the interval can be expressed as either $x_2 - x_0$ or $x_2 - (-x_2)$

$$x_2 - x_0 = x_2 - (-x_2)$$

$$x_2 - x_0 = 2x_2$$

$$x_2 = \frac{x_2 - x_0}{2}$$

These final two substitutions for parameter c and x_2 in Equation (5.18) define one application of Simpson's rule, A_S.

$$\int_{x_0}^{x_2} f(x)\, dx \approx \int_{x_0}^{x_2} f_2(x)\, dx = A_S = \frac{x_2 - x_0}{6}(f(x_0) + 4f(x_1) + f(x_2)) \tag{5.19}$$

Example 5.7. Use Simpson's rule to estimate the area between $f(x) = 13/(x^3 + 2)$ and the x-axis over the interval $[1.7, 5.4]$.

Solution. Using three equally spaced points, given $x_0 = 1.7$ and $x_2 = 5.4$, means $x_1 = 3.55$, the midpoint of the interval. According to Equation (5.19)

$$\int_{1.7}^{5.4} \frac{13}{x^3 + 2}\, dx \approx A_S = \frac{5.4 - 1.7}{6}(f(1.7) + 4f(3.55) + f(5.4))$$

$$= \frac{3.7}{6}(1.881 + 4(0.278) + 0.082) = 1.896$$

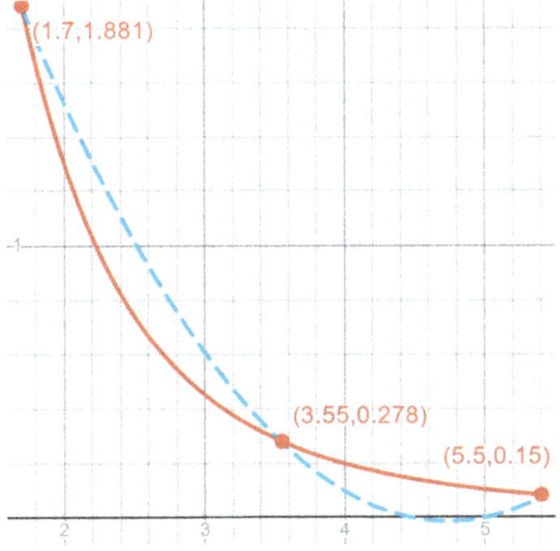

Figure 5.8: The blue parabolic second-degree polynomial of Simpson's rule that estimates the red $f(x) = 13/(x^3 + 2)$.

Figure 5.8 displays the graph of $f(x) = 13/(x^3 + 2)$ in red and the parabolic-second-degree approximation $f_2(x)$ in blue using Simpson's rule of Example 5.7. The actual quadratic function through the three points is $f_2(x) = 0.205x^2 - 1.944x + 4.592$. Integrated analytically, $f_2(x)$ does estimate the definite integral. But Simpson's rule precisely removes the need to know $f_2(x)$ explicitly by eliminating parameters a, b, and c and expressing the estimation in terms of the three known points.

Figure 5.9 shows the true area in red, while Figure 5.10 highlights, on two separate intervals, both an overestimation and underestimation of the true area based on $f_2(x)$ of Example 5.7. The green area from $x_0 = 1.7$ to $x_1 = 3.55$ is an overestimation because $f_2(x) > f(x)$ there, while the blue area from $x_1 = 3.55$ to $x_2 = 5.4$ is an underestimation because $f_2(x) < f(x)$ there.

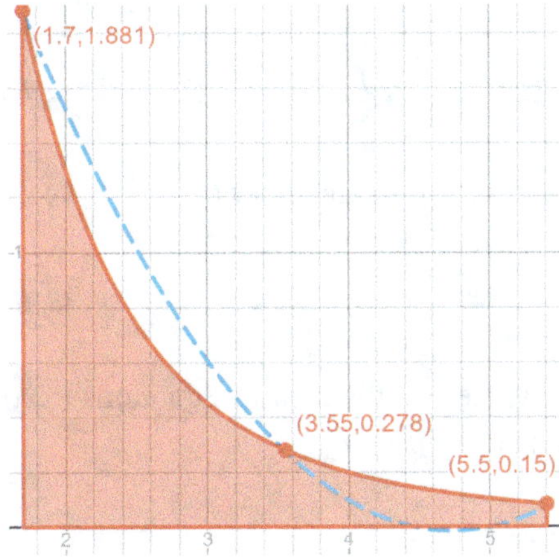

Figure 5.9: The red area is the true integral value $\int_{1.7}^{5.4} \frac{13}{x^3+2}\, dx.$

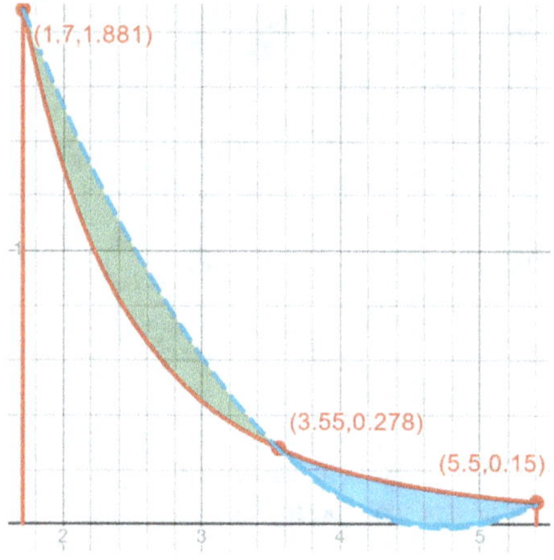

Figure 5.10: The green area is an overestimation and the blue area is an underestimation, using dashed $f_2(x)$ for red $f(x)$.

5.2.2 Single application, truncation error adjusted

Determining the truncation error adjustment for Simpson's rule mimics the approach used for the trapezoidal rule in Section 5.1.2. Altered slightly by beginning with two additional terms of the Taylor series representation of a function $f(x)$ centered at x_0

$$f(x) = f(x_0) + f'(x_0)(x - x_0) + \frac{f''(x_0)(x - x_0)^2}{2}$$
$$+ \frac{f^{(3)}(x_0)(x - x_0)^3}{3!} + \frac{f^{(4)}(c)(x - x_0)^4}{4!} \tag{5.20}$$

The first four of the five terms of Equation (5.20) comprise the third-degree Taylor polynomial approximation $T_3(x)$.

$$T_3(x) = f(x_0) + f'(x_0)(x - x_0) + \frac{f''(x_0)(x - x_0)^2}{2} + \frac{f^{(3)}(x_0)(x - x_0)^3}{3!} \tag{5.21}$$

The last term of Equation (5.20) is the Lagrange form of the remainder $R_3(x)$.

$$R_3(x) = \frac{f^{(4)}(c)(x - x_0)^4}{4!} \tag{5.22}$$

Thus the right side of Equation (5.20) can be written as the sum of Equations (5.21) and (5.22), which equals the left side of Equation (5.20) $f(x)$.

$$f(x) = T_3(x) + R_3(x).$$

Integrating both sides from x_0 to x_2, the definite integral $\int_{x_0}^{x_2} f(x)\, dx$ is

$$\int_{x_0}^{x_2} f(x)\, dx = \int_{x_0}^{x_2} T_3(x)\, dx + \int_{x_0}^{x_2} R_3(x)\, dx. \tag{5.23}$$

Note that Simpson's rule involves integrating a second-degree polynomial fit through three points to approximate the integration of a second-degree Taylor polynomial $T_2(x)$, not $T_3(x)$. However, by employing a shifted interval of $[-x_2, x_2]$ with Simpson's rule, integrating the third-degree term of the Taylor polynomial $T_3(x)$ equals zero, effectively enabling Simpson's rule to approximate the integration of $T_3(x)$ exactly as $T_2(x)$.

Theorem 5.2. *Integrating the third-degree term of the Taylor polynomial $T_3(x)$ on the shifted integral $[-x_2, x_2]$ equals zero.*

Proof. As in Equation (5.2), not centered at one point at x_0 but approximated by two points at x_0 and x_1, Equation (5.21) is, by extension, approximated around three points, an additional one at x_2. A similar extended approximation is thus used.

$$(x - x_0)^3 \approx (x - x_0)(x - x_1)(x - x_2)$$

Substituting this into the third-degree term of Equation (5.21) and integrating

$$\int_{x_0}^{x_2} \frac{f^{(3)}(x_0)(x - x_0)^3}{3!}\, dx \approx \int_{x_0}^{x_2} \frac{f^{(3)}(x_0)(x - x_0)(x - x_1)(x - x_2)}{6}\, dx.$$

Using the shifted interval means $x_0 = -x_2$ and $x_1 = 0$, so that

$$\int_{x_0}^{x_2} \frac{f^{(3)}(x_0)(x - x_0)(x - x_1)(x - x_2)}{6}\, dx = \int_{-x_2}^{x_2} \frac{f^{(3)}(-x_2)(x - (-x_2))(x - 0)(x - x_2)}{6}\, dx.$$

Multiplying the factors and factoring out the constant, $\frac{f^{(3)}(-x_2)}{6}$

$$= \frac{f^{(3)}(-x_2)}{6} \int_{-x_2}^{x_2} (x + x_2)(x)(x - x_2)\, dx = \frac{f^{(3)}(-x_2)}{6} \int_{-x_2}^{x_2} (x^3 - x_2^2 x)\, dx$$

Evaluating the integral,

$$= \left(\frac{x^4}{4} - \frac{x_2^2 x^2}{2} \right)\Big|_{-x_2}^{x_2} = \frac{x_2^4}{4} - \frac{x_2^2 x_2^2}{2} - \left(\frac{(-x_2)^4}{4} - \frac{x_2^2(-x_2)^2}{2} \right)$$

$$= \frac{x_2^4}{4} - \frac{x_2^2 x_2^2}{2} - \frac{x_2^4}{4} + \frac{x_2^2 x_2^2}{2} = 0. \qquad \square$$

With Simpson's rule of Equation (5.19) now proved to approximate $T_3(x)$ the same as $T_2(x)$, Equation (5.19) can be substituted in Equation (5.23) such that

$$\int_{x_0}^{x_2} f(x)\, dx \approx \frac{x_2 - x_0}{6}\left(f(x_0) + 4f(x_1) + f(x_2)\right) + \int_{x_0}^{x_2} R_3(x)\, dx.$$

What remains is evaluating $\int_{x_0}^{x_2} R_3(x)\, dx$, the truncation error adjustment $E_{t,S}$ defined below. Although $T_2(x)$ and $T_3(x)$ are approximated identically using Simpson's rule, using $T_3(x)$ and $R_3(x)$ instead of $T_2(x)$ and $R_2(x)$ in Equation (5.23) is significant. The truncation error adjustment produced from $R_3(x)$ is based on the fourth-degree term of the Taylor series rather than the third-degree term of $R_2(x)$, and the fourth-degree term results in a smaller magnitude of error.

Derivation of the truncation error adjustment $E_{t,S}$

Theorem 5.2 referenced the approximation of $(x - x_0)^3$ using three factors, each containing one of the known points at x_0, x_1, and x_2. $R_3(x)$ has a fourth-degree product, similarly approximated using the same points

$$(x - x_0)^4 \approx (x - x_0)(x - x_1)^2(x - x_2). \tag{5.24}$$

Because x values are equally spaced, using x_1 twice is justified because it is centrally located in the interval $[x_0, x_2]$. Also, using the equivalent shifted interval $[-x_2, x_2]$ means $x_1 = 0$, which simplifies the derivation of the truncation error adjustment. Substituting Equation (5.24) in Equation (5.22), $x_0 = -x_2$, and $x_1 = 0$

$$
\begin{aligned}
R_3(x) &= \frac{f^{(4)}(c)(x - x_0)^4}{4!} \approx \frac{f^{(4)}(c)(x - x_0)(x - x_1)^2(x - x_2)}{24} \\
&= \frac{f^{(4)}(c)(x - (-x_2))(x - 0)^2(x - x_2)}{24} = \frac{f^{(4)}(c)}{24}(x + x_2)x^2(x - x_2) \\
&= \frac{f^{(4)}(c)}{24}\left(x^4 - x_2^2 x^2\right).
\end{aligned}
$$

Integrating the last expression over the shifted interval

$$
\begin{aligned}
\int_{-x_2}^{x_2} \frac{f^{(4)}(c)}{24}\left(x^4 - x_2^2 x^2\right) dx &= \frac{f^{(4)}(c)}{24}\left(\frac{x^5}{5} - \frac{x_2^2 x^3}{3}\right)\Bigg|_{-x_2}^{x_2} \\
&= \frac{f^{(4)}(c)}{24}\left(\frac{x_2^5}{5} - \frac{x_2^2 x_2^3}{3} - \left(\frac{(-x_2)^5}{5} - \frac{x_2^2(-x_2)^3}{3}\right)\right) \\
&= \frac{f^{(4)}(c)}{24}\left(\frac{x_2^5}{5} - \frac{x_2^5}{3} + \frac{x_2^5}{5} - \frac{x_2^5}{3}\right)
\end{aligned}
$$

$$= \frac{f^{(4)}(c)}{24}\left(\frac{-4x_2^{\,5}}{15}\right) = -\frac{f^{(4)}(c)x_2^{\,5}}{90}$$

As in the final step of the derivation of Simpson's rule of Equation (5.19), the substitution $x_2 = \frac{x_2 - x_0}{2}$ is made in preparation for extending the truncation error adjustment for multiple applications of Simpson's rule. Substituting this expression for x_2 on the right side above produces the initial definition for the truncation error adjustment of Simpson's rule.

$$E_{t,S} = \int_{x_0}^{x_2} R_3(x)\,dx \approx -\frac{f^{(4)}(c)x_2^{\,5}}{90} = -\frac{f^{(4)}(c)}{90}\left(\frac{x_2 - x_0}{2}\right)^5$$

$$= -\frac{f^{(4)}(c)}{90}\frac{(x_2 - x_0)^5}{32} = -\frac{f^{(4)}(c)(x_2 - x_0)^5}{2880}. \tag{5.25}$$

One last substitution mirrors the derivation of the trapezoidal rule truncation error adjustment, using the mean for $f^{(4)}(c)$ in Equation (5.25).

$$f^{(4)}(c) \approx \overline{f^{(4)}(x)} = \frac{\int_{x_0}^{x_2} f^{(4)}(x)\,dx}{x_2 - x_0} = \frac{f^{(3)}(x)\big|_{x_0}^{x_2}}{x_2 - x_0} = \frac{f^{(3)}(x_2) - f^{(3)}(x_0)}{x_2 - x_0} \tag{5.26}$$

Substituting Equation (5.26) into Equation (5.25) and simplifying produces the final form of the truncation error adjustment for a single application of Simpson's rule $E_{t,S}$.

$$E_{t,S} = -\frac{f^{(4)}(c)(x_2 - x_0)^5}{2880} = -\frac{(f^{(3)}(x_2) - f^{(3)}(x_0))}{x_2 - x_0}\frac{(x_2 - x_0)^5}{2880}$$

$$E_{t,S} = -\frac{(f^{(3)}(x_2) - f^{(3)}(x_0))(x_2 - x_0)^4}{2880} \tag{5.27}$$

Equation (5.23) now represents the basis for the improved Simpson's rule $A_{S,t}$, substituting Simpson's rule alone, Equation (5.19), and the truncation error adjustment, Equation (5.27), respectively, on the right side.

$$\int_{x_0}^{x_2} f(x)\,dx = \int_{x_0}^{x_2} T_3(x)\,dx + \int_{x_0}^{x_2} R_3(x)\,dx$$

$$\int_{x_0}^{x_2} f(x)\,dx \approx A_{S,t} = A_S + E_{t,S} \tag{5.28}$$

Example 5.8. Calculate $E_{t,S}$ for the definite integral of Example 5.7. Apply $E_{t,S}$ for the improved Simpson's rule estimation $A_{S,t}$. Use $E_{t,S}$ to determine whether Simpson's rule alone A_S overestimates or underestimates the true integral value.

Solution. Using $f^{(3)}(x) = \frac{-780x^6 + 2496x^3 - 312}{(x^3+2)^4}$ and Equation (5.27)

$$E_{t,S} = -\frac{(f^{(3)}(5.4) - f^{(3)}(1.7))(5.4 - 1.7)^4}{2880} = -\frac{(-0.029 - 3.011)(3.7)^4}{2880} = -0.198$$

$E_{t,S} < 0$ indicates Simpson's rule alone overestimated the true integral value and requires adding this *negative* $E_{t,S}$ value to improve the integral estimation of Example 5.7. The adjusted estimation applying Equation (5.28) is

$$A_{S,t} = A_S + E_{t,S} = 1.896 + (-0.198) = 1.698$$

Reexamining Figure 5.10 of Example 5.7, the greater amount of green area versus the smaller amount of blue area suggests an overall overestimation of the integral value. The algebra of Example 5.8 confirms this conjecture. When the graphical difference between overestimated and underestimated area is distinguishably difficult, and Figure 5.10 is a reasonable example, the algebraic sign of $E_{t,S}$ can convey valuable information.

Computer routines maintaining a high degree of precision estimate the value of the definite integral of Example 5.7 as $I = 1.733$. Although the truncation error adjustment of Example 5.8 overcorrected and now produces an underestimation, the absolute error indeed improved. Using Simpson's rule alone and the resulting integral value $A_S = 1.896$, the absolute error is

$$|I - A_S| = |1.733 - 1.896| = 0.163.$$

Adjusting Simpson's rule with the truncation error adjustment for the integral value $A_{S,t} = 1.698$, the absolute error shrinks to

$$|I - A_{S,t}| = |1.733 - 1.698| = 0.035.$$

An exact relative percentage error, akin to the one used in Section 5.1.5 for the various trapezoidal rules, is examined in Section 5.2.5.

A bonus feature of Simpson's rule

If the underlying function $f(x)$ is a second-degree polynomial function through three known points, using Simpson's rule on $f(x)$ at the three points produces the exact integral value $\int_{x_0}^{x_2} f(x)\,dx$. This is apparent; Simpson's rule fits the unique second-degree polynomial function through the three given points, and those three points satisfy the unique underlying second-degree polynomial function $f(x)$.

What may be surprising is using Simpson's rule on three points produces the exact integral value if the underlying function $f(x)$ is a third-degree polynomial function $f(x)$ as well. The following Theorem 5.3 demonstrates this additional perk of Simpson's rule.

Theorem 5.3. *If $f(x)$ is a third-degree polynomial through points at increasing values of x_0, x_1, and x_2, Simpson's rule exactly evaluates the following definite integral so that*

$$\int_{x_0}^{x_2} f(x)\, dx = \frac{x_2 - x_0}{6}\left(f(x_0) + 4f(x_1) + f(x_2)\right).$$

Proof. Given three ordered, equally spaced points at x_0, x_1, and x_2, and a fourth point at $x_j \neq x_1$ and $x_0 < x_j < x_2$, all of which satisfy a unique third-degree polynomial function $f(x)$. There exists a unique third-degree Taylor polynomial function $T_3(x)$ through those four points such that $f(x) = T_3(x)$ and so their definite integral values over $[x_0, x_2]$ are equal as well.

$$\int_{x_0}^{x_2} f(x)\, dx = \int_{x_0}^{x_2} T_3(x)\, dx$$

There also exists a unique second-degree Taylor polynomial function $T_2(x)$ through x_0, x_1, and x_2. Since Simpson's rule equals the definite integral of a second-degree polynomial function fit through these same three points, the integral value is both unique and exact.

$$\int_{x_0}^{x_2} T_2(x)\, dx = \frac{x_2 - x_0}{6}\left(f(x_0) + 4f(x_1) + f(x_2)\right)$$

Theorem 5.2 established integrating the third-degree term of the Taylor polynomial function $T_3(x)$ on the shifted interval $[-x_2, x_2]$ equals zero. In turn, integrating $T_3(x)$ is equivalent to integrating $T_2(x)$ on both the shifted interval $[-x_2, x_2]$ and its equivalent original interval $[x_0, x_2]$, so that

$$\int_{x_0}^{x_2} T_3(x)\, dx = \int_{x_0}^{x_2} T_2(x)\, dx.$$

Combining all three above equalities in the order of first, third, and second

$$\int_{x_0}^{x_2} f(x)\, dx = \int_{x_0}^{x_2} T_3(x)\, dx = \int_{x_0}^{x_2} T_2(x)\, dx = \frac{x_2 - x_0}{6}\left(f(x_0) + 4f(x_1) + f(x_2)\right)$$

Thus, the exact integral value of a third-degree polynomial function $f(x)$ over $[x_0, x_2]$ equals Simpson's rule of Equation (5.19). $\qquad\square$

Example 5.9. Use Simpson's rule to estimate $\int_1^{13}(0.1x^3 - 0.14x^2 - 2.6x + 31)\,dx$. Analytically solve the integral and explain how both results are related.

Solution. The limits of integration correspond to $x_0 = 1$ and $x_2 = 13$, with an equally spaced midpoint of $x_1 = 7$. Using Equation (5.19) for Simpson's rule

$$\int_1^{13}\left(0.1x^3 - 0.14x^2 - 2.6x + 31\right)dx = \frac{13-1}{6}\left(f(1) + 4f(7) + f(13)\right)$$

$$= 2\left(28.36 + 4(40.24) + 193.24\right) = 765.12$$

Solving analytically with an antiderivative,

$$\int_1^{13}\left(0.1x^3 - 0.14x^2 - 2.6x + 31\right)dx = \left(0.025x^4 - \frac{7x^3}{150} - 1.3x^2 + 31x\right)\Bigg|_1^{13}$$

$$= 794.79\overline{83} - 29.67\overline{83} = 765.12$$

By Theorem 5.3, the answers should be equal because the integrand is a third-degree polynomial function.

The result of Theorem 5.3 is even more intriguing because, while using Simpson's rule and its corresponding unique second-degree polynomial function $f_2(x)$ produces the same integral value as the underlying third-degree polynomial function $f(x)$, $f_2(x)$ and $f(x)$ are not the same functions over the interval $[x_0, x_2]$. Thus, two different antiderivatives over the same interval yield equivalent answers.

Using Example 5.9 as an illustration, the three points $f(1) = 28.36$, $f(7) = 40.24$, and $f(13) = 193.24$ satisfy the unique second-degree polynomial function $f_2(x) = ax^2 + bx + c$ such that

$$f_2(1) = a(1)^2 + b(1) + c = a + b + c = 28.36$$

$$f_2(7) = a(7)^2 + b(7) + c = 49a + 7b + c = 40.24$$

$$f_2(13) = a(13)^2 + b(13) + c = 169a + 13b + c = 193.24$$

Representing the system of equations on the right in an augmented matrix and using Gauss–Jordan elimination to produce RREF

$$\begin{bmatrix} 1 & 1 & 1 & \vdots & 28.36 \\ 49 & 7 & 1 & \vdots & 40.24 \\ 169 & 13 & 1 & \vdots & 193.24 \end{bmatrix} \longrightarrow \begin{bmatrix} 1 & 0 & 0 & \vdots & 1.96 \\ 0 & 1 & 0 & \vdots & -13.7 \\ 0 & 0 & 1 & \vdots & 40.1 \end{bmatrix}$$

Thus, $a = 1.96$, $b = -13.7$, and $c = 40.1$, so $f_2(x) = 1.96x^2 - 13.7x + 40.1$. Integrating $f_2(x)$ over $[1, 13]$,

$$\int_{1}^{13} (1.96x^2 - 13.7x + 40.1)\, dx = \left(\frac{49}{75}x^3 - 6.85x^2 + 40.1x \right)\Big|_{1}^{13}$$

$$= 799.02\overline{3} - 33.90\overline{3} = 765.12$$

This matches the result generated by the fourth-degree antiderivative of Example 5.9.

Figure 5.11 displays the graphs of $f(x)$ in red and $f_2(x)$ in black, overlapping and clearly not identical on the interval $[1, 13]$. With respect to the x-axis, the blue area represents an underestimation of $f(x)$ by $f_2(x)$ because $f_2(x) < f(x)$ on $[1, 7]$. Conversely, the green area represents an overestimation of $f(x)$ by $f_2(x)$ since $f_2(x) > f(x)$ on $[7, 13]$. Theorem 5.3 guarantees the magnitudes of both areas are equal.

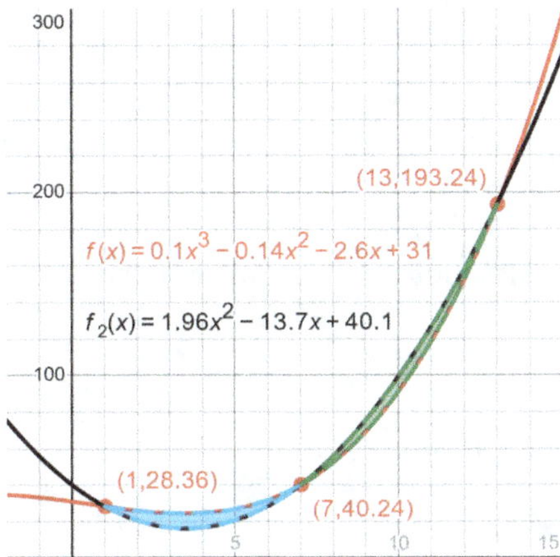

Figure 5.11: The green area is an overestimation of $f(x)$ by $f_2(x)$ and the blue area is an underestimation of $f(x)$ by $f_2(x)$. Both areas have equal magnitude.

5.2.3 Multiple applications

Similar to extending the trapezoidal rule from one to multiple applications, Simpson's rule can be applied to each group of *three* consecutive x values for an interval of equally spaced x values $x_0, x_1, x_2, \ldots, x_{n-1}, x_n$. The right endpoint of one group equals the left endpoint of the next group for all applications, save the last group ending at x_n. Such grouping requires an even n for x_n, stemming from the following.

Starting at x_0, one application of Simpson's rule, which uses three points, requires two additional points at x_1 and x_2, so that $n = 2$ for x_n. For a second application, beginning at x_2, two more x values at x_3 and x_4 are needed, and $n = 4$. For a third application, x_5 and x_6 are needed, $n = 6$, and so forth. It follows easily that $n/2$ applications of Simpson's rule mandates an even n when using an interval of equally spaced x values $x_0, x_1, x_2, \ldots, x_{n-1}, x_n$.

Recall when developing Equation (5.19) for one application of Simpson's rule in Section 5.2.1, the width of the single interval used was expressed as $x_2 - x_0$, in preparation of extending the estimation to the current multiple applications scenario. For $n/2$ applications over the interval $[x_0, x_n]$, the width of each subinterval is

$$\frac{x_n - x_0}{n/2} = \frac{2(x_n - x_0)}{n}.$$

Substituting this single subinterval width for the single width $x_2 - x_0$ in Equation (5.19) results in an equivalent leading constant for one application as

$$\frac{x_2 - x_0}{6} = \frac{2(x_n - x_0)/n}{6} = \frac{x_n - x_0}{3n}$$

Now, applying Simpson's rule of Equation (5.19) to one subinterval $[x_{i-1}, x_{i+1}]$, where i is odd and using the above constant

$$\int_{x_{i-1}}^{x_{i+1}} f(x)\, dx \approx \frac{x_n - x_0}{3n}\left(f(x_{i-1}) + 4f(x_i) + f(x_{i+1})\right).$$

Summing all applications of Simpson's rule over all subintervals, the integral estimation of $f(x)$ over the entire interval is

$$\int_{x_0}^{x_n} f(x)\, dx \approx \sum_{i=1,3,5}^{n-1} \frac{x_n - x_0}{3n}\left(f(x_{i-1}) + 4f(x_i) + f(x_{i+1})\right)$$

$$= \frac{x_n - x_0}{3n} \sum_{i=1,3,5}^{n-1} \left(f(x_{i-1}) + 4f(x_i) + f(x_{i+1})\right). \tag{5.29}$$

Expanding the summation of each group of three function values at x_{i-1}, x_i, and x_{i+1} highlights the pattern that emerges when adding all groups.

$$\left(f(x_0) + 4(x_1) + f(x_2)\right) + \left(f(x_2) + 4f(x_3) + f(x_4)\right)$$
$$+ \left(f(x_4) + 4(x_5) + f(x_6)\right) + \left(f(x_6) + 4f(x_7) + f(x_8)\right)$$
$$+ \cdots + \left(f(x_{n-4}) + 4f(x_{n-3}) + f(x_{n-2})\right) + \left(f(x_{n-2}) + 4f(x_{n-1}) + f(x_n)\right)$$

Note that the subintervals end and start at x_k values, where k is even so that those function values are represented twice each with the exception of the endpoints, which occur once each. When k is odd, the middle point of each interval is multiplied by four but never duplicated in subsequent applications. Combining terms in this manner, the above sum equals

$$f(x_0) + 4f(x_1) + 2f(x_2) + 4f(x_3) + 2f(x_4) + \cdots + 4f(x_{n-1}) + f(x_n). \tag{5.30}$$

Observe the pattern of coefficients is $1, 4, 2, 4, 2, \ldots, 4, 1$. Substituting Equation (5.30) for the summation in Equation (5.29), the integral estimation of $f(x)$ over $[x_0, x_n]$ using n subintervals or $n/2$ applications of Simpson's rule $A_S(n)$ is

$$\int_{x_0}^{x_n} f(x)\,dx \approx A_S(n) = \frac{x_n - x_0}{3n}\left(f(x_0) + \sum_{i=1,3,5}^{n-1} 4f(x_i) + \sum_{i=2,4,6}^{n-2} 2f(x_i) + f(x_n) \right) \tag{5.31}$$

Example 5.10. Use Simpson's rule with $n = 4$ subintervals to estimate $\int_{1.7}^{5.4} 13/(x^3 + 2)\,dx$.
Solution. The width of each of the $n = 4$ equal subintervals on $[1.7, 5.4]$ is

$$h = \frac{5.4 - 1.7}{4} = 0.925.$$

The left endpoint is x_0, and repeatedly adding $h = 0.925$ to each successive x value generates

$$x_0 = 1.7, \; x_1 = 2.625, \; x_2 = 3.55, \; x_3 = 4.475, \text{ and } x_4 = 5.4$$

Using Equation (5.31),

$$\int_{1.7}^{5.4} \frac{13}{x^3 + 2}\,dx \approx A_S(4) = \frac{x_4 - x_0}{3n}\left(f(x_0) + 4f(x_1) + 2f(x_2) + 4f(x_3) + f(x_4) \right)$$

$$= \frac{5.4 - 1.7}{3(4)}\left(f(1.7) + 4f(2.625) + 2f(3.55) + 4f(4.475) + f(5.4) \right)$$

$$= \frac{3.7}{12}\left(1.881 + 4(0.647) + 2(0.278) + 4(0.142) + 0.082 \right) = 1.750.$$

The explanation following Example 5.8 used $I = 1.733$ as the true integral value. For $n = 4$ only $n/2$ or two applications of Simpson's rule was used in Example 5.10. In this case, the absolute error for $A_S(4) = 1.750$ is

$$\left| I - A_S(4) \right| = |1.733 - 1.750| = 0.017.$$

So, for this function on the given interval, two applications of Simpson's rule without any truncation error adjustment is superior to one application of Simpson's rule with a truncation error adjustment (Example 5.8); the latter has an absolute error of 0.035.

The next example presents a problem in terms of the number of times to apply Simpson's rule rather than by using n subintervals as in the previous example.

Example 5.11. Use three applications of Simpson's rule to estimate $\int_{1}^{4} e^{(2/x)}\,dx$.
Solution. Three applications of Simpson's rule means $n/2 = 3$ or $n = 6$. Thus each x value is equally spaced by a width of $h = \frac{4-1}{6} = 0.5$. The x values start at $x_0 = 1$, end at $x_6 = 4$, and the values between are

$$x_1 = 1.5, \; x_2 = 2, \; x_3 = 2.5, \; x_4 = 3, \text{ and } x_5 = 3.5$$

Using Equation (5.31),

$$\int_{x_0}^{x_6} e^{(2/x)}\, dx \approx A_S(6)$$

$$= \frac{x_6 - x_0}{3n}\left(f(x_0) + 4f(x_1) + 2f(x_2) + 4f(x_3) + 2f(x_4) + 4f(x_5) + f(x_6)\right)$$

$$= \frac{4-1}{3(6)}\left(f(1) + 4f(1.5) + 2f(2) + 4f(2.5) + 2f(3) + 4f(3.5) + f(4)\right)$$

$$= \frac{1}{6}\left(7.389 + 4(3.794) + 2(2.718) + 4(2.226) + 2(1.948) + 4(1.771) + 1.649\right) = 8.256$$

5.2.4 Multiple applications, truncation error adjusted

For a single application of Simpson's rule, the truncation error adjustment of Equation (5.25) is one term based on the entire interval $[x_0, x_2]$ and containing the corresponding factor $(x_2 - x_0)^5$. When using multiple applications of Simpson's rule, the truncation error adjustment of Equation (5.25) is applied to every subinterval $[x_{i-2}, x_i]$, with each subinterval factor $(x_i - x_{i-2})^5$ replacing the single $(x_2 - x_0)^5$ factor. All the subinterval terms are then summed.

$$E_{t,S}(n) = \sum_{i=2,4,6}^{n} -\frac{f^{(4)}(c_i)(x_i - x_{i-2})^5}{2880} \tag{5.32}$$

Recall in Section 5.2.3 for n even subintervals, there are $n/2$ applications of Simpson's rule. The width of the interval for each application is therefore

$$h = \frac{x_n - x_0}{n/2} = \frac{2(x_n - x_0)}{n}$$

Substituting this width h for $x_i - x_{i-2}$ in Equation (5.32)

$$E_{t,S}(n) = \sum_{i=2,4,6}^{n} -\frac{f^{(4)}(c_i)\left(\frac{2(x_n - x_0)}{n}\right)^5}{2880} = -\frac{\frac{2^5(x_n - x_0)^5}{n^5}}{2880} \sum_{i=2,4,6}^{n} f^{(4)}(c_i)$$

$$= -\frac{(x_n - x_0)^5}{90n^5} \sum_{i=2,4,6}^{n} f^{(4)}(c_i). \tag{5.33}$$

As with multiple applications of the trapezoidal rule the mean value is used for each $f^{(4)}(c_i)$ in Equation (5.33).

$$\sum_{i=2,4,6}^{n} f^{(4)}(c_i) \approx \sum_{i=2,4,6}^{n} \frac{\int_{x_{i-2}}^{x_i} f^{(4)}(x)\, dx}{x_i - x_{i-2}}. \tag{5.34}$$

Expanding the right side and substituting the width h of each subinterval for each application in the denominator

$$\frac{\int_{x_0}^{x_2} f^{(4)}(x)\,dx}{\frac{2(x_n-x_0)}{n}} + \frac{\int_{x_2}^{x_4} f^{(4)}(x)\,dx}{\frac{2(x_n-x_0)}{n}} + \cdots + \frac{\int_{x_{n-4}}^{x_{n-2}} f^{(4)}(x)\,dx}{\frac{2(x_n-x_0)}{n}} + \frac{\int_{x_{n-2}}^{x_n} f^{(4)}(x)\,dx}{\frac{2(x_n-x_0)}{n}} \tag{5.35}$$

The common denominator can be factored out of each term and written as its reciprocal, multiplied by the sum of the integrals. Because each upper limit of integration equals each lower limit of integration of the next integral, the sum of all integrals is the integral of $f^{(4)}(x)$ over the entire interval $[x_0, x_n]$. Rewriting Equation (5.35) as such and then substituting for the right side of Equation (5.34)

$$\sum_{i=2,4,6}^{n} f^{(4)}(c_i) \approx \frac{n}{2(x_n - x_0)} \int_{x_0}^{x_n} f^{(4)}(x)\,dx$$

Integrating the fourth derivative is equivalent to the third derivative, so that

$$\sum_{i=2,4,6}^{n} f^{(4)}(c_i) \approx \frac{n\int_{x_0}^{x_n} f^{(4)}(x)\,dx}{2(x_n - x_0)} = \frac{nf^{(3)}(x)|_{x_0}^{x_n}}{2(x_n - x_0)} = \frac{n(f^{(3)}(x_n) - f^{(3)}(x_0))}{2(x_n - x_0)}. \tag{5.36}$$

Substituting the right side of Equation (5.36) for the summation in Equation (5.33) and reducing produces the final form of the truncation error adjustment for n subintervals or $n/2$ applications of Simpson's rule $E_{t,S}(n)$.

$$\begin{aligned} E_{t,S}(n) &= -\frac{(x_n - x_0)^5}{90n^5} \sum_{i=2,4,6}^{n} f^{(4)}(c_i) \approx -\frac{(x_n - x_0)^5}{90n^5} \cdot \frac{n(f^{(3)}(x_n) - f^{(3)}(x_0))}{2(x_n - x_0)} \\ &= -\frac{(x_n - x_0)^4 (f^{(3)}(x_n) - f^{(3)}(x_0))}{180n^4} \end{aligned} \tag{5.37}$$

The truncation error adjustment for a single application of Simpson's rule of Equation (5.27) is now viewed as simply a special case of Equation (5.37). Using Equation (5.37) for one application, $1 = n/2$ so $n = 2$ and $x_n = x_2$. Making these two substitutions along with the denominator $180(2)^4 = 2880$ in Equation (5.37) make it identical to Equation (5.27).

The adjusted estimation $A_{S,t}(n)$ for n subintervals or $n/2$ applications of Simpson's rule using Equation (5.31) along with the truncation error adjustment of Equation (5.37) can be expressed as

$$A_{S,t}(n) = A_S(n) + E_{t,S}(n). \tag{5.38}$$

Example 5.12. Calculate $E_{t,S}(n)$ for the integral of Example 5.10. Apply $E_{t,S}(n)$ for the improved Simpson's rule estimation $A_{S,t}(n)$. Use $E_{t,S}(n)$ to determine whether multiple applications of Simpson's rule alone $A_S(n)$ in this example underestimates or overestimates the true integral value.

Solution. Using $f^{(3)}(x) = \frac{-780x^6 + 2496x^3 - 312}{(x^3+2)^4}$ and Equation (5.37)

$$E_{t,S}(4) = -\frac{(5.4 - 1.7)^4 (f^{(3)}(5.4) - f^{(3)}(1.7))}{180(4)^4} = -\frac{(3.7)^4(-0.029 - 3.011)}{46080} = -0.012$$

$E_{t,S}(4) < 0$ indicates multiple applications of Simpson's rule alone of Example 5.10 overestimated the true integral value. The adjusted estimation is

$$A_{S,t}(4) = A_S(4) + E_{t,S}(4) = 1.750 + (-0.012) = 1.738$$

Like the trapezoidal rule, multiple applications of Simpson's rule with the truncation error adjustment $E_{t,S}(n)$ can produce very reasonable results even for a small number of applications. Using $I = 1.733$ as the true integral value, as explained after Example 5.10, the absolute error of Example 5.12 is just

$$\left|I - A_{S,t}(4)\right| = |1.733 - 1.738| = 0.005.$$

This is an improvement over the absolute error of 0.017 of Example 5.10, prior to applying the truncation error adjustment.

5.2.5 Simpson's rule variations, evaluated by ε_S

The four Simpson's rule estimations of Sections 5.2.1 through 5.2.4 are now considered in conjunction with an example where the exact definite integral value I can be solved analytically. Analogous to the exact relative percentage error used for the trapezoidal rules in Section 5.1.5, here the exact relative percentage error for Simpson's rules ε_S is defined by A_{Sim} instead of A_{Trap}.

$$\varepsilon_S = \left|\frac{I - A_{\text{Sim}}}{I}\right| \times 100\,\% \tag{5.39}$$

A_{Sim} is an estimation type using either single or multiple applications of Simpson's rule of Sections 5.2.1 or 5.2.3 or a truncation error-adjusted estimation of Sections 5.2.2 or 5.2.4. Table 5.3 lists each variation of Equation (5.39) for ε_S.

Table 5.3: Estimations for the integral A_{Sim} used for ε_S of Equation (5.39) can be defined by any of the four rows, each a variation on Simpson's rule of Sections 5.2.1 through 5.2.4.

Type	Applications	Error Adjusted?	Equation	Uses Equations
A_S	single	no	(5.19)	
$A_{S,t}$	single	yes, by $E_{t,S}$	(5.28)	(5.19) and (5.27)
$A_S(n)$	multiple	no	(5.31)	
$A_{S,t}(n)$	multiple	yes, by $E_{t,S}(n)$	(5.38)	(5.31) and (5.37)

Example 5.13. Analytically solve $I = \int_{2.5}^{16.9} 10xe^{-0.1x}\, dx$. Solve each of the four Simpson's rule variations of Table 5.3 to estimate I. Calculate the exact relative percentage error ε_S for each estimation.

Solution. The antiderivative $F(x)$ for $f(x) = 10xe^{-0.1x}$ is found using integration by parts, where

$$u = 10x \text{ and } dv = e^{-0.1x}\, dx$$

For $F(x) = uv - \int v\, du$, $F(x) = -100e^{-0.1x}(x + 10)$ and

$$I = \int_{2.5}^{16.9} 10xe^{-0.1x}\, dx = -100e^{-0.1x}(x + 10)\big|_{2.5}^{16.9}$$

$$= -100e^{-0.1(16.9)}(16.9 + 10) - \left(-100e^{-0.1(2.5)}(2.5 + 10)\right)$$

$$= -496.358 - (-973.501) = 477.143$$

Case 1. Using Simpson's rule once with no adjustment, $x_0 = 2.5$, $x_2 = 16.9$, and the midpoint is at $x_1 = \frac{16.9+2.5}{2} = 9.7$. The function values $f(x)$ at these three points are

$$f(x_0) = f(2.5) = 10(2.5)e^{-0.1(2.5)} = 19.470$$

$$f(x_1) = f(9.7) = 10(9.7)e^{-0.1(9.7)} = 36.771$$

$$f(x_2) = f(16.9) = 10(16.9)e^{-0.1(16.9)} = 31.184$$

According to Equation (5.19),

$$\int_{2.5}^{16.9} f(x)\, dx \approx A_S = \frac{16.9 - 2.5}{6}\left(f(2.5) + 4f(9.7) + f(16.9)\right)$$

$$= (2.4)\left(19.470 + 4(36.771) + 31.184\right) = 474.571.$$

The exact relative percentage error of Equation (5.39) is

$$\varepsilon_S = \left|\frac{I - A_S}{I}\right| = \left|\frac{477.143 - 474.571}{477.143}\right| \times 100\,\% = 0.54\,\%.$$

Case 2. The truncation error adjustment for one application of Simpson's rule requires the third derivative of $f(x)$. Using a product rule for each derivative up to and including the third, and combining terms, $f^{(3)}(x) = (0.3 - 0.01x)e^{-0.1x}$. At the two endpoints, the third derivative is

$$f^{(3)}(x_0) = f^{(3)}(2.5) = \left(0.3 - 0.01(2.5)\right)e^{-0.1(2.5)} = 0.214$$

$$f^{(3)}(x_2) = f^{(3)}(16.9) = \left(0.3 - 0.01(16.9)\right)e^{-0.1(16.9)} = 0.024.$$

By Equation (5.27), the truncation error adjustment is

$$E_{t,S} = -\frac{(f^{(3)}(16.9) - f^{(3)}(2.5))(16.9 - 2.5)^4}{2880} = -\frac{(0.024 - 0.214)(14.4)^4}{2880} = 2.837.$$

The adjusted estimation $A_{S,t}$ using Equation (5.28) is then

$$A_{S,t} = A_S + E_{t,S} = 474.571 + 2.837 = 477.408.$$

The exact relative percentage error according to Equation (5.39) has improved.

$$\varepsilon_S = \left| \frac{I - A_{S,t}}{I} \right| = \left| \frac{477.143 - 477.408}{477.143} \right| \times 100\,\% = 0.06\,\%$$

Case 3. An even $n = 6$ is selected for applying multiple applications of Simpson's rule. The endpoints are now labeled $x_0 = 2.5$ and $x_6 = 16.9$, with interior points spaced an equal distance of $h = \frac{16.9 - 2.5}{6} = 2.4$, so that

$$x_1 = 4.9,\ x_2 = 7.3,\ x_3 = 9.7,\ x_4 = 12.1,\ \text{and}\ x_5 = 14.5.$$

The function values for each of these x values are

$$f(x_0) = f(2.5) = 10(2.5)e^{-0.1(2.5)} = 19.470$$
$$f(x_1) = f(4.9) = 10(4.9)e^{-0.1(4.9)} = 30.019$$
$$f(x_2) = f(7.3) = 10(7.3)e^{-0.1(7.3)} = 35.179$$
$$f(x_3) = f(9.7) = 10(9.7)e^{-0.1(9.7)} = 36.771$$
$$f(x_4) = f(12.1) = 10(12.1)e^{-0.1(12.1)} = 36.082$$
$$f(x_5) = f(14.5) = 10(14.5)e^{-0.1(14.5)} = 34.013$$
$$f(x_6) = f(16.9) = 10(16.9)e^{-0.1(16.9)} = 31.184$$

By Equation (5.31),

$$\int_{2.5}^{16.9} f(x)\,dx \approx A_S(6) = \frac{x_6 - x_0}{3(6)}\left(f(x_0) + \sum_{i=1,3,5}^{5} 4f(x_i) + \sum_{i=2,4}^{4} 2f(x_i) + f(x_6) \right)$$

$$= \frac{16.9 - 2.5}{18}\big(f(x_0) + 4f(x_1) + 2f(x_2) + 4f(x_3) + 2f(x_4) + 4f(x_5) + f(x_6)\big)$$

$$= \frac{14.4}{18}\big(f(2.5) + 4f(4.9) + 2f(7.3) + 4f(9.7) + 2f(12.1) + 4f(14.5) + f(16.9)\big)$$

$$= 0.8\big(19.470 + 4(30.019) + 2(35.179) + 4(36.771)$$

$$+ 2(36.082) + 4(34.013) + 31.184\big) = 477.110$$

The exact relative percentage error of Equation (5.39) continues to improve.

$$\varepsilon_S = \left| \frac{I - A_S(6)}{I} \right| = \left| \frac{477.143 - 477.110}{477.143} \right| \times 100\,\% = 0.01\,\%$$

Case 4. The truncation error adjustment for multiple applications of Simpson's rule uses the same third derivative values at the limits of integration as in Case 2. The upper limit here is relabeled x_n from x_2, and since in this example $n = 6$, $x_n = x_6 = x_2 = 16.9$. Correspondingly, $f^{(3)}(x_6) = f^{(3)}(x_2) = 0.024$. Using Equation (5.37) for the truncation error adjustment

$$E_{t,S}(6) = -\frac{(x_6 - x_0)^4(f^{(3)}(x_6) - f^{(3)}(x_0))}{180n^4} = -\frac{(16.9 - 2.5)^4(0.024 - 0.214)}{180(6)^4} = 0.035.$$

The adjusted estimation $A_{S,t}(6)$ by Equation (5.38) is thus

$$A_{S,t}(6) = A_S(6) + E_{t,S}(6) = 477.110 + 0.035 = 477.145.$$

This is the true integral value I accurate to the tenths place, and the exact relative percentage error is virtually 0 %.

$$\varepsilon_S = \left| \frac{I - A_{S,t}(6)}{I} \right| = \left| \frac{477.143 - 477.145}{477.143} \right| \times 100\,\%$$

$$= \frac{0.002}{477.143} \times 100\,\% \approx 0.000004 \times 100\,\% = 0.00\,\%$$

The exact relative percentage errors using the Simpson's rule variations consistently decreased from Case 1 to Case 4, from 0.54 % to 0.06 % to 0.01 % to virtually 0 %, identical to the progression of the correspondingly constructed trapezoidal rule variations of Example 5.6, summarized in Table 5.2. The same three conclusions following Example 5.6 are drawn now about the variations of Simpson's rule. First, applying a truncation error adjustment in the single or multiple applications case can offer improvement. Second, multiple applications are superior to a single application. Third, multiple applications without a truncation error adjustment are an improvement over a single application with a truncation error adjustment.

5.2.6 What could go wrong?

Examples 5.4 and 5.5 produced the results of Table 5.4 for $\int_{0.5}^{3.5} \ln(6x - x^2)\, dx$. The true integral value used for the ε_T values in Table 5.4 is $I = 5.817$, as noted after Example 5.5.

Table 5.4: Trapezoidal rule estimation data for $\int_{0.5}^{3.5} \ln(6x - x^2)\, dx$ with $n = 4$ of Examples 5.4 and 5.5.

Type	Error adjusted?	Estimation	ε_T, %
$A_T(4)$	no	5.730	1.50
$A_{T,t}(4)$	yes, by $E_{t,T}(4)$	5.821	0.07

If Simpson's rule of Equation (5.31), without $E_{t,S}(4)$, is applied with $n = 4$

$$A_S(4) = \frac{3.5 - 0.5}{3(4)}\left(f(0.5) + 4f(1.25) + 2f(2) + 4f(2.75) + f(3.5) \right)$$

$$= 0.25\left(1.012 + 4(1.781) + 2(2.079) + 4(2.190) + 2.169 \right) = 5.806$$

The corresponding exact relative percentage error is

$$\varepsilon_S = \left| \frac{I - A_{\text{Sim}}}{I} \right| \times 100\,\% = \left| \frac{5.817 - 5.806}{5.817} \right| \times 100\,\% = 0.19\,\%.$$

As expected, Simpson's rule $A_S(4)$ without $E_{t,S}(4)$ is a better estimation than the trapezoidal rule $A_T(4)$ without $E_{t,T}(4)$, but not better than the trapezoidal rule $A_{T,t}(4)$ which includes $E_{t,T}(4)$.

So, attempting further refinement using Simpson's rule with $E_{t,S}(4)$, the third derivative $f^{(3)}(x) = \frac{432-216x+36x^2-4x^3}{(6x-x^2)^3}$ of $f(x) = \ln(6x - x^2)$ is evaluated at the endpoints of the integral.

$$f^{(3)}(0.5) = 15.988$$
$$f^{(3)}(3.5) = -0.081$$

By Equation (5.37) the Simpson's rule truncation error adjustment is

$$E_{t,S}(4) = -\frac{(3.5 - 0.5)^4(-0.081 - 15.988)}{180(4)^4} = 0.028.$$

The adjusted estimation $A_{S,t}(4)$ by Equation (5.38) is then

$$A_{S,t}(4) = A_S(4) + E_{t,S}(4) = 5.806 + 0.028 = 5.834.$$

This produces an exact relative percentage error of

$$\varepsilon_S = \left|\frac{I - A_{\text{Sim}}}{I}\right| \times 100\,\% = \left|\frac{5.817 - 5.834}{5.817}\right| \times 100\,\% = 0.29\,\%.$$

What happened? The Simpson's rule truncation error adjustment $E_{t,S}(4)$ made the estimation $A_{S,t}(4)$ worse than the estimation $A_S(4)$ without $E_{t,S}(4)$. $A_{S,t}(4)$ is also inferior to the trapezoidal rule estimation $A_{T,t}(4)$, which includes $E_{t,T}(4)$. In these examples, the derivatives are culpable.

In the process of defining $E_{t,T}(n)$ and $E_{t,S}(n)$ for the trapezoidal and Simpson's rules respectively, Equations (5.9) and (5.34) substituted a single mean value for all derivative values $f''(c_i)$ and $f^{(4)}(c_i)$, respectively, later substituted for all derivative values $f'(c_i)$ and $f^{(3)}(c_i)$, respectively, due to integration. Extended from $E_{t,T}$ of Equation (5.4) of Section 5.1.2 was the assumption that substituting a single mean derivative value for all derivative values $f'(c_i)$ and $f^{(3)}(c_i)$ is appropriate because the derivative behaves constantly on a small interval.

Figure 5.12 displays the first and third derivatives of $f(x) = \ln(6x - x^2)$ on the interval $[0.5, 3.5]$. The first derivative $f'(x)$ in red is fairly constant on $[0.5, 3.5]$, thus $E_{t,T}(4)$ defined by $f'(x)$ for the trapezoidal rule improves the estimation, verified by Table 5.4. However, the third derivative $f^{(3)}(x)$ in blue which defines $E_{t,S}(4)$ for Simpson's rule clearly is not constant on $[0.5, 3.5]$; it decreases severely on the first third of the interval $[0.5, 1.5]$ before maintaining an almost constant value for the second third of the interval $[1.5, 3.5]$. Using a single mean value for $f^{(3)}(x)$ over the entire interval $[0.5, 3.5]$ which is not fairly constant overall is decidedly not appropriate. Consequently, the $E_{t,S}(4)$ for Simpson's rule detrimentally affects the initial estimation $A_S(4)$, making $A_{S,t}(4)$ worse.

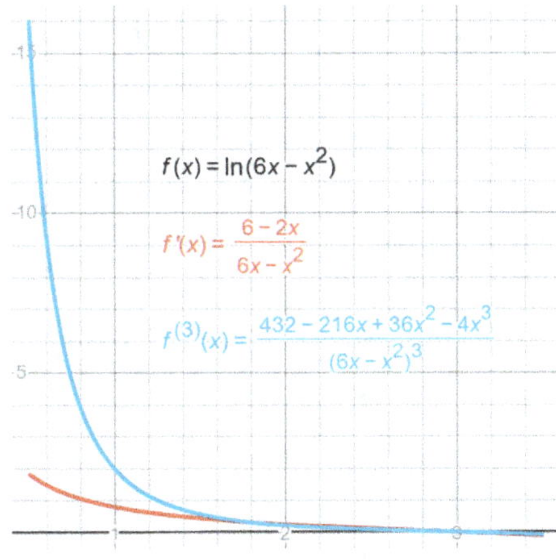

Figure 5.12: Using the mean value to represent the red first derivative is viable because $f'(x)$ is fairly constant. The mean value is not a good substitution for the blue third derivative because $f^{(3)}(x)$ varies significantly.

Avoiding pitfalls

The example in this section highlights why simply invoking an algorithm without understanding its definition, assumptions in its derivation, or the function or data on which it is applied, can produce poor results. Conversely, such knowledge is key in rectifying lurking problems.

As observed in Figure 5.12, the third derivative $f^{(3)}(x)$ behaves much differently on the subintervals $[0.5, 1.5]$ and $[1.5, 3.5]$. Therefore, rather than employing a single mean value over the entire interval $[0.5, 3.5]$ that poorly represents $f^{(3)}(x)$, two mean values are used, one for each subinterval. Each of these mean values offers a better characterization of $f^{(3)}(x)$ on their respective subintervals versus the single mean value over the entire interval. Note the mean values themselves are not found directly but embedded in the truncation error adjustments $E_{t,S}$ and $E_{t,S}(n)$, and implemented by specifying the endpoints of the subintervals.

The example in this section used $n = 4$, corresponding to $n/2 = 4/2 = 2$ applications of Simpson's rule. Two applications are used again, a single application on each subinterval $[0.5, 1.5]$ and $[1.5, 3.5]$.

On $[0.5, 1.5]$, $x_0 = 0.5$, $x_1 = 1$, and $x_2 = 1.5$. Using Equation (5.19), the first Simpson's rule estimation is

$$A_{S_1} = \frac{1.5 - 0.5}{6}\left(f(0.5) + 4f(1) + f(1.5)\right) = \frac{1.012 + 4(1.609) + 1.910}{6} = 1.560.$$

The first truncation error adjustment needs $f^{(3)}(0.5) = 15.988$, $f^{(3)}(1.5) = 0.571$, and Equation (5.27).

$$E_{t,S_1} = -\frac{(f^{(3)}(1.5) - f^{(3)}(0.5))(1.5 - 0.5)^4}{2880} = -\frac{0.571 - 15.988}{2880} = 0.005.$$

Including the truncation error adjustment, the integral estimation on the first subinterval is

$$A_{S_1,t} = A_{S_1} + E_{t,S_1} = 1.560 + 0.005 = 1.565.$$

Repeating these steps for the second subinterval $[1.5, 3.5]$, using $x_0 = 1.5, x_1 = 2.5, x_2 = 3.5$, and Equation (5.19), the second Simpson's rule estimation is

$$A_{S_2} = \frac{3.5 - 1.5}{6}(f(1.5) + 4f(2.5) + f(3.5)) = \frac{1.910 + 4(2.169) + 2.169}{3} = 4.252$$

The second truncation error adjustment uses $f^{(3)}(1.5) = 0.571, f^{(3)}(3.5) = -0.081$, and Equation (5.27).

$$E_{t,S_2} = -\frac{(f^{(3)}(3.5) - f^{(3)}(1.5))(3.5 - 1.5)^4}{2880} = -\frac{16(-0.081 - 0.571)}{2880} = 0.004.$$

With the truncation error adjustment, the integral estimation on the second subinterval is

$$A_{S_2,t} = A_{S_2} + E_{t,S_2} = 4.252 + 0.004 = 4.256.$$

Adding the integral estimations of each subinterval to estimate the entire interval $[0.5, 3.5]$, including the truncation error adjustment

$$A_{S,t} = A_{S_1,t} + A_{S_2,t} = 1.565 + 4.256 = 5.821.$$

The associated exact relative percentage error is

$$\varepsilon_S = \left| \frac{I - A_{\text{Sim}}}{I} \right| \times 100\,\% = \left| \frac{5.817 - 5.821}{5.817} \right| \times 100\,\% = 0.07\,\%.$$

Lastly, for comparison, the integral estimation without the truncation error adjustment over the entire interval using the above subintervals is

$$A_S = A_{S_1} + A_{S_2} = 1.560 + 4.252 = 5.812$$

This exact relative percentage error is

$$\varepsilon_S = \left| \frac{I - A_{\text{Sim}}}{I} \right| \times 100\,\% = \left| \frac{5.817 - 5.812}{5.817} \right| \times 100\,\% = 0.09\,\%$$

Table 5.5 combines the results of Table 5.4 with the revised Simpson's rule estimations, and all new estimations are intuitively consistent. Simpson's rule $A_{S,t}$ is now an improvement over A_S. Furthermore, Simpson's rule with the truncation error adjustment, $A_{S,t}$, is identical to the trapezoidal rule with the truncation error adjustment, $A_{T,t}(4)$, to

Table 5.5: Trapezoidal rule data of Table 5.4 and Simpson's rule data from this section for $\int_{0.5}^{3.5} \ln(6x - x^2)\, dx$.

Type	Error Adjusted?	Estimation	Error, %
$A_T(4)$	no	5.730	$\varepsilon_T = 1.50$
$A_{T,t}(4)$	yes, by $E_{t,T}(4)$	5.821	$\varepsilon_T = 0.07$
A_S	no	5.812	$\varepsilon_S = 0.09$
$A_{S,t}$	yes, by $E_{t,S}$	5.821	$\varepsilon_S = 0.07$

the thousandths place. Additionally note the identical results are achieved using half as many applications of Simpson's rule as the trapezoidal rule. Recall $n = 4$ applications of the trapezoidal rules corresponds to $n/2 = 4/2 = 2$ applications of Simpson's rule, implemented in this example by using one application of Simpson's rule on two separate subintervals and summing.

Erratic behavior of third derivatives is not uncommon, especially for initially complicated functions. Analogously, if a first derivative $f'(x)$ fluctuates significantly, the truncation error adjustments $E_{t,T}$ and $E_{t,T}(n)$ for the trapezoidal rules, defined with the mean value of $f'(x)$, warrant careful inspection before application.

5.3 Romberg integration

Romberg integration is a two-tiered recursive algorithm which uses two successive estimations in conjunction for an improved estimation. In proportion, the better estimation is weighted more and positively while the worse is weighted less and negatively. The proportion depends on how many applications are applied in the recursive process. The more applications the more accurate and the larger the proportion.

1. First tier estimations $I_{0,k}$

The first tier estimations for Romberg integration $I_{0,k}$ are defined.

$$I_{0,k} = \frac{x_{2^k} - x_0}{2^k}\left(\frac{f(x_0) + f(x_{2^k})}{2} + \sum_{i=1}^{2^k - 1} f(x_i)\right) \tag{5.40}$$

Equation (5.40) is simply multiple applications of the trapezoidal rule as defined by Equation (5.7), but using slightly different notation with 2^k replacing n. Just like n, 2^k represents the number of equal subintervals and $n + 1 = 2^k + 1$ is the number of points used, with x values indexed $x_0, x_1, x_2, \ldots, x_{2^k-1}, x_{2^k}$. Table 5.6 shows the relationship between $I_{0,k}$, the number of subintervals used, and the list of required indexed x values, for the first five k values.

Table 5.6: Each first tier estimation $I_{0,k}$ uses 2^k equal subintervals and $2^k + 1$ points for a trapezoidal rule estimation using multiple applications.

k	$I_{0,k}$	Number of equal subintervals, 2^k	Required x values indexed, $x_0, \ldots, x_{2^k}$
0	$I_{0,0}$	1	x_0, x_1
1	$I_{0,1}$	2	x_0, x_1, x_2
2	$I_{0,2}$	4	x_0, x_1, x_2, x_3, x_4
3	$I_{0,3}$	8	$x_0, x_1, x_2, \ldots, x_7, x_8$
4	$I_{0,4}$	16	$x_0, x_1, x_2, \ldots, x_{15}, x_{16}$

2. Second tier estimations $I_{m,k}$, $m > 0$

The second tier estimations for Romberg integration $I_{m,k}$ are found recursively, proportionally weighted for $m = 1, 2, 3, \ldots$ and defined.

$$I_{m,k} = \frac{4^m \cdot I_{m-1,k+1} - I_{m-1,k}}{4^m - 1} \tag{5.41}$$

The recursion occurs according to the m values. For m on the left side, $I_{m,k}$ uses two previous estimations for $m - 1$ on the right side, i.e., $I_{m-1,k+1}$ and $I_{m-1,k}$. Of these, the former uses 2^{k+1} subintervals and the latter uses 2^k subintervals, according to Table 5.6. More subintervals results in a better estimation, so $I_{m-1,k+1}$ is weighted more and positively, by a factor of 4^m. The worse estimation $I_{m-1,k}$ is weighted always once, negatively. The sum of these weighted coefficients equals the denominator divisor. Intuitively understanding the weighted tiers suffices here; Ralston and Rabinowitz [1] prove algebraically. Table 5.7 lists the coefficient weights for each m value up to $m = 4$ with the corresponding denominator.

Table 5.7: Each second tier estimation $I_{m,k}$ uses two previous estimations for $m - 1$, with the $k + 1$ estimation weighted more and positively according to Equation (5.41).

m	$I_{m,k}$	Weight of $I_{m-1,k+1}$	Weight of $I_{m-1,k}$	Denominator
1	$I_{1,k}$	4	−1	3
2	$I_{2,k}$	16	−1	15
3	$I_{3,k}$	64	−1	63
4	$I_{4,k}$	256	−1	255

3. Second tier recursion table

A final estimation for Romberg integration can occur for any m value when $k = 0$, that is, any $I_{m,0}$ value. For a particular m, the number of first tier estimations $I_{0,k}$ required are

$k = 0, \ldots, m$. For example, if $I_{2,0}$ represents the final estimation, then $m = 2$ and three first tier estimations for $k = 0, 1, 2$ are needed by Table 5.6, along with $m = 2$ recursive iterations of second tier estimations.

Now suppose the final estimation is at $I_{3,0}$ for $m = 3$. An additional first tier estimation $I_{0,3}$ for $k = m = 3$ is required. Correspondingly, additional recursive second tier estimations are calculated, one each for $m = 1, 2, 3$. Table 5.9 highlights the additional values calculated in red that distinguish $I_{2,0}$ of Table 5.8 from $I_{3,0}$.

Table 5.8: For $m = 2$, three first tier estimations $I_{0,k}$ for $k = 0, 1, 2$ are needed. Two iterations of recursive second tier estimations $I_{m,k}$ are needed for $m = 1, 2$. The final estimation is $I_{2,0}$ in the yellow cell.

k	First tier $m = 0$ $I_{0,k}$	Second tier $m = 1$ $I_{1,k}$	Second tier $m = 2$ $I_{2,k}$
0	$I_{0,0}$	$I_{1,0}$	$I_{2,0}$
1	$I_{0,1}$	$I_{1,1}$	
2	$I_{0,2}$		

Table 5.9: For $m = 3$, the four red values on the diagonal are the additional estimations appended to Table 5.8. The final estimation is now $I_{3,0}$ in the yellow cell.

k	First tier $m = 0$ $I_{0,k}$	Second tier $m = 1$ $I_{1,k}$	Second tier $m = 2$ $I_{2,k}$	Second tier $m = 3$ $I_{3,k}$
0	$I_{0,0}$	$I_{1,0}$	$I_{2,0}$	$I_{3,0}$
1	$I_{0,1}$	$I_{1,1}$	$I_{2,1}$	
2	$I_{0,2}$	$I_{1,2}$		
3	$I_{0,3}$			

Example 5.14. Use Romberg integration to estimate $\int_{1.7}^{5.4} 13/(x^3 + 2)\, dx$ up to $I_{3,0}$.

Solution. For $m = 3$, first tier estimations $I_{0,k}$ for $k = 0, 1, 2, 3$ are needed as described in Table 5.9. According to Table 5.6, the maximum number of equally spaced intervals is eight for $I_{0,3}$, so the width h between x values is

$$h = \frac{5.4 - 1.7}{8} = 0.4625$$

Eight intervals produces a set of nine h-spaced points at x values starting at the lower bound of 1.7 and ending at the upper bound of 5.4.

$$\{1.7,\ 2.1625,\ 2.625,\ 3.0875,\ 3.55,\ 4.0125,\ 4.475,\ 4.9375,\ 5.4\}$$

Subsets of these nine x values can be used for each $k = 0, 1, 2$, described below.

For $k = 0$ and by Table 5.6, a single subinterval is used, so only two of the nine points are needed, the endpoints, so that

$$x_0 = 1.7 \text{ and } x_1 = 5.4$$

Using the trapezoidal rule first tier estimation of Equation (5.40),

$$I_{0,0} = \frac{x_{2^0} - x_0}{2^0}\left(\frac{f(x_0) + f(x_{2^0})}{2} + \sum_{i=1}^{2^0-1} f(x_i) \right)$$

$$= \frac{x_1 - x_0}{1}\left(\frac{f(x_0) + f(x_1)}{2} + \sum_{i=1}^{0} f(x_i) \right).$$

There are no terms in the summation because the upper bound of zero is less than the starting index $i = 1$, so this reduces to

$$I_{0,0} = (5.4 - 1.7)\left(\frac{f(1.7) + f(5.4)}{2} \right) = 3.7\left(\frac{1.881 + 0.082}{2} \right) = 3.632$$

For $k = 1$ and by Table 5.6, two subintervals are used with three equally spaced points from the complete set of the above nine, corresponding to the two endpoints and their midpoint, so that

$$x_0 = 1.7,\ x_1 = 3.55,\ \text{and } x_2 = 5.4.$$

Using Equation (5.40), again,

$$I_{0,1} = \frac{x_{2^1} - x_0}{2^1}\left(\frac{f(x_0) + f(x_{2^1})}{2} + \sum_{i=1}^{2^1-1} f(x_i) \right)$$

$$= \frac{x_2 - x_0}{2}\left(\frac{f(x_0) + f(x_2)}{2} + \sum_{i=1}^{1} f(x_i) \right)$$

$$= \frac{5.4 - 1.7}{2}\left(\frac{f(1.7) + f(5.4)}{2} + f(3.55) \right)$$

$$= 1.85\left(\frac{1.881 + 0.082}{2} + 0.278 \right) = 2.330$$

For $k = 2$ and by Table 5.6, the breakdown is four subintervals and five equally spaced points from the set of nine.

$$x_0 = 1.7,\ x_1 = 2.625,\ x_2 = 3.55,\ x_3 = 4.475,\ \text{and } x_4 = 5.4$$

Using Equation (5.40),

$$I_{0,2} = \frac{x_{2^2} - x_0}{2^2}\left(\frac{f(x_0) + f(x_{2^2})}{2} + \sum_{i=1}^{2^2-1} f(x_i) \right)$$

$$= \frac{x_4 - x_0}{4}\left(\frac{f(x_0) + f(x_4)}{2} + \sum_{i=1}^{3} f(x_i)\right)$$

$$= \frac{5.4 - 1.7}{4}\left(\frac{f(1.7) + f(5.4)}{2} + f(2.625) + f(3.55) + f(4.475)\right)$$

$$= 0.925\left(\frac{1.881 + 0.082}{2} + 0.647 + 0.278 + 0.142\right) = 1.895$$

Lastly, $k = 3$ uses all nine points and Equation (5.40) yields

$$I_{0,3} = \frac{x_8 - x_0}{8}\left(\frac{f(x_0) + f(x_8)}{2} + \sum_{i=1}^{7} f(x_i)\right)$$

$$= \frac{5.4 - 1.7}{8}\left(\frac{f(1.7) + f(5.4)}{2} + f(2.1625) + f(2.625) + f(3.0875) + f(3.55)\right.$$

$$\left. + f(4.0125) + f(4.475) + f(4.9375)\right)$$

$$= 0.4625\left(\frac{1.881 + 0.082}{2} + 1.073 + 0.647 + 0.414 + 0.278 + 0.195 + 0.142 + 0.106\right) = 1.775$$

Now, the second tier estimations using proportional weighting are applied recursively until $I_{3,0}$ is reached. For $m = 1$, $I_{1,0}$, $I_{1,1}$, and $I_{1,2}$ are needed, as depicted in Table 5.9. Using Equation (5.41) repeatedly and the weights 4 and -1 of Table 5.7

$$I_{1,0} = \frac{4^1 \cdot I_{0,1} - I_{0,0}}{4^1 - 1} = \frac{4(2.330) - 3.632}{3} = 1.896$$

$$I_{1,1} = \frac{4^1 \cdot I_{0,2} - I_{0,1}}{4^1 - 1} = \frac{4(1.895) - 2.330}{3} = 1.750$$

$$I_{1,2} = \frac{4^1 \cdot I_{0,3} - I_{0,2}}{4^1 - 1} = \frac{4(1.775) - 1.895}{3} = 1.735$$

For $m = 2$, $I_{2,0}$ and $I_{2,1}$ are required, illustrated in Table 5.9. Using Equation (5.41) twice and the weights 16 and -1 of Table 5.7

$$I_{2,0} = \frac{4^2 \cdot I_{1,1} - I_{1,0}}{4^2 - 1} = \frac{16(1.750) - 1.896}{15} = 1.740$$

$$I_{2,1} = \frac{4^2 \cdot I_{1,2} - I_{1,1}}{4^2 - 1} = \frac{16(1.735) - 1.750}{15} = 1.734$$

Lastly, using Equation (5.41) for $m = 3$ and Table 5.7 for the weights 64 and -1

$$I_{3,0} = \frac{4^3 \cdot I_{2,1} - I_{2,0}}{4^3 - 1} = \frac{64(1.734) - 1.740}{63} = 1.734$$

Table 5.10 organizes all the values of Example 5.14 in a recursion table akin to Table 5.9. The yellow cells of the first row indicate final the Romberg integration estimation for a particular m value.

Table 5.10: A recursion table organization of all values of Example 5.14. Yellow cells are each potential final estimations using Romberg integration.

k	$m = 0$	$m = 1$	$m = 2$	$m = 3$
0	$I_{0,0} = 3.632$	$I_{1,0} = 1.896$	$I_{2,0} = 1.740$	$I_{3,0} = 1.734$
1	$I_{0,1} = 2.330$	$I_{1,1} = 1.750$	$I_{2,1} = 1.734$	
2	$I_{0,2} = 1.895$	$I_{1,2} = 1.735$		
3	$I_{0,3} = 1.775$			

5.3.1 Error analysis

As previously stated, a potential final estimation occurs at any $k = 0$ or $I_{m,0}$ value, all of which comprise the first row of the recursion table. When the last two estimations in the first row differ by less than a given tolerance δ Romberg integration stops, i. e., when $|I_{m,0} - I_{m-1,0}| < \delta$.

An associated estimation relative change error for Romberg integration $E_{c,r}$ can also be defined.

$$E_{c,r} = \left| \frac{I_{m,0} - I_{m-1,0}}{I_{m,0}} \right| \times 100\,\% \tag{5.42}$$

Consider why $I_{m,0}$ is used in the denominator of $E_{c,r}$ in place of the unknown true integral value. The estimation $I_{m,0}$ is closer than $I_{m-1,0}$ is to the true integral value because $I_{m,0}$ includes an extra first tier estimation for $k = m$, whereas $I_{m-1,0}$ has a maximum first tier estimation for $k = m - 1$. The extra first tier estimation incorporates an additional trapezoidal rule estimation that has double the number of intervals than the maximum number of intervals of the previous trapezoidal rule first tier estimation.

Example 5.15. Find the smallest $I_{m,0}$ that satisfies the tolerance $\delta = 0.01$ for the data of Table 5.10. Calculate the estimation relative change error $E_{c,r}$.

Solution. Using $I_{2,0}$ is not accurate enough to stop Romberg integration since

$$|I_{2,0} - I_{1,0}| = |1.740 - 1.896| = 0.156 > \delta$$

However, using $I_{3,0}$ suffices.

$$|I_{3,0} - I_{2,0}| = |1.734 - 1.740| = 0.006 < \delta$$

Therefore, $I_{3,0} = 1.734$ is the final Romberg integration estimation. The associated estimation relative change error $E_{c,r}$ using Equation (5.42) is

$$E_{c,r} = \left| \frac{1.734 - 1.740}{1.734} \right| \times 100\,\% = 0.35\,\%$$

As explained after Example 5.8, the true integral value is 1.733. With a relatively small number of iterations the Romberg integration estimation is accurate to the hundredths place, supporting the use of $E_{c,r}$ of Example 5.15 as a reasonable stopping criterion.

5.3.2 Proof of Romberg integration convergence

In Section 5.1.3, it was proved n applications of the trapezoidal rule converges to the true definite integral value as $n \to \infty$. Convergence is now proved for Romberg integration.

Theorem 5.4. *For $f(x)$ continuous on the interval $[x_0, x_n]$ and $\int_{x_0}^{x_n} f(x)\, dx = G$*

$$\lim_{\substack{m \to \infty \\ k \to \infty}} I_{m,k} = G$$

Proof. A proof by induction is employed.

Base case: For $m = 0$, $I_{0,k}$ is the first tier estimation of Equation (5.40), equivalent to n applications of the trapezoid rule for $n = 2^k$. If $k \to \infty$, then $n \to \infty$. Thus $I_{0,k}$ as $k \to \infty$ is equivalent to n applications of the trapezoidal rule as $n \to \infty$, the latter proved to converge to the true integral value G by Theorem 5.1.

Induction step: For some $j > 0$ assume $I_{j,k}$, the second tier estimation of Equation (5.41), converges to the true integral value G as $k \to \infty$, i. e.,

$$\lim_{k \to \infty} I_{j,k} = \lim_{k \to \infty} \frac{4^j \cdot I_{j-1,k+1} - I_{j-1,k}}{4^j - 1} = G.$$

It must be shown $I_{j+1,k}$ also converges to G, or

$$\lim_{k \to \infty} I_{j+1,k} = \lim_{k \to \infty} \frac{4^{j+1} \cdot I_{j,k+1} - I_{j,k}}{4^{j+1} - 1} = G$$

When $k \to \infty$, then $k + 1 \to \infty$ as well so that their corresponding limits of $I_{j,k}$ and $I_{j,k+1}$ are equivalent.

$$\lim_{k+1 \to \infty} I_{j,k+1} = \lim_{k \to \infty} I_{j,k} = G$$

Substituting G for each of these in the induction step completes the proof.

$$\lim_{k \to \infty} I_{j+1,k} = \lim_{k \to \infty} = \frac{4^{j+1} \cdot G - G}{4^{j+1} - 1} = \frac{G(4^{j+1} - 1)}{4^{j+1} - 1} = G \qquad \square$$

5.3.3 Advantages of Romberg integration

While using only multiple applications of the trapezoidal rule and using them within Romberg integration both converge to the true integral value, Romberg integration does so more efficiently, and the improved accuracy does not rely on a truncation error adjustment and the first derivative. A more pronounced advantage manifests over Simpson's rule with a truncation error adjustment because that requires a typically algebraically complicated third derivative. Furthermore, as Section 5.2.6 cautioned, the truncation error adjustment for Simpson's rule could actually worsen the estimation depending on the third derivative.

Following the construction of ε_T for trapezoidal rules and ε_S for Simpson's rules, the exact relative percentage error for Romberg integration ε_R is defined. For the exact definite integral value I solved analytically and an $m \geq 0$

$$\varepsilon_R = \left| \frac{I - I_{m,0}}{I} \right| \times 100\,\% \tag{5.43}$$

Consider a final estimation for the definite integral of Example 5.14 stopping at $I_{2,0} = 1.740$. Using the true value of $I = 1.733$ per the explanation following Example 5.8, the exact relative percentage error is

$$\varepsilon_R = \left| \frac{1.733 - 1.740}{1.733} \right| \times 100\,\% = 0.40\,\%$$

Note that $I_{2,0}$ and Simpson's rule with a truncation error adjustment of Example 5.12 both use the same five points. Although Example 5.12 produced a slightly better corresponding exact relative percentage error ε_S, it needed an algebraically intense third derivative.

$$\varepsilon_S = \left| \frac{1.733 - 1.738}{1.733} \right| \times 100\,\% = 0.29\,\%$$

Nevertheless, even with that truncation error adjustment, Simpson's rule was not better than $I_{3,0} = 1.734$ of Example 5.14, which used four additional points but required no derivative.

$$\varepsilon_R = \left| \frac{1.733 - 1.734}{1.733} \right| \times 100\,\% = 0.06\,\%$$

Even when an analytic solution is available, Romberg integration may be a suitably quicker compromise, as the next example demonstrates.

Example 5.16. Use Simpson's rule $A_{S,t}(4)$ and Romberg integration up to $I_{2,0}$ to estimate $\int_{1.2}^{2.2} (2 - x^7 \cos x)\,dx$. Compare the results.

Solution. For $n = 4$, consecutive x values on $[1.2, 2.2]$ are equally spaced by a width of $h = \frac{2.2-1.2}{4} = 0.25$ so that

$$x_0 = 1.2,\ x_1 = 1.45,\ x_2 = 1.7,\ x_3 = 1.95,\ \text{and } x_4 = 2.2$$

Using Equation (5.31) for Simpson's rule with $f(x) = 2 - x^7 \cos x$,

$$A_S(4) = \frac{x_4 - x_0}{3(4)}\left(f(x_0) + 4f(x_1) + 2f(x_2) + 4f(x_3) + f(x_4)\right)$$

$$= \frac{2.2 - 1.2}{12}\left(f(1.2) + 4f(1.45) + 2f(1.7) + 4f(1.95) + f(2.2)\right)$$

$$= \frac{1}{12}\left(0.702 + 4(0.376) + 2(7.287) + 4(41.688) + 148.793\right) = 27.694$$

Using the third derivative and Equation (5.37) for the truncation error adjustment with $f^{(3)}(x) = -x^7 \sin x + 21x^6 \cos x + 126x^5 \sin x - 210x^4 \cos x$

$$E_{t,S}(4) = -\frac{(2.2 - 1.2)^4 (f^{(3)}(2.2) - f^{(3)}(1.2))}{180(4)^4} = -\frac{(6542.215 - 153.812)}{46080} = -0.139$$

The adjusted estimation by Equation (5.38) is

$$A_{S,t}(4) = A_S(4) + E_{t,S}(4) = 27.694 + (-0.139) = 27.555$$

Romberg integration up to $I_{2,0}$ begins with trapezoidal rule first tier estimations $I_{0,k}$ for $k = 0, 1, 2$. For $k = 0$ and by Table 5.6, only two equally spaced points are needed for $I_{0,0}$, the endpoints $x_0 = 1.2$ and $x_1 = 2.2$. Using Equation (5.40)

$$I_{0,0} = \frac{x_1 - x_0}{1}\left(\frac{f(x_0) + f(x_1)}{2} + \sum_{i=1}^{0} f(x_i)\right)$$

Recall there are no terms in the summation because the upper bound of zero is less than the starting index $i = 1$, reducing the above equation to

$$I_{0,0} = (2.2 - 1.2)\left(\frac{f(1.2) + f(2.2)}{2}\right) = \left(\frac{0.702 + 148.793}{2}\right) = 74.748$$

For $k = 1$ and by Table 5.6, three equally spaced points on the interval are required for $I_{0,1}$, the endpoints and the midpoint. Thus $x_0 = 1.2$, $x_1 = 1.7$, and $x_2 = 2.2$. Using Equation (5.40), again,

$$I_{0,1} = \frac{x_2 - x_0}{2}\left(\frac{f(x_0) + f(x_2)}{2} + \sum_{i=1}^{1} f(x_i)\right)$$

$$= \frac{2.2 - 1.2}{2}\left(\frac{f(1.2) + f(2.2)}{2} + f(1.7)\right)$$

$$= 0.5\left(\frac{0.702 + 148.793}{2} + 7.287\right) = 41.017$$

For $k = 2$ and by Table 5.6, $I_{0,2}$ uses five equally spaced points, the same used and indexed for Simpson's rule above. Using Equation (5.40) one last time,

$$I_{0,2} = \frac{x_4 - x_0}{4}\left(\frac{f(x_0) + f(x_4)}{2} + \sum_{i=1}^{3} f(x_i)\right)$$

$$I_{0,2} = \frac{2.2 - 1.2}{4}\left(\frac{f(1.2) + f(2.2)}{2} + f(1.45) + f(1.7) + f(1.95)\right)$$

$$= 0.25\left(\frac{0.702 + 148.793}{2} + 0.376 + 7.287 + 41.688\right) = 31.025$$

Now, second tier estimations with proportional weighting are applied recursively in two iterations until $I_{2,0}$ is reached. In the first iteration, for $m = 1$, Equation (5.41) is used twice, and in conjunction with the weights of Table 5.7.

$$I_{1,0} = \frac{4^1 \cdot I_{0,1} - I_{0,0}}{4^1 - 1} = \frac{4(41.017) - 74.748}{3} = 29.773$$

$$I_{1,1} = \frac{4^1 \cdot I_{0,2} - I_{0,1}}{4^1 - 1} = \frac{4(31.025) - 41.017}{3} = 27.694$$

In the second iteration, for $m = 2$, Equation (5.41) is used only once, coupled with the weights of Table 5.7.

$$I_{2,0} = \frac{4^2 \cdot I_{1,1} - I_{1,0}}{4^2 - 1} = \frac{16(27.694) - 29.773}{15} = 27.555$$

The Romberg integration values are summarized in a recursion table using Table 5.8 as a template. The final estimation is in the yellow cell.

k	$m = 0$	$m = 1$	$m = 2$
0	$I_{0,0} = 74.748$	$I_{1,0} = 29.773$	$I_{2,0} = 27.555$
1	$I_{0,1} = 41.017$	$I_{1,1} = 27.694$	
2	$I_{0,2} = 31.025$		

The estimations using Simpson's rule with the truncation error adjustment $A_{S,t}(4)$ and Romberg integration $I_{2,0}$ are equal to the thousandths place.

Note the integral of Example 5.16 *does* have an antiderivative, but it requires integration by parts seven times. It produces an exact answer accurate to the thousandths place of 27.554. Both Simpson's rule with the truncation error adjustment and Romberg integration estimations yield the same corresponding exact relative percentage error of virtually 0 %.

$$\varepsilon_S = \varepsilon_R = \left|\frac{27.554 - 27.555}{27.554}\right| \times 100\,\% = \frac{0.001}{27.554} \times 100\,\% = 0.0036\,\%$$

However, the truncation error adjustment $E_{t,S}(4)$ of Equation (5.37) with Simpson's rule requires a relatively complicated third derivative, whereas Romberg integration only needs minimal trapezoidal rule estimations, sans adjustment.

Furthermore, because Theorem 5.4 verified Romberg integration converges to the true definite integral value, any accuracy is possible. Successive iterations $I_{m-1,0}$ and $I_{m,0}$ identical to a specified decimal place equal the true integral value to the same decimal place.

5.4 Monte Carlo integration

An entirely different approach to estimating a definite integral involves running a Monte Carlo simulation. However, a salient underlying characteristic of the integrand $f(x)$ now needs consideration. The previous methods in this chapter, although visually motivated by trapezoids and parabolas, are not dependent on the sign of $f(x)$. For instance, a graph deviates from the shape of a trapezoid and appears as two triangles when $f(x_i)$ and $f(x_{i+1})$ differ in sign on interval $[x_i, x_{i+1}]$ which crosses the x-axis. But the trapezoidal rules still apply because they were constructed algebraically without any restrictions on $f(x)$, just like Simpson's rules and Romberg integration. In contrast, the sign of $f(x)$ is directly involved in Monte Carlo integration and thus different cases must be addressed.

5.4.1 A nonnegative integrand, $f(x) \geq 0$

Consider the integral $\int_a^b f(x)\,dx$, where $f(x)$ is nonnegative and has an upper bound of M such that $0 \leq f(x) \leq M$ for all $a \leq x \leq b$. Define the exact integral area I.

$$I = \int_a^b f(x)\,dx$$

The area of the rectangle A encompassing the entire view of the definite integral equals length, or the distance between the limits of integration, multiplied by the maximum height M.

$$A = (b - a)M \tag{5.44}$$

The area of the definite integral I comprises a decimal portion p of the area of the entire rectangle A such that

$$p = \frac{I}{A} \tag{5.45}$$

The definite integral therefore can be expressed by rewriting Equation (5.45).

$$I = pA$$

The Monte Carlo simulation generates $\hat{p}$, an estimation for p, which in turn estimates the above integral I and defines positive Monte Carlo integration.

$$I \approx \hat{p}A = A^+_{MC} \tag{5.46}$$

A set of N random points (x_i, y_i) within the bounds of the entire rectangle A is generated. The x_i-coordinates can be randomly spread as long as they represent the entire rectangle sufficiently, but for ease can be equally spaced so that for $x_1 = a$ and $i = 2, \ldots, N$

$$x_i = x_{i-1} + \frac{b-a}{N}$$

For each of the x_i-coordinates, an associated random y_i-coordinate is generated where $0 \leq y_i \leq M$. Every y_i-coordinate is then compared to the true y-coordinate $f(x_i)$. If $y_i \leq f(x_i)$ the point is below or on $y = f(x)$ and if $y_i > f(x_i)$ the point is above $y = f(x)$. Define n as the number of points where $y_i \leq f(x_i)$.

$$n = \sum_{i=1}^{N} a_i \text{ where } a_i = \begin{cases} 0, & y_i > f(x_i) \\ 1, & y_i \leq f(x_i) \end{cases}$$

The ratio of the number of points n below or on $y = f(x)$ to the total number of points N estimates the ratio of *all* points below or on $y = f(x)$ to the total number of points on the *entire* rectangle.

$$\frac{n}{N} \approx \frac{\text{all points below or on } y = f(x)}{\text{total points on the entire rectangle}}$$

All points below or on $y = f(x)$ precisely represents the area of the definite integral I, and the total number of points on the entire rectangle corresponds to the area of the rectangle itself A. Thus, the above approximation can be written

$$\frac{n}{N} \approx \frac{I}{A} \tag{5.47}$$

Define

$$\hat{p} = \frac{n}{N} \tag{5.48}$$

Combining Equations (5.48), (5.47), and (5.45) justifies $\hat{p}$ estimates p.

$$\hat{p} = \frac{n}{N} \approx \frac{I}{A} = p$$
$$\hat{p} \approx p$$

Lastly, substituting Equations (5.48) and (5.44) in Equation (5.46) defines positive Monte Carlo integration A_{MC}^{+} as

$$A_{MC}^{+} = \hat{p}A = \frac{n}{N}(b-a)M = \frac{nM(b-a)}{N} \tag{5.49}$$

A larger total number of points N generally produces a $\hat{p}$ closer to the true p and in turn, a better estimation A_{MC}^{+}.

Example 5.17. Use positive Monte Carlo integration with 1000 points to estimate $\int_{0.25}^{2.75}(2-4x+4x^2-x^3)\,dx$.

Solution. The upper bound M can be found exactly in this case. The extreme value theorem (EVT) from calculus guarantees a maximum value on the closed interval $[0.25, 2.75]$ for the continuous function $f(x) = 2-4x+4x^2-x^3$. Recall the maximum must occur at either endpoint or where $f'(x) = -4+8x-3x^2 = 0$, i. e., $x = 2$ or $x = 2/3$. $f(2) = 2$ is larger than $f(0.25) = 1.234$, $f(2.75) = 0.453$, and $f(2/3) = 0.815$, so $M = 2$.

1000 random y_i-coordinates in the interval $[0, 2]$ are generated and assigned to one of the x_i-coordinates from x_1 to x_{1000} inclusive. Figure 5.13 displays all the points and Table 5.11 lists a subset of these points.

There are 693 points below or on $y = f(x)$ in blue and 307 points above $y = f(x)$ in green. The fraction $n/N = 693/1000 = \hat{p}$, the estimation for the true ratio p of all area below or on $y = f(x)$, divided by the area of the entire rectangle $[0.25, 2.75] \times [0, 2]$. Using Equation (5.49), the estimated integral value is

$$A_{MC}^{+} = \frac{nM(b-a)}{N} = \frac{693(2)(2.75-0.25)}{1000} = 3.465$$

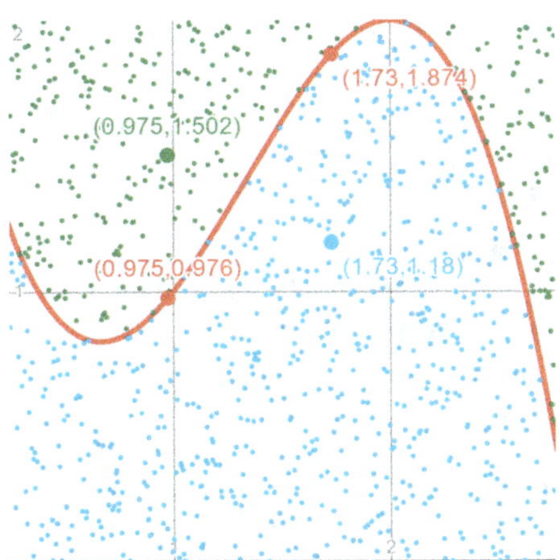

Figure 5.13: 1000 randomly generated points in the rectangle where $0.25 \leq x \leq 2.75$ and $0 \leq y \leq 2$. The red points satisfy the red $y = f(x)$ exactly. The blue points are below or on $y = f(x)$ and the green points are above $y = f(x)$. The number of blue points divided by 1000 is the estimated fraction of the area of the definite integral $\int_{0.25}^{2.75} f(x)\,dx$ relative to the entire rectangle.

The function of Example 5.17 was chosen because it can be solved analytically and compared to the estimated value. Using an antiderivative, the exact integral value I is

$$I = \int_{0.25}^{2.75} (2 - 4x + 4x^2 - x^3)\,dx = \left(2x - 2x^2 + \frac{4}{3}x^3 - \frac{x^4}{4}\right)\Bigg|_{0.25}^{2.75} = 3.411$$

An exact relative percentage error for Monte Carlo integration ε_{MC} is defined just as it was for the trapezoidal rules, Simpson's rules, and Romberg integration, using A_{MC} in place of A_{Trap}, A_{Sim}, and $I_{m,0}$ in Equations (5.16), (5.39), and (5.43), respectively.

$$\varepsilon_{MC} = \left|\frac{I - A_{MC}}{I}\right| \times 100\,\% \tag{5.50}$$

Table 5.11: 1000 random y_i-coordinates between 0 and 2 are assigned to each equally spaced x_i-coordinate and compared to the true y-coordinate $f(x_i)$. The green and blue cells are highlighted points on Figure 5.13.

x_i	y_i-coordinate random in $[0, 2]$	$f(x_i)$ true y-coordinate	y_i-coordinate $\leq f(x_i)$?
0.2500	0.870	1.2344	true
0.2525	1.650	1.2289	false
0.2550	1.108	1.2235	true
$\vdots$	$\vdots$	$\vdots$	$\vdots$
0.9750	1.502	0.9756	false
$\vdots$	$\vdots$	$\vdots$	$\vdots$
1.7250	1.180	1.8739	true
$\vdots$	$\vdots$	$\vdots$	$\vdots$
2.7425	0.519	0.4880	false
2.7450	0.259	0.4765	true
2.7475	1.939	0.4648	false
			Number of true = 639

A_{MC} can represent A_{MC}^+ of Equation (5.49), A_{MC}^- of Equation (5.51) defined in Section 5.4.2, or the sum of both if the integrand $f(x)$ changes sign once between the limits of integration, as Example 5.19 and the explanation that follows will expound.

In Example 5.17, the exact relative percentage error is

$$\varepsilon_{MC} = \left| \frac{I - A_{MC}^+}{I} \right| \times 100\,\% = \left| \frac{3.411 - 3.465}{3.411} \right| \times 100\,\% = 1.58\,\%$$

Because A_{MC} changes according to the total number of random points generated and the points themselves, ε_{MC} varies correspondingly.

Although the exact upper bound of $M = 2$ for the function of Example 5.17 was easily calculated, an exact upper bound M is not required. It is only necessary that $f(x) < M$ because A_{MC}^+ depends on the product of both A and $\hat{p}$. Increasing M also increases of the area of the total rectangle $A = (b - a)M$, which simultaneously decreases the decimal portion p of the area covered by the integral I, per Equation (5.45). This corresponds to the decrease also in the estimation $\hat{p}$ of Equation (5.48). Out of the total number of N random points, there is an increased probability more points will lie above $y = f(x)$. In turn, less points will lie below or on $y = f(x)$, decreasing the number n and $\hat{p}$.

Equation (5.49) explicitly shows this inverse relationship by the factors nM in the numerator. Increasing M decreases n, maintaining the balance of the integral estimation A_{MC}^+. This is illustrated in Example 5.18.

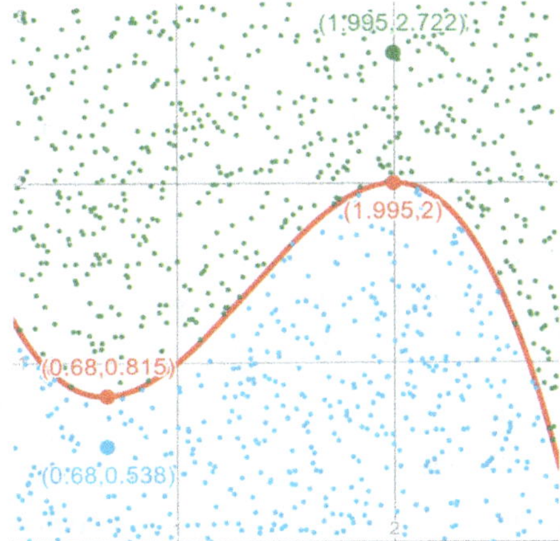

Figure 5.14: 1000 randomly generated points in the rectangle where $0.25 \leq x \leq 2.75$ and $0 \leq y \leq 3$. The red points satisfy the red $y = f(x)$ exactly, the blue points are below or on $y = f(x)$, and the green points are above $y = f(x)$. The number n of blue points divided by 1000 is the estimated fraction of the integral area, relative to the entire rectangle. This fraction is less than the one of Figure 5.13 but offset by the area of the larger rectangle here.

Example 5.18. Repeat the positive Monte Carlo integration of Example 5.17 but use an upper bound of $M = 3$ instead of $M = 2$.

Solution. 1000 random y_i-coordinates, now in the interval $[0, 3]$, are generated and assigned to one of the x_i-coordinates from x_1 to x_{1000} inclusive. Figure 5.14 displays all these points.

There are 446 points below or on $y = f(x)$ in blue and 554 points above $y = f(x)$ in green. The fraction $n/N = 446/1000 = \hat{p}$, the estimation of the true ratio p of Equation (5.45), the integral area I below or on $y = f(x)$ divided by the area of the entire rectangle A, $[0.25, 2.75] \times [0, 3]$. Note this $\hat{p} = 0.446$ is less than the $\hat{p} = 0.693$ of Example 5.17 which used a smaller rectangle. But here the points generated in the additional area outside the integral area between $y = 2$ and $y = 3$ simultaneously decreased the number of points n inside the integral area out of the same total number of $N = 1000$ random points. Thus, $\hat{p}$ decreased because n decreased and $\hat{p} = n/N$.

Using Equation (5.49), while n has decreased from 693 of Example 5.17 to 446, M has correspondingly increased from 2 to 3 and altered the product nM from 1386 to 1338. The estimated integral value here is

$$A_{MC}^+ = \frac{nM(b - a)}{N} = \frac{446(3)(2.75 - 0.25)}{1000} = 3.345$$

The exact relative percentage error using Equation (5.50) differs from Example 5.17 since A_{MC}^+ is different.

$$\varepsilon_{MC} = \left| \frac{3.411 - 3.345}{3.411} \right| \times 100\,\% = 1.93\,\%$$

5.4.2 A nonnegative and nonpositive integrand $f(x)$

The initial stipulation that $f(x)$ is nonnegative on $a \leq x \leq b$ of Section 5.4.1 is not an absolute requirement for Monte Carlo integration. But an $f(x)$ that changes sign on $a \leq x \leq b$ must be handled differently. Suppose $a < c < b$, $f(c) = 0$, $f(x) > 0$ on $[a, c)$, and $f(x) < 0$ on $(c, b]$. The summation property of definite integrals from calculus states

$$\int_a^b f(x)\, dx = \int_a^c f(x)\, dx + \int_c^b f(x)\, dx.$$

The first integrand on the right side of the equation is positive, estimated by A_{MC}^+. The second integrand on the right side of the equation is negative, necessitating two adjustments to positive Monte Carlo integration. First, $f(x)$ now requires a *lower* bound m where $m \leq f(x) \leq 0$ for all $c < x \leq b$. Second, the fraction of the area here is represented by points *above* or on $y = f(x)$.

Area on the subinterval $(c, b]$ is negative because $f(x) < 0$. Since the lower bound $m < 0$, this is handled automatically for m simply replacing the positive upper bound M in Equation (5.49). Negative Monte Carlo integration A_{MC}^- is correspondingly defined.

$$A_{MC}^- = \frac{nm(b - a)}{N} \tag{5.51}$$

Example 5.19 details Monte Carlo integration for an integrand which is positive and negative between the limits of integration.

Example 5.19. Use Monte Carlo integration to estimate $\int_4^{8.5} (1 - 2\sin x)\, dx$.

Solution. Set the integrand $f(x) = 0$ for possible locations where the sign of $f(x)$ changes.

$$f(x) = 1 - 2\sin x = 0$$

$$\sin x = \frac{1}{2}$$

On the given interval $[4, 8.5]$ there is one solution at

$$x = \frac{\pi}{6} + 2\pi = \frac{13\pi}{6} = 6.807$$

Since $-1 \leq \sin x \leq 1$, the upper bound of $f(x)$ is $M = 1 - 2(-1) = 3$ and the lower bound is $m = 1 - 2(1) = -1$.

$f(x) > 0$ on the subinterval $[4, 6.807)$. 1000 random y_i-coordinates in the interval $[0, M] = [0, 3]$ are generated and assigned to one of the x_i-coordinates from x_1 to x_{1000} inclusive. Figure 5.15 displays all these points to the left of the black dashed vertical line $x = 6.807$. There are 687 points below or on $y = f(x)$ in blue and 313 points above $y = f(x)$ in green. The fraction $n/N = 687/1000 = \hat{p}$ estimates the true ratio p of the integral area below or on $y = f(x)$ divided by the area of the entire rectangle $[4, 6.807] \times [0, 3]$. Using Equation (5.49), the estimated integral value on this subinterval is

$$A_{MC}^+ = \frac{nM(b - a)}{N} = \frac{687(3)(6.807 - 4)}{1000} = 5.785$$

$f(x) < 0$ on the subinterval $(6.807, 8.5]$. The area of the entire rectangle used here is nearly one-fifth the size of the rectangle used for $f(x) > 0$. For visual similarity, 200 random y_i-coordinates in the interval $[m, 0] = [-1, 0]$ are generated and assigned to one of the x_i-coordinates on this new subinterval from x_1 to x_{200} inclusive. Figure 5.15 displays all these points to the right of the black dashed vertical line $x = 6.807$. There are 148 points *above* or on $y = f(x)$ in blue and 52 points *below* $y = f(x)$ in green. The fraction $n/N = 148/200 = \hat{p}$ estimates the true ratio p of the integral area above or on $y = f(x)$ divided by the area of the entire rectangle $[6.807, 8.5] \times [-1, 0]$. Using Equation (5.51), the estimated integral value on this subinterval is

$$A_{MC}^- = \frac{nm(b - a)}{N} = \frac{148(-1)(8.5 - 6.807)}{200} = -1.253$$

The estimated value of the entire integral between 4 and 8.5 is

$$A_{MC}^+ + A_{MC}^- = 5.785 + (-1.253) = 4.532$$

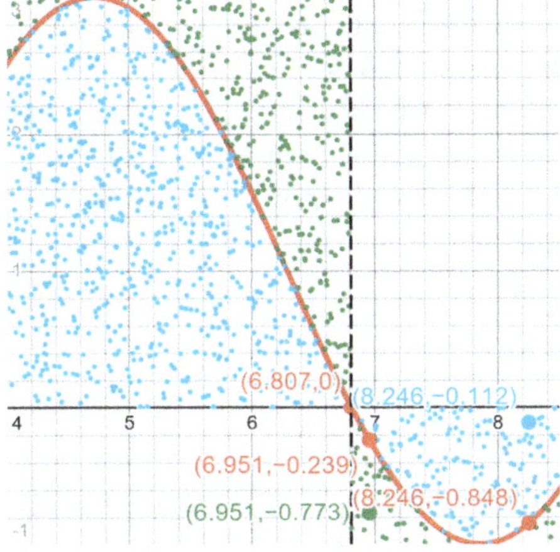

Figure 5.15: 1000 randomly generated points in the left rectangle $4 \le x < 6.807$ by $0 \le y \le 3$ and 200 randomly generated points in the right rectangle $6.807 < x \le 8.5$ by $-1 \le x \le 0$. The red points satisfy the red $y = f(x)$ exactly. For each rectangle, the number of blue points divided by the total points in that rectangle is the estimated fraction of the area of the definite integral relative to that rectangle. The left rectangle has blue points below or on $y = f(x)$ for $f(x) > 0$, while the right rectangle has blue points above or on $y = f(x)$ for $f(x) < 0$.

Like Example 5.17, the function $f(x) = 1 - 2\sin x$ of Example 5.19 was chosen because it can be solved analytically and compared with the estimation. Additionally, the exact zero of $f(x)$ could be found and used to divide the entire definite integral into positive and negative areas. Lastly, the exact upper and lower bounds are known. But as with the upper bound M, the only requirement for the lower bound is $m < f(x)$; the exact value is not necessary.

Using the antiderivative of $f(x)$, the true integral value is

$$I = \int\limits_{4}^{8.5} (1 - 2\sin x)\, dx = (x + 2\cos x)\big|_4^{8.5} = 4.603$$

The exact relative percentage error according to Equation (5.50) for this particular Monte Carlo integration is

$$\varepsilon_{MC} = \left| \frac{4.603 - 4.532}{4.603} \right| \times 100\,\% = 1.54\,\%$$

A proportionate total number of points was used for visual similarity in Figure 5.15 of Example 5.19, but for simplicity in practice the same N can be used for each rectangle, regardless of size. Note in such cases, smaller rectangles will tend to yield more accurate results than larger rectangles.

The function $f(x)$ of Example 5.19 changed sign once on the given interval and needed two separate simulations, one each for approximating the positive area and negative area. In general, if $f(c_i) = 0$ and changes sign around c_i for $i = 1, \ldots, k$ on the interval $[a, b]$, $k + 1$ simulations are required. Each simulation corresponds to one definite subintegral on the right side of the equation.

$$\int\limits_{a}^{b} f(x)\, dx = \int\limits_{a}^{c_1} f(x)\, dx + \int\limits_{c_1}^{c_2} f(x)\, dx + \cdots + \int\limits_{c_{k-1}}^{c_k} f(x)\, dx + \int\limits_{c_k}^{b} f(x)\, dx$$

For multiple subintegrals where $f(x) > 0$, it is not necessary to find an upper bound for each one. Example 5.18 demonstrated an exact upper bound is never required. Thus, a single upper bound where $f(x) \leq M$ for all subintegrals where $f(x) > 0$ is sufficient. Likewise, a single lower bound $m \leq f(x)$ suffices for all subintegrals where $f(x) < 0$.

If exact zeros where a function $f(x)$ changes sign are unknown through analytic methods, any of the zero-finding algorithms of Chapter 1 could be employed to estimate zeros.

5.4.3 Is Monte Carlo integration practical?

Computational efficiency can be defined as the total number of calculations an algorithm requires for a specified accuracy. The concept was referenced in Chapter 1, in which the less efficient bisection and false position algorithms typically required more iterations and computations than other algorithms when finding a solution. Monte Carlo integration certainly holds conceptual value, but otherwise lacks computational efficiency compared to other methods in this chapter.

For example, for a simulation using 1000 points, 1000 random function values must be generated. Then 1000 true function values need computation. Another 1000 evaluations are required comparing the random and true function values. A final percentage is necessary for the definite integral estimation.

Now, consider multiple applications of either the trapezoidal rule or Simpson's rule, without truncation error adjustments. Even relatively few applications for small n produced some reasonable results in Sections 5.1 and 5.2. Estimations improve for larger n, and in fact were proved to converge to the true value as n increases. For $n = 100$, 100 function evaluations are necessary, plus the final trapezoidal or Simpson's rule formula calculation, which includes less than 100 weightings of those evaluations, their addition, and determination of the leading coefficient.

With much less computational effort, the trapezoidal or Simpson's rule definite integral estimations for $n = 100$ will most likely exceed the accuracy of a Monte Carlo integration using 1000 points. In theory, Monte Carlo integration is an interesting idea, but in practice is easily surpassed.

Integration concepts summary

1. Numerical integration is often necessary; many functions have no closed-form antiderivatives consisting of elementary functions.
2. Numerical integration also could be an acceptable solution for functions in which finding the antiderivative is possible but algebraically intense.
3. The trapezoidal rule uses linear functions to approximate the curve between two points.

4. As the number of applications of the trapezoidal rule approaches infinity, the definite integral converges to its exact value.
5. Simpson's rule uses quadratic functions to approximate the curve through three equally spaced consecutive points.
6. Simpson's rule directly uses given points; the coefficients of the quadratic functions are not found.
7. There are $n/2$ applications of Simpson's rule on points at $x_0, \ldots, x_n$; n must be even.
8. Simpson's rule exactly evaluates definite integrals in which the underlying data is from a polynomial function up to degree three, inclusive.
9. Truncation error adjustments for the trapezoidal and Simpson's rules can improve an initial estimation and indicate whether the initial estimation was an underestimation or overestimation.
10. Multiple applications of the trapezoidal and Simpson's rules generally yield better results than single applications with a truncation error adjustment.
11. If the first or third derivative is not fairly constant, the truncation error adjustments can, respectively, adversely affect the trapezoidal and Simpson's rule estimations.
12. Romberg integration recursively uses weighted trapezoidal estimations and converges to the exact value of a definite integral.
13. Each successive recursion of Romberg integration uses a greater number of points within the entire interval and a greater proportional weighting of the better estimation.
14. Romberg integration attains any level of accuracy by comparing $I_{m,0}$ and $I_{m-1,0}$ estimations for any $m > 0$.
15. Romberg integration refinement does not rely on the truncation error adjustments using first and third derivatives like the trapezoidal and Simpson's rules, respectively.
16. Monte Carlo integration generates random points to estimate the area represented by a definite integral in a rectangle.
17. The Monte Carlo integration rectangle must enclose the limits of integration of the definite integral and the maximum and minimum function values; exact upper and lower bounds are not necessary.
18. Separate Monte Carlo integrations are necessary for functions which cross the x-axis; trapezoidal rules, Simpson's rules, and Romberg integration require no separation for such functions.
19. Monte Carlo integration is valuable conceptually but computationally inefficient practically.

Exercises

Section 5.1.1

Use a single application of the trapezoidal rule A_T to estimate the area between $f(x)$ and the x-axis over the interval $[a, b]$.

1. $f(x) = e^{(1-x^2)}$ $[0, 1.5]$
2. $f(x) = 3^{-\sqrt{x}}$ $[0.5, 1.75]$
3. $f(x) = 5\log_3(x)$ $[2, 4]$
4. $f(x) = 3\log_8(x)$ $[3, 9]$
5. $f(x) = 2^x - 3x$ $[0.5, 2.5]$
6. $f(x) = 1 - e^{x/3}$ $[1, 3]$
7. $f(x) = 6\tan^{-1}(1 - \sqrt{x})$ $[0.25, 1.75]$
8. $f(x) = \tan^{-1}(\tfrac{1}{4}x^2 - 1)$ $[1.5, 4]$

Section 5.1.2

Use a single application of the trapezoidal rule A_T to estimate the area between $f(x)$ and the x-axis over the interval $[a, b]$. Calculate the truncation error adjustment $E_{t,T}$, the adjusted estimation $A_{T,t}$, and whether A_T underestimates or overestimates the true integral value.

9. $f(x) = 3^{1/x}$ $[1, 2.5]$

10. $f(x) = e^{(x^2 - x)}$ $[0.75, 2]$

11. $f(x) = \ln(10x^2 + 3)$ $[0.5, 3.5]$

12. $f(x) = \log(36x + 15)$ $[0, 2.25]$

13. $f(x) = \log(3 - \frac{x}{2})$ $[4, 5.25]$

14. $f(x) = \ln(2 - \frac{1}{8}x^2)$ $[3, 3.75]$

15. $f(x) = \sin(4\sqrt{x})$ $[1, 2.75]$

16. $f(x) = \cos(0.25x^2)$ $[2, 3]$

Section 5.1.3

Use n applications of the trapezoidal rule $A_T(n)$ to estimate the integral.

17. $\int_0^2 2e^{0.25x^2}\, dx$ $n = 4$ 18. $\int_{-1}^1 10^{x^3}\, dx$ $n = 4$

19. $\int_{-2}^3 \ln(0.5x^3 + 5)\, dx$ $n = 5$ 20. $\int_1^{2.25} \log_2(x^4 + 2)\, dx$ $n = 5$

21. $\int_1^{1.75} 2^x \log_5 x\, dx$ $n = 5$ 22. $\int_0^4 x^2 \ln(6 - x)\, dx$ $n = 4$

23. $\int_2^{4.4} \tan(0.1x^2 + 11)\, dx$ $n = 6$ 24. $\int_{-0.2}^1 -\sin e^x\, dx$ $n = 6$

Section 5.1.4

Use n applications of the trapezoidal rule $A_T(n)$ to estimate the integral. Calculate the truncation error adjustment $E_{t,T}(n)$ and the adjusted estimation $A_{T,t}(n)$.

25. $\int_0^{1.6} \ln(8 + x - x^2 - x^3)\, dx$ $n = 4$ 26. $\int_{0.25}^{2.25} \ln(3x^{2/3} + x)\, dx$ $n = 4$

27. $\int_{1.75}^3 e^{4/x^3}\, dx$ $n = 5$ 28. $\int_2^3 5^{6/(x+1)}\, dx$ $n = 5$

29. $\int_1^8 \frac{20x}{x^2 + 4x + 8}\, dx$ $n = 7$ 30. $\int_{0.5}^{3.5} \frac{16\sqrt{x}}{x^3 + 2}\, dx$ $n = 6$

31. $\int_{1.4}^2 10\cos(x^3 - 5x)\, dx$ $n = 6$ 32. $\int_{3.5}^7 \sin\left(\frac{7}{1-x}\right)\, dx$ $n = 7$

Section 5.1.5

Analytically solve the integral. Use A_T, $A_{T,t}$, $A_T(n)$, and $A_{T,t}(n)$ to estimate the integral. Calculate the exact relative percentage error ε_T for each estimation.

33. $\int_2^6 (0.4x^3 + 3\sin 2x)\, dx$ $n = 4$ 34. $\int_{0.7}^{1.6} \left(\frac{39}{x^2} - \cos 3x\right) dx$ $n = 4$

35. $\int_{0.35}^{2.6} \frac{30x^2}{x^3 + 3}\, dx$ $n = 5$ 36. $\int_{4.6}^{6.7} (2x - 9)\sqrt{x^2 - 9x + 21}\, dx$ $n = 6$

Section 5.2.1

Use a single application of Simpson's rule A_S to estimate the area between $f(x)$ and the x-axis over the interval $[a, b]$.

37. $f(x) = e^{(8/x^2)}$ $\quad\quad$ [4, 8]

38. $f(x) = 1.25^{x^2}$ $\quad\quad$ [0, 2]

39. $f(x) = \ln(x^3 - 1)$ $\quad\quad$ [2, 4]

40. $f(x) = \log_3(2x - 7)$ $\quad\quad$ [5, 9]

41. $f(x) = \cos(x^{2/3} + 2)$ $\quad\quad$ [0.25, 3]

42. $f(x) = \sin(4 - x^3)$ $\quad\quad$ [−0.5, 0.25]

43. $f(x) = 3.5\log_2(x + 4)$ $\quad\quad$ [−2.8, 0.4]

44. $f(x) = \ln(9x + \sqrt{x})$ $\quad\quad$ [0.3, 2.1]

Section 5.2.2

Use a single application of Simpson's rule A_S to estimate the area between $f(x)$ and the x-axis over the interval $[a, b]$. Calculate the truncation error adjustment $E_{t,S}$, the adjusted estimation $A_{S,t}$, and whether A_S underestimates or overestimates the true integral value.

45. $f(x) = 5e^{-x^2}$ $\quad\quad$ [−1, 1]

46. $f(x) = e^{-0.5x^2}$ $\quad\quad$ [−1.5, 1.5]

47. $f(x) = \frac{1}{x^5 + 2}$ $\quad\quad$ [−0.6, 1.7]

48. $f(x) = \frac{4}{x^3 + 1}$ $\quad\quad$ [1.8, 4.2]

49. $f(x) = \tan^{-1} x^2$ $\quad\quad$ [0.25, 3.75]

50. $f(x) = \tan^{-1}(2x - 3)$ $\quad\quad$ [1.25, 3.25]

51. $f(x) = \ln(x^3 - x)$ $\quad\quad$ [4, 10]

52. $f(x) = 1 + \ln(3 - x^3)$ $\quad\quad$ [−1.8, 0]

Use a single application of Simpson's rule A_S to estimate the integral. Analytically solve the integral to verify Simpson's rule produces the exact value for integrations of polynomials up to and including degree three.

53. $\int_1^8 (-\frac{1}{4}x + 7)\, dx$ $\quad\quad$ 54. $\int_1^5 (\frac{4}{5}x - 6)\, dx$

55. $\int_2^{11} (0.12x^2 - 2x + 9)\, dx$ $\quad\quad$ 56. $\int_{-3}^0 (4 - 6x - 0.9x^2)\, dx$

57. $\int_1^3 (2x^3 - 3x^2 - 8x - 5)\, dx$ $\quad\quad$ 58. $\int_{-1}^2 (4x^3 - 6x^2 - 5x + 2)\, dx$

59. $\int_{-3}^3 (6 + 2x + x^2 - 4x^3)\, dx$ $\quad\quad$ 60. $\int_1^4 (1 + 4x + 3x^2 - x^3)\, dx$

Section 5.2.3

Use Simpson's rule $A_S(n)$ to estimate the integral.

61. $\int_{-1}^{3}(4x^2 + 3)(0.75)^{x^3}\,dx \quad n = 4$

62. $\int_{-1.5}^{0.5}(2e^{x^3} - 1)\,dx \quad n = 4$

63. $\int_{3}^{10.2}\log_2(x^2 + x - 3)\,dx \quad n = 6$

64. $\int_{0}^{6}\ln(8x^2 + 15)\,dx \quad n = 6$

Use k applications of Simpson's rule $A_S(n)$ to estimate the integral.

65. $\int_{3.6}^{6} e^{(10/x)}\,dx \quad k = 3$

66. $\int_{0.5}^{2} 7^{\frac{2x}{x^2+1}}\,dx \quad k = 3$

67. $\int_{7.2}^{8} \sin\frac{20}{x}\,dx \quad k = 4$

68. $\int_{0.6}^{3} 2\cos(\log_5 x)\,dx \quad k = 4$

Section 5.2.4

Use Simpson's rule $A_S(n)$ to estimate the integral. Calculate the truncation error adjustment $E_{t,S}(n)$, the adjusted estimation $A_{S,t}(n)$, and whether $A_S(n)$ underestimates or overestimates the true integral value.

69. $\int_{-3}^{4} 4(1.05)^{-x^2}\,dx \quad n = 4$

70. $\int_{-1.25}^{3.75}(5 - (0.85)^{-0.5x^2})\,dx \quad n = 4$

71. $\int_{-1}^{8}\ln(x^4 + 3)\,dx \quad n = 6$

72. $\int_{-2.2}^{8.6} 3\ln(e^x + 3)\,dx \quad n = 6$

Use k applications of Simpson's rule $A_S(n)$ to estimate the integral. Calculate the truncation error adjustment $E_{t,S}(n)$, the adjusted estimation $A_{S,t}(n)$, and whether $A_S(n)$ underestimates or overestimates the true integral value.

73. $\int_{1.2}^{5.1}\frac{36}{x^{3.5}+8}\,dx \quad k = 3$

74. $\int_{-0.7}^{1.1}(2 + x^2\tan x)\,dx \quad k = 3$

75. $\int_{6.3}^{14.3}(\ln x)(\cos x)\,dx \quad k = 4$

76. $\int_{5.6}^{15.2}(\log_2 x)(\sin x)\,dx \quad k = 4$

Section 5.2.5

Analytically solve the integral. Use A_S, $A_{S,t}$, $A_S(n)$, and $A_{S,t}(n)$ to estimate the integral. Calculate the exact relative percentage error ε_S for each estimation.

77. $\int_{1}^{6.4}(0.2x^4 - 4x - x^{-1} + 7)\,dx \quad n = 6$

78. $\int_{0.7}^{3.4}(-x + \sin 2x)\,dx \quad n = 6$

79. $\int_{14}^{17} 5x\cos x\,dx \quad n = 4$

80. $\int_{5.9}^{8.3}\frac{48}{4-x}\,dx \quad n = 4$

Section 5.3

Use Romberg integration up to $I_{n,0}$ to estimate the integral. Use $I_{n-1,0}$ and $I_{n,0}$, rounded to the nearest thousandth, to calculate the estimation relative change error $E_{c,r}$.

81. $I_{2,0} \quad \int_{0}^{6} e^{x^{0.25}}\,dx$

82. $I_{2,0} \quad \int_{-1.5}^{1.5} e^{0.1x^3}\,dx$

83. $I_{2,0} \quad \int_{1.5}^{2.5}\tan^{-1}(6x^2 - 15x)\,dx$

84. $I_{2,0} \quad \int_{0.5}^{4.5}\tan^{-1}(\frac{1}{4}x - x^{1.25})\,dx$

85. $I_{3,0} \quad \int_{1.05}^{9.05}\log(\frac{512}{x-1} + 4)\,dx$

86. $I_{3,0} \quad \int_{0.15}^{4.55}\log_3(x^2 + 1/x)\,dx$

87. $I_{3,0}$ $\int_{-0.75}^{5.65} \sqrt{\frac{5}{2}x^3 + \frac{3}{2}x^2 + \frac{1}{2}}\, dx$ 88. $I_{3,0}$ $\int_{-1.4}^{1.8} \sqrt[3]{12 + 5x - x^4}\, dx$

89. $I_{4,0}$ $\int_0^{11.2} 3 \cdot 2^{-x} \sqrt[3]{x}\, dx$ 90. $I_{4,0}$ $\int_{0.2}^{11.4} 10\sqrt{x}\, e^{-0.4x}\, dx$

91. $I_{4,0}$ $\int_{8.7}^{13.5} \tan(3\ln x)\, dx$ 92. $I_{4,0}$ $\int_{1.5}^{14.3} \ln(x^{2.5} - 2)\, dx$

Sections 5.2.4 and 5.3

Use Simpson's rules $A_S(n)$ and $A_{S,t}(n)$ and Romberg integration up to $I_{k,0}$ to estimate the integral. Show the latter two estimations are sufficiently similar.

93. $\int_{1.25}^{5.25} \log_{1.25}(3x^2 + 1)\, dx$ $n = 4$ $I_{2,0}$

94. $\int_{0.4}^{5.8} \frac{28}{x^{1.5}+6}\, dx$ $n = 4$ $I_{2,0}$

95. $\int_{0.75}^{3.75} (e^x/x)\, dx$ $n = 4$ $I_{2,0}$

96. $\int_{1.8}^{5.8} (2^x/x^2)\, dx$ $n = 4$ $I_{2,0}$

97. $\int_{-0.6}^{1} e^{2x^3}\, dx$ $n = 8$ $I_{3,0}$

98. $\int_3^9 e^{7/x}\, dx$ $n = 8$ $I_{3,0}$

99. $\int_{-1.5}^{2.5} e^x \tan^{-1} x\, dx$ $n = 8$ $I_{3,0}$

100. $\int_{0.4}^{4.4} e^x \ln x\, dx$ $n = 8$ $I_{3,0}$

Sections 5.4.1 and 5.4.2

Use Monte Carlo integration to estimate the integral using 1000 points when the integrand is strictly nonnegative. Otherwise, use one simulation for all nonnegative integrand values and one for all nonpositive integrand values. (A proportionate total number of points for each rectangle is not necessary; one is suggested in the answers in the Appendix.) Analytically solve the integral using calculus techniques and verify the estimation is reasonable by calculating the exact relative percentage error ε_{MC}. Answers will vary according to the simulation.

101. $\int_0^2 \frac{12x}{x^2+4}\, dx$ 102. $\int_{0.5}^{2.5} \frac{20x^3}{(x^4+2)^2}\, dx$ 103. $\int_1^5 6xe^{-0.5x}\, dx$

104. $\int_7^{8.5} x\sin x\, dx$ 105. $\int_{0.5}^{7.5} \frac{x+12}{x^2+4x}\, dx$ 106. $\int_3^9 \frac{7x+13}{x^2+2x-3}\, dx$

107. $\int_{-0.75}^{1.75} 8xe^{-x^2}\, dx$ 108. $\int_{0.5}^{1.5} (2\ln x)/x\, dx$ 109. $\int_{0.25}^{1.75} 4x\ln x\, dx$

110. $\int_0^{2.5} (2x\tan^{-1} x - \frac{\pi}{2})\, dx$ 111. $\int_{-1.5}^{3.5} \frac{-10x-5}{x^2+7x+10}\, dx$ 112. $\int_{0.5}^{3.5} \frac{8-4x}{x^2+2x}\, dx$

6 Differentiation

Given a closed-form function where $f(x)$ equals an expression of x in terms of algebraic, trigonometric, exponential, logarithmic, or inverse trigonometric functions, there exists an $f'(x)$ as an expression of x in those terms as well. However, calculating $f'(x)$ may be complex or difficult to implement, and a numerical method may provide a sufficiently accurate result with much less effort. Furthermore, numerical derivatives are still necessary for non-closed form functions, such as a set of points or a table of values. Writing $f(x)$ as the Taylor polynomial $T_n(x)$ plus the remainder term $R_n(x)$, both centered at $x = x_i$, is the foundation for all methods described in Sections 6.1 through 6.5 of this chapter.

$$f(x) = T_n(x) + R_n(x)$$

$$f(x) = \sum_{k=0}^{n} \frac{f^{(k)}(x_i)(x - x_i)^k}{k!} + \frac{f^{(n+1)}(c)(x - x_i)^{n+1}}{(n+1)!} \tag{6.1}$$

Specifically, the Taylor polynomial $T_n(x)$ is the sum of all Taylor series terms up to and including the nth power and $R_n(x)$ is the $(n+1)$th power term for a value c between x and x_i.

The first three sections examine first derivative approximations using $T_1(x)$, $T_2(x)$, and $T_4(x)$, corresponding to the first (linear), second (quadratic), and fourth (quartic) degree Taylor polynomials, respectively. The next two sections detail second derivative approximations. The following section presents an idea similar to Romberg integration of Section 5.3, in which multiple derivative methods are weighted and combined for an improved approximation. The chapter concludes with a technique used for non-equally spaced points.

6.1 Approximations for $f'(x)$ using $T_1(x)$

Suppose M points are equally spaced so that $x_{i+1} - x_i = h$ for $i = 0, \ldots, M - 1$. Using Equation (6.1) with $n = 1$ and centered at x_i to evaluate the point at x_{i+1}

$$f(x) = T_1(x) + R_1(x)$$

$$f(x_{i+1}) = f(x_i) + f'(x_i)(x_{i+1} - x_i) + \frac{f''(c)(x_{i+1} - x_i)^2}{2}$$

$$f(x_{i+1}) = f(x_i) + f'(x_i)h + \frac{f''(c)h^2}{2}$$

Solving for $f'(x_i)$,

$$f(x_{i+1}) - f(x_i) - \frac{f''(c)h^2}{2} = f'(x_i)h$$

https://doi.org/10.1515/9783112221051-006

$$f'(x_i) = \frac{f(x_{i+1}) - f(x_i)}{h} - \frac{f''(c)h}{2} \tag{6.2}$$

Using just the first fraction of Equation (6.2)

$$f'(x_i) \approx \frac{f(x_{i+1}) - f(x_i)}{h}$$

$$\mathrm{FDD}_1'(x_i) = \frac{f(x_{i+1}) - f(x_i)}{h} \tag{6.3}$$

Equation (6.3) defines the forward divided difference approximation for the first derivative $\mathrm{FDD}_1'(x_i)$. This approximation uses x_{i+1}, the next x value to the right of, or forward from, the center at x_i. The subscript indicates it was derived using $T_1(x)$ to distinguish it from later approximations.

The reason Equation (6.3) is an approximation of the derivative is because it disregards the second fraction of Equation (6.2). Since that second fraction is the difference between the approximation and the true value of the derivative, it describes the truncation error for the $\mathrm{FDD}_1'(x_i)$.

$$\varepsilon_{t,1} = -\frac{f''(c)h}{2} \tag{6.4}$$

If a maximum value for $f''(c)$ can be found for some c in the interval $[x_i, x_{i+1}]$, Equation (6.4) provides an upper bound for the truncation error. Because it is multiplied by h, the error is described as order h, or $O(h)$. Order was briefly explained in Section 4.1. Here $O(h)$ indicates changing the size of h correspondingly changes the error by the same magnitude. For example, if the size of the interval h is halved, the truncation error is halved because substituting $h/2$ in Equation (6.4) is

$$\varepsilon_{t,1} = \frac{-f''(c)(h/2)}{2} = \frac{-f''(c)h}{4}$$

Note that the right side is half the size of Equation (6.4). The order of the truncation error will be revisited later in this section for comparison purposes.

Example 6.1. Use the $\mathrm{FDD}_1'(x_i)$ to approximate $f'(1)$ for the data and $h = 0.5$.

time x_i in s	0.0	0.25	0.5	0.75	1.0	1.25	1.5	1.75	2.0
position $f(x_i)$ in m	0.0	1.5	4.2	6.4	7.7	9.3	8.9	8.3	7.9

Solution. According to Equation (6.3) with $x_i = 1$ and $x_{i+1} = x_i + h = 1.5$

$$f'(1) \approx \mathrm{FDD}_1'(1) = \frac{f(1.5) - f(1)}{0.5} = \frac{8.9 - 7.7}{0.5} = 2.4\,\mathrm{m/s}$$

A backward divided difference approximation for the first derivative $\text{BDD}'_1(x_i)$ using $T_1(x)$ is found similarly using the x value at x_{i-1} to the left of, or backward from, the center at x_i, instead of the right or forward x value at x_{i+1}. Using Equation (6.1) again with $n = 1$ and centered at x_i

$$f(x_{i-1}) = f(x_i) + f'(x_i)(x_{i-1} - x_i) + \frac{f''(c)(x_{i-1} - x_i)^2}{2}$$

$$f(x_{i-1}) = f(x_i) + f'(x_i)(-h) + \frac{f''(c)(-h)^2}{2}$$

$$f'(x_i) = \frac{f(x_i) - f(x_{i-1})}{h} + \frac{f''(c)h}{2} \tag{6.5}$$

Using the first fraction only of Equation (6.5) defines the $\text{BDD}'_1(x_i)$.

$$f'(x_i) \approx \frac{f(x_i) - f(x_{i-1})}{h}$$

$$\text{BDD}'_1(x_i) = \frac{f(x_i) - f(x_{i-1})}{h} \tag{6.6}$$

The truncation error is the second fraction of Equation (6.5), equivalent to the negative of Equation (6.4) and again with $O(h)$.

Example 6.2. Use the $\text{BDD}'_1(x_i)$ to approximate $f'(1)$ for the data of Example 6.1 and $h = 0.25$.
Solution. Using Equation (6.6) with $x_i = 1$ and $x_{i-1} = x_i - h = 0.75$

$$f'(1) \approx \text{BDD}'_1(1) = \frac{f(1) - f(0.75)}{0.25} = \frac{7.7 - 6.4}{0.25} = 5.2\,\text{m/s}$$

6.2 Approximations for $f'(x)$ using $T_2(x)$

6.2.1 A centered divided difference

The $\text{FDD}'_1(x_i)$ and $\text{BDD}'_1(x_i)$ use an x value to the right or left of the center at x_i, respectively, to approximate $f'(x_i)$. A centered divided difference approximation for $f'(x_i)$ uses one x value to the left and one x value to the right of the center at x_i, but not the center at x_i itself. Furthermore, it uses a higher-degree Taylor polynomial to improve the approximation, because including more terms captures more of the underlying characteristics of $f(x)$. Equation (6.1) with $n = 1$ was used in the previous section for evaluating $f(x_{i+1})$ and $f(x_{i-1})$ en route to deriving the $\text{FDD}'_1(x_i)$ and $\text{BDD}'_1(x_i)$. Now, Equation (6.1) with $n = 2$, and correspondingly $T_2(x) + R_2(x)$, is used.

For the point to the right of the center at x_i,

$$f(x_{i+1}) = f(x_i) + f'(x_i)(x_{i+1} - x_i) + \frac{f''(x_i)(x_{i+1} - x_i)^2}{2} + \frac{f^{(3)}(c_1)(x_{i+1} - x_i)^3}{3!}$$

Since the points are equally spaced, $x_{i+1} - x_i = h$, so that

$$f(x_{i+1}) = f(x_i) + f'(x_i)h + \frac{f''(x_i)h^2}{2} + \frac{f^{(3)}(c_1)h^3}{6}. \tag{6.7}$$

For the point to the left of the center at x_i,

$$f(x_{i-1}) = f(x_i) + f'(x_i)(x_{i-1} - x_i) + \frac{f''(x_i)(x_{i-1} - x_i)^2}{2} + \frac{f^{(3)}(c_2)(x_{i-1} - x_i)^3}{3!}$$

Here $x_{i-1} - x_i = -h$, so substituting,

$$f(x_{i-1}) = f(x_i) + f'(x_i)(-h) + \frac{f''(x_i)(-h)^2}{2} + \frac{f^{(3)}(c_2)(-h)^3}{6}$$

$$f(x_{i-1}) = f(x_i) - f'(x_i)h + \frac{f''(x_i)h^2}{2} - \frac{f^{(3)}(c_2)h^3}{6}. \tag{6.8}$$

The function values $f(x_{i-1})$, $f(x_i)$, and $f(x_{i+1})$ are known, and the two remainder terms at c_1 and c_2 are constants, so Equations (6.7) and (6.8) form a system of equations in terms of the two unknowns, the first and second derivatives. The goal is finding an expression for the first derivative, so the second derivative is eliminated by subtracting Equation (6.8) from Equation (6.7).

$$f(x_{i+1}) - f(x_{i-1}) = 2f'(x_i)h + \frac{f^{(3)}(c_1)h^3}{6} + \frac{f^{(3)}(c_2)h^3}{6}$$

Solving for $f'(x_i)$,

$$f'(x_i) = \frac{f(x_{i+1}) - f(x_{i-1})}{2h} - \frac{h^2}{12}(f^{(3)}(c_1) + f^{(3)}(c_2)) \tag{6.9}$$

The first fraction of Equation (6.9) is the centered divided difference approximation for the first derivative $\mathrm{CDD}'_2(x_i)$ using $T_2(x)$.

$$f'(x_i) \approx \frac{f(x_{i+1}) - f(x_{i-1})}{2h}$$

$$\mathrm{CDD}'_2(x_i) = \frac{f(x_{i+1}) - f(x_{i-1})}{2h} \tag{6.10}$$

The second term of Equation (6.9) defines the truncation error for the $\mathrm{CDD}'_2(x_i)$ and has order h^2 or $O(h^2)$, an improvement over the truncation error of order $O(h)$ for the $\mathrm{FDD}'_1(x_i)$ and $\mathrm{BDD}'_1(x_i)$. $O(h^2)$ is superior to $O(h)$ because if, for example, the size of the interval is halved, the error is not halved as with $O(h)$, but quartered. Observe substituting $h/2$ for h in the second term of Equation (6.9).

$$-\frac{(h/2)^2}{12}(f^{(3)}(c_1) + f^{(3)}(c_2)) = -\frac{h^2}{48}(f^{(3)}(c_1) + f^{(3)}(c_2))$$

The divisor of the truncation error changed from originally 12 in Equation (6.9) to 48, so halving the interval reduced the error by one half squared, or one quarter.

Example 6.3. Use the $CDD'_2(x_i)$ to approximate $f'(1)$ for the data of Example 6.1 and $h = 0.25$.
Solution. Using Equation (6.10) with $x_i = 1, x_{i-1} = x_i - h = 0.75$, and $x_{i+1} = x_i + h = 1.25$

$$f'(1) \approx CDD'_2(1) = \frac{f(1.25) - f(0.75)}{2(0.25)} = \frac{9.3 - 6.4}{0.5} = 5.8 \, \text{m/s}$$

Figures 6.1 through 6.3 provide further evidence that the $CDD'_2(x_i)$ is superior to both the $FDD'_1(x_i)$ and $BDD'_1(x_i)$ for a given spacing h of x values. In each figure, the red tangent line slope represents the true derivative value $f'(x_i)$ at x_i. Using one x value on each side of x_i avoids an overreliance on how the function is shaped in just one direction from x_i. Figure 6.1 demonstrates this, employing the $CDD'_2(x_i)$ for the orange dashed secant line slope approximation of the tangent line slope $f'(x_i)$. Conversely, an overreliance in one direction occurs using only one x value on either side of x_i. Approximations for $f'(x_i)$ exhibit this behavior in Figures 6.2 and 6.3, using the $FDD'_1(x_i)$ for the green dashed secant line slope and the $BDD'_1(x_i)$ for the blue dashed secant line slope, respectively.

The conclusion of the calculus mean value theorem (MVT) for derivatives supports the graphical evidence that the $CDD'_2(x_i)$ is an improvement over the $FDD'_1(x_i)$ and $BDD'_1(x_i)$. The MVT states for a continuous function $f(x)$ on $[a,b]$ and differentiable on (a,b) there exists an $x = c$ such that $a < c < b$ and

$$f'(c) = \frac{f(b) - f(a)}{b - a} \tag{6.11}$$

Or, the secant line slope through the endpoints, the right side of Equation (6.11), equals the tangent line slope, the left side of Equation (6.11), at some $x = c$ between, but not including, the endpoints at $x = a$ and $x = b$.

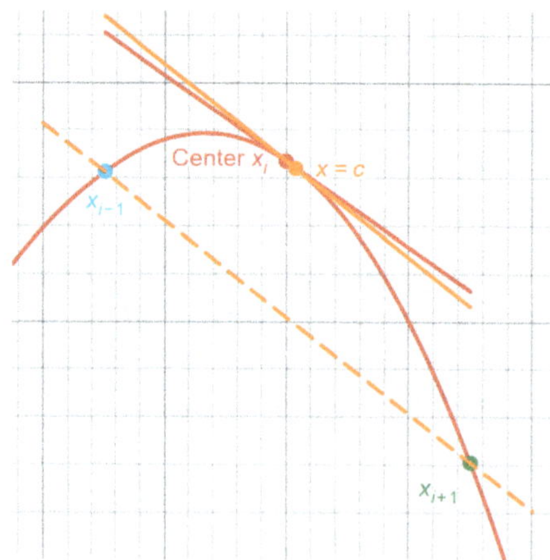

Figure 6.1: A $CDD'_2(x_i)$ approximation of $f'(x_i)$ using the dashed orange secant line slope between an x value to the left and right of x_i at x_{i-1} and x_{i+1}. The slope of the red line is the true value of $f'(x_i)$. The solid orange tangent line slope $f'(c)$ equals the dashed orange secant line slope by the MVT. Since $c \approx x_i, f'(c) \approx f'(x_i)$.

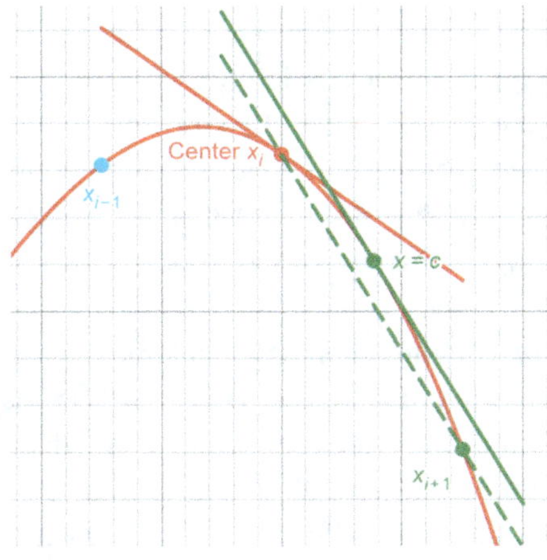

Figure 6.2: A FDD$_1'(x_i)$ approximation of $f'(x_i)$ using the dashed green secant line slope from x_i to an x value to the right of x_i at x_{i+1}. The slope of the red line is the true value of $f'(x_i)$. The solid green tangent line slope $f'(c)$ equals the dashed green secant line slope by the MVT. Unlike Figure 6.1, $f'(c) \neq f'(x_i)$ because $c \neq x_i$.

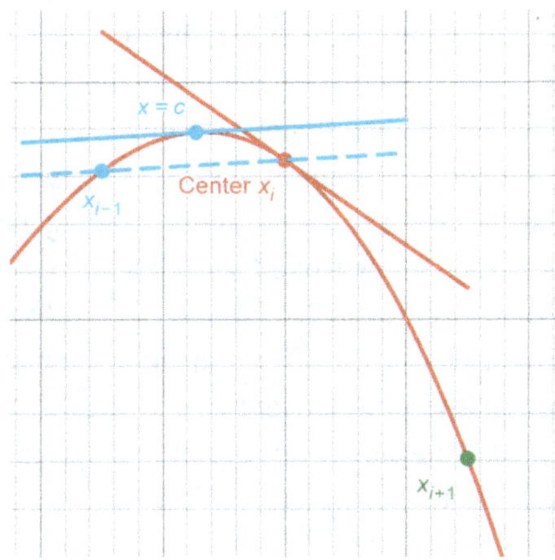

Figure 6.3: A BDD$_1'(x_i)$ approximation of $f'(x_i)$ using the dashed blue secant line slope from x_i to an x value to the left of x_i at x_{i-1}. The slope of the red line is the true value of $f'(x_i)$. The solid blue tangent line slope $f'(c)$ equals the dashed blue secant line slope by the MVT. Similar to Figure 6.2 and unlike Figure 6.1, $f'(c) \neq f'(x_i)$ because $c \neq x_i$.

Suppose the unknown function $f(x)$ is continuous on $[x_{i-1}, x_{i+1}]$ and differentiable on (x_{i-1}, x_{i+1}), thereby satisfying the MVT. Substituting these endpoints in Equation (6.11) and h for the equally spaced $x_{i+1} - x_i = h$ and $x_i - x_{i-1} = h$

$$f'(c) = \frac{f(x_{i+1}) - f(x_{i-1})}{x_{i+1} - x_{i-1}} = \frac{f(x_{i+1}) - f(x_{i-1})}{x_{i+1} - x_i + (x_i - x_{i-1})}$$

$$f'(c) = \frac{f(x_{i+1}) - f(x_{i-1})}{2h} \tag{6.12}$$

The right side of Equation (6.12) is precisely the definition of the CDD$_2'(x_i)$, Equation (6.10). But because Equation (6.10) only approximates $f'(x_i)$, it cannot be combined with Equation (6.12) to conclude $f'(c) = f'(x_i)$ or $c = x_i$. Nevertheless, the MVT guarantees an $x = c$ does exist where $x_{i-1} < c < x_{i+1}$. Furthermore, the center at x_i is by definition in that same open interval because it is midway between x_{i-1} and x_{i+1}. So although $c = x_i$ is not true in general, there always exists the possibility they are equal.

In contrast, approximations for $f'(x_i)$ using the $\mathrm{FDD}'_1(x_i)$ and $\mathrm{BDD}'_1(x_i)$ are calculated on the intervals $[x_i, x_{i+1}]$ and $[x_{i-1}, x_i]$, respectively. The secant line slopes through the endpoints of both of these intervals equals the tangent line slope $f'(c)$ at an $x = c$ in the corresponding *open* intervals (x_i, x_{i+1}) and (x_{i-1}, x_i), neither of which contain x_i, so $c \neq x_i$ always when using the $\mathrm{FDD}'_1(x_i)$ or $\mathrm{BDD}'_1(x_i)$.

Certainly, whether it is possible for $c = x_i$ or not, the center at x_i could be close to $x = c$ so that $f'(x_i) \approx f'(c)$, regardless of whether the $\mathrm{CDD}'_2(x_i)$, $\mathrm{FDD}'_1(x_i)$, or $\mathrm{BDD}'_1(x_i)$ is used. Note when using the $\mathrm{CDD}'_2(x_i)$, the center at x_i is located exactly in the middle of the interval between x_{i-1} and x_{i+1}. But when employing the $\mathrm{FDD}'_1(x_i)$ and $\mathrm{BDD}'_1(x_i)$, the center at x_i is located at one of the endpoints. For so-called well-behaved functions on a small interval, such as ones that do not change concavity and are relatively symmetric, the $x = c$ that satisfies the MVT will tend toward the middle of the interval rather than either end. Thus it is more likely $x = c$ will be closer to the center at x_i if the $\mathrm{CDD}'_2(x_i)$ is applied in lieu of the $\mathrm{FDD}'_1(x_i)$ or $\mathrm{BDD}'_1(x_i)$.

Reexamining Figures 6.1 through 6.3, each dashed secant line slope through the endpoints equals the correspondingly colored solid tangent line slope at the $x = c$ satisfying the MVT. In Figure 6.1 $x = c$ is almost identical to the center at x_i, so that the red tangent line slope $f'(x_i)$ is nearly identical to the orange tangent line slope $f'(c)$, which equals the orange secant line slope through the endpoints used by the $\mathrm{CDD}'_2(x_i)$; $f'(x_i) \approx f'(c)$.

But in Figures 6.2 and 6.3 $x = c$ is farther away significantly from the center at x_i. In turn, $f'(x_i)$ differs noticeably from the green and blue tangent line slopes $f'(c)$, correspondingly equal to the green and blue secant line slopes through the endpoints used, respectively, by the $\mathrm{FDD}'_1(x_i)$ and $\mathrm{BDD}'_1(x_i)$. In both cases, $f'(x_i) \neq f'(c)$.

6.2.2 Forward and backward divided differences

Similar to the development of the $\mathrm{CDD}'_2(x_i)$ of Section 6.2.1, improved forward and backward divided difference approximations for the first derivative are found using two points to form a system of equations in terms of the first and second derivatives. Eliminating the second derivative terms, an equation can be solved in terms of the first derivative $f'(x_i)$ only, and removing the remainder term from that equation leaves $f'(x_i)$ as an approximation of the derivative, derived from $T_2(x)$. But unlike using one x value to the left and one x value to the right of x_i in the $\mathrm{CDD}'_2(x_i)$ approximation, two x values to the right or two x values to the left of x_i are used for the $\mathrm{FDD}'_2(x_i)$ and $\mathrm{BDD}'_2(x_i)$ approximations, respectively.

For the forward divided difference, consider the additional function value at x_{i+2}. Using Equation (6.1) with $n = 2$

$$f(x_{i+2}) = f(x_i) + f'(x_i)(x_{i+2} - x_i) + \frac{f''(x_i)(x_{i+2} - x_i)^2}{2} + \frac{f^{(3)}(c_1)(x_{i+2} - x_i)^3}{3!}$$

Since the points are equally spaced,

$$x_{i+2} - x_i = (x_{i+2} - x_{i+1}) + (x_{i+1} - x_i) = h + h = 2h$$

Substituting $2h$ into the above function evaluation for $f(x_{i+2})$

$$f(x_{i+2}) = f(x_i) + f'(x_i)2h + \frac{f''(x_i)(2h)^2}{2} + \frac{f^{(3)}(c_2)(2h)^3}{6}$$

$$f(x_{i+2}) = f(x_i) + 2f'(x_i)h + 2f''(x_i)h^2 + \frac{4f^{(3)}(c_2)h^3}{3} \tag{6.13}$$

The function value at the second point at x_{i+1} with $x_{i+1} - x_i = h$ is Equation (6.7), restated here.

$$f(x_{i+1}) = f(x_i) + f'(x_i)h + \frac{f''(x_i)h^2}{2} + \frac{f^{(3)}(c_1)h^3}{6}$$

Equations (6.13) and (6.7) form the previously noted system of equations in terms of the first and second derivatives. To eliminate $f''(x_i)$, Equation (6.7) is multiplied by negative four and added to Equation (6.13).

$$-4f(x_{i+1}) = -4f(x_i) - 4f'(x_i)h - 2f''(x_i)h^2 - \frac{2f^{(3)}(c_1)h^3}{3}$$

$$f(x_{i+2}) - 4f(x_{i+1}) = -3f(x_i) - 2f'(x_i)h + \frac{4f^{(3)}(c_2)h^3}{3} - \frac{2f^{(3)}(c_1)h^3}{3}$$

Solving for $f'(x_i)$,

$$2f'(x_i)h = -f(x_{i+2}) + 4f(x_{i+1}) - 3f(x_i) + \frac{4f^{(3)}(c_2)h^3}{3} - \frac{2f^{(3)}(c_1)h^3}{3}$$

$$f'(x_i) = \frac{-f(x_{i+2}) + 4f(x_{i+1}) - 3f(x_i)}{2h} + \frac{h^2}{3}(2f^{(3)}(c_2) - f^{(3)}(c_1)) \tag{6.14}$$

The first fraction of Equation (6.14) is the forward divided difference approximation for the first derivative $\mathrm{FDD}_2'(x_i)$, using $T_2(x)$.

$$f'(x_i) \approx \frac{-f(x_{i+2}) + 4f(x_{i+1}) - 3f(x_i)}{2h}$$

$$\mathrm{FDD}_2'(x_i) = \frac{-f(x_{i+2}) + 4f(x_{i+1}) - 3f(x_i)}{2h} \tag{6.15}$$

The second fraction of Equation (6.14) is the truncation error.

$$\varepsilon_{t,2} = \frac{h^2}{3}(2f^{(3)}(c_2) - f^{(3)}(c_1)) \tag{6.16}$$

Equation (6.16) shows the $\text{FDD}_2'(x_i)$ has error $O(h^2)$, an improvement over the $\text{FDD}_1'(x_i)$ which has error $O(h)$ as demonstrated in Section 6.1.

Example 6.4. Use the $\text{FDD}_2'(x_i)$ to approximate $f'(0.75)$ for the data of Example 6.1 and $h = 0.25$.
Solution. Using Equation (6.15) with $x_i = 0.75$, $x_{i+1} = 1$, and $x_{i+2} = 1.25$

$$f'(0.75) \approx \text{FDD}_2'(0.75) = \frac{-f(1.25) + 4f(1) - 3f(0.75)}{2(0.25)}$$

$$= \frac{-9.3 + 4(7.7) - 3(6.4)}{0.5} = 4.6 \text{ m/s}$$

The corresponding $\text{BDD}_2'(x_i)$ is easily derived from Equation (6.14). Since two x values to the left of x_i are used instead of two points to the right, $-h$ replaces h. Also, the order of x value evaluations in the numerator is reversed, from decreasing x_{i+2}, x_{i+1}, x_i to increasing x_{i-2}, x_{i-1}, x_i.

$$f'(x_i) = \frac{-f(x_{i-2}) + 4f(x_{i-1}) - 3f(x_i)}{-2h} + \frac{(-h)^2}{3}(2f^{(3)}(c_2) - f^{(3)}(c_1))$$

$$f'(x_i) = \frac{f(x_{i-2}) - 4f(x_{i-1}) + 3f(x_i)}{2h} + \frac{h^2}{3}(2f^{(3)}(c_2) - f^{(3)}(c_1)) \tag{6.17}$$

The first fraction of Equation (6.17) is the backward divided difference approximation for the first derivative $\text{BDD}_2'(x_i)$, using $T_2(x)$.

$$f'(x_i) \approx \frac{f(x_{i-2}) - 4f(x_{i-1}) + 3f(x_i)}{2h}$$

$$\text{BDD}_2'(x_i) = \frac{f(x_{i-2}) - 4f(x_{i-1}) + 3f(x_i)}{2h} \tag{6.18}$$

The second fraction of Equation (6.17) is the truncation error, equivalent to Equation (6.16) with $O(h^2)$.

Example 6.5. Use the $\text{BDD}_2'(x_i)$ to approximate $f'(0.75)$ for the data of Example 6.1 and $h = 0.25$.
Solution. Using Equation (6.18) with $x_i = 0.75$, $x_{i-2} = 0.25$, and $x_{i-1} = 0.5$

$$f'(0.75) \approx \text{BDD}_2'(0.75) = \frac{f(0.25) - 4f(0.5) + 3f(0.75)}{2(0.25)}$$

$$= \frac{1.5 - 4(4.2) + 3(6.4)}{0.5} = 7.8 \text{ m/s}$$

6.3 An approximation for $f'(x)$ using $T_4(x)$

Deriving a centered divided difference approximation for the first derivative using $T_4(x)$, $\text{CDD}_4'(x_i)$, is not as intuitive as it was for the $\text{CDD}_2'(x_i)$. Nevertheless, the same approach applies of forming a system of equations using Equation (6.1) and combining them in a way that eliminates all the derivative terms except $f'(x_i)$. The algebraically intense details are omitted, but an outline of the procedure is presented which should suffice as a guide to insert those details if interested; see Yew [2].

Equation (6.1) with $n = 4$ is used to evaluate two x values to the left and two x values to the right of x_i.

$$f(x_{i-2}) = T_4(x_{i-2}) + R_4(x_{x-2}) \tag{6.19}$$

$$f(x_{i-1}) = T_4(x_{i-1}) + R_4(x_{i-1}) \tag{6.20}$$

$$f(x_{i+1}) = T_4(x_{i+1}) + R_4(x_{i+1}) \tag{6.21}$$

$$f(x_{i+2}) = T_4(x_{i+2}) + R_4(x_{i+2}) \tag{6.22}$$

Equations (6.19) through (6.22) are then combined in the following manner

$$8f(x_{i+1}) - 8f(x_{i-1}) - f(x_{i+2}) + f(x_{i-2})$$
$$= 8T_4(x_{i+1}) + 8R_4(x_{i+1}) - 8T_4(x_{i-1}) - 8R_4(x_{i-1})$$
$$- T_4(x_{i+2}) - R_4(x_{i+2}) + T_4(x_{i-2}) + R_4(x_{i-2}) \tag{6.23}$$

Three groupings motivate the simplification of the right side of Equation (6.23). $8T_4(x_{i+1}) - 8T_4(x_{i-1})$ and $-T_4(x_{i+2}) + T_4(x_{i-2})$ eliminate the even derivative terms $f''(x_i)$ and $f^{(4)}(x_i)$. The third derivative terms for $-T_4(x_{i+2}) + T_4(x_{i-2})$ use an interval of $2h$, so when raised to the third power $(2h)^3 = 8h^3$. The third derivative terms for $8T_4(x_{i+1}) - 8T_4(x_{i-1})$ use an interval of h so when raised to the third power it equals just h^3. But the multiple of eight and addition to $-T_4(x_{i+2}) + T_4(x_{i-2})$ eliminates the third derivative terms $f^{(3)}(x_i)$. What remains is only a multiple of the first derivative term $12hf'(x_i)$ and the truncation error term as a sum of the four $R_4(x_k)$ remainder terms.

$$8f(x_{i+1}) - 8f(x_{i-1}) - f(x_{i+2}) + f(x_{i-2})$$
$$= 12hf'(x_i) + 8R_4(x_{i+1}) - 8R_4(x_{i-1}) - R_4(x_{i+2}) + R_4(x_{i-2})$$

Solving for $f'(x_i)$ and reordering the points right to left

$$f'(x_i) = \frac{-f(x_{i+2}) + 8f(x_{i+1}) - 8f(x_{i-1}) + f(x_{i-2})}{12h}$$
$$+ \frac{R_4(x_{i+2}) - 8R_4(x_{i+1}) + 8R_4(x_{i-1}) - R_4(x_{i-2})}{12h} \tag{6.24}$$

The first fraction of Equation (6.24) is the centered divided difference approximation for the first derivative $\mathrm{CDD}_4'(x_i)$ using $T_4(x)$.

$$f'(x_i) \approx \frac{-f(x_{i+2}) + 8f(x_{i+1}) - 8f(x_{i-1}) + f(x_{i-2})}{12h}$$

$$\mathrm{CDD}_4'(x_i) = \frac{-f(x_{i+2}) + 8f(x_{i+1}) - 8f(x_{i-1}) + f(x_{i-2})}{12h} \tag{6.25}$$

Note that the terms $f(x_{i+1})$ and $f(x_{i-1})$ each contribute eight times as much to the approximation as each of the terms $f(x_{i+2})$ and $f(x_{i-2})$, which makes sense intuitively since both x values at x_{i+1} and x_{i-1} are closer to x_i than both x values at x_{i+2} and x_{i-2}.

The second fraction of Equation (6.24) is the truncation error. Since each remainder term $R_4(x_k)$ is a fifth power term of h and the fraction is divided by h, the resulting error is $O(h^4)$, an appreciable improvement over the $\mathrm{CDD}_2'(x_i)$ with error $O(h^2)$. For example, if h is halved, the error using the $\mathrm{CDD}_2'(x_i)$ is reduced from h to $(h/2)^2 = 0.25h$. Using the $\mathrm{CDD}_4'(x_i)$, the error is reduced from h to $(h/2)^4 = 0.0625h$.

Example 6.6. Use the $\mathrm{CDD}_4'(x_i)$ to approximate $f'(0.75)$ for the data of Example 6.1 and $h = 0.25$.
Solution. Using Equation (6.25) with $x_i = 0.75$, $x_{i-2} = 0.25$, $x_{i-1} = 0.5$, $x_{i+1} = 1$, and $x_{i+2} = 1.25$

$$f'(0.75) \approx \mathrm{CDD}_4'(0.75) = \frac{-f(1.25) + 8f(1) - 8f(0.5) + f(0.25)}{12(0.25)}$$

$$= \frac{-9.3 + 8(7.7) - 8(4.2) + 1.5}{3} = 6.733 \, \mathrm{m/s}$$

More accurate (and complex) approximation formulae exist, which use more terms in the Taylor series and higher-degree Taylor polynomials. But those described using $T_1(x)$ in Section 6.1, $T_2(x)$ in Section 6.2, and $T_4(x)$ in this section sufficiently demonstrate how such extensions could be derived.

6.4 Approximations for $f''(x)$ using $T_2(x)$

In finding the $\mathrm{FDD}_2'(x_i)$ in Section 6.2.2, Equations (6.13) and (6.7) formed a system of equations in terms of the first and second derivatives, and the $f''(x_i)$ terms were eliminated in order to find a first derivative approximation. That system is reused here, but now to eliminate the $f'(x_i)$ terms and solve for a second derivative approximation. Multiplying Equation (6.7) by negative two

$$-2f(x_{i+1}) = -2f(x_i) - 2f'(x_i)h - f''(x_i)h^2 - \frac{f^{(3)}(c_1)h^3}{3}$$

Adding this multiple to Equation (6.13),

$$f(x_{i+2}) - 2f(x_{i+1}) = -f(x_i) + f''(x_i)h^2 + \frac{4f^{(3)}(c_2)h^3}{3} - \frac{f^{(3)}(c_1)h^3}{3}$$

Lastly, solving for $f''(x_i)$,

$$f''(x_i)h^2 = f(x_{i+2}) - 2f(x_{i+1}) + f(x_i) + \frac{f^{(3)}(c_1)h^3}{3} - \frac{4f^{(3)}(c_2)h^3}{3}$$

$$f''(x_i) = \frac{f(x_{i+2}) - 2f(x_{i+1}) + f(x_i)}{h^2} + \frac{h}{3}(f^{(3)}(c_1) - 4f^{(3)}(c_2)) \qquad (6.26)$$

The first fraction of Equation (6.26) is the forward divided difference approximation for the second derivative $\mathrm{FDD}_2''(x_i)$ using $T_2(x)$.

$$f''(x_i) \approx \frac{f(x_{i+2}) - 2f(x_{i+1}) + f(x_i)}{h^2}$$

$$\mathrm{FDD}_2''(x_i) = \frac{f(x_{i+2}) - 2f(x_{i+1}) + f(x_i)}{h^2} \tag{6.27}$$

The second fraction of Equation (6.26) is the truncation error of $O(h)$.

Example 6.7. Use the $\mathrm{FDD}_2''(x_i)$ to approximate $f''(0.25)$ for the data of Example 6.1 and $h = 0.5$.
Solution. Using Equation (6.27) with $x_i = 0.25$, $x_{i+1} = 0.75$, and $x_{i+2} = 1.25$

$$f''(0.25) \approx \mathrm{FDD}_2''(0.25) = \frac{f(1.25) - 2f(0.75) + f(0.25)}{(0.5)^2}$$

$$= \frac{9.3 - 2(6.4) + 1.5}{0.25} = -8 \,\mathrm{m/s^2}$$

The corresponding $\mathrm{BDD}_2''(x_i)$ follows easily from Equation (6.27). Since two x values to the left of x_i are used instead of two x values to the right, $-h$ replaces h but since it is squared there is no change. The order of x value evaluations in the numerator is reversed as well, from decreasing x_{i+2}, x_{i+1}, x_i to increasing x_{i-2}, x_{i-1}, x_i. Reordering is not algebraically necessary because the largest and smallest function values are positively weighted once; reordering merely preserves the symmetry of the $\mathrm{FDD}_2''(x_i)$ and $\mathrm{BDD}_2''(x_i)$ approximation formulae. The backward divided difference approximation for the second derivative $\mathrm{BDD}_2''(x_i)$ using $T_2(x)$ is

$$f''(x_i) \approx \frac{f(x_{i-2}) - 2f(x_{i-1}) + f(x_i)}{h^2}$$

$$\mathrm{BDD}_2''(x_i) = \frac{f(x_{i-2}) - 2f(x_{i-1}) + f(x_i)}{h^2} \tag{6.28}$$

Although replacing $-h$ for h in the second fraction of Equation (6.26) changes the sign of the two truncation error terms, the overall order of the truncation error for the $\mathrm{BDD}_2''(x_i)$ is still $O(h)$, the same as for the $\mathrm{FDD}_2''(x_i)$.

Example 6.8. Use the $\mathrm{BDD}_2''(x_i)$ to approximate $f''(1.75)$ for the data of Example 6.1 and $h = 0.25$.
Solution. Using Equation (6.28) with $x_i = 1.75$, $x_{i-2} = 1.25$, and $x_{i-1} = 1.5$

$$f''(1.75) \approx \mathrm{BDD}_2''(1.75) = \frac{f(1.25) - 2f(1.5) + f(1.75)}{(0.25)^2}$$

$$= \frac{9.3 - 2(8.9) + 8.3}{0.0625} = -3.2 \,\mathrm{m/s^2}$$

6.5 An approximation for $f''(x)$ using $T_3(x)$

As with the $\mathrm{CDD}_2'(x_i)$, the centered divided difference approximation for the second derivative uses an x value to the right and left of the center at x_i and Equation (6.1). But now $n = 3$ is used instead of $n = 2$, increasing the two expansions of Equation (6.1) used for the $\mathrm{CDD}_2'(x_i)$ by one term each. Equation (6.7) extends to

$$f(x_{i+1}) = f(x_i) + f'(x_i)h + \frac{f''(x_i)h^2}{2} + \frac{f^{(3)}(x_i)h^3}{6} + \frac{f^{(4)}(c_1)h^4}{4!} \tag{6.29}$$

While Equation (6.8) adds the extra term to produce

$$f(x_{i-1}) = f(x_i) - f'(x_i)h + \frac{f''(x_i)h^2}{2} - \frac{f^{(3)}(x_i)h^3}{6} + \frac{f^{(4)}(c_2)h^4}{4!} \tag{6.30}$$

Adding Equations (6.29) and (6.30),

$$f(x_{i+1}) + f(x_{i-1}) = 2f(x_i) + \frac{2f''(x_i)h^2}{2} + \frac{f^{(4)}(c_1)h^4}{4!} + \frac{f^{(4)}(c_2)h^4}{4!}$$

Notice both the first and third derivative terms were eliminated, leaving just the second derivative term. Solving for $f''(x_i)$

$$f''(x_i)h^2 = f(x_{i+1}) - 2f(x_i) + f(x_{i-1}) - \frac{f^{(4)}(c_1)h^4}{24} - \frac{f^{(4)}(c_2)h^4}{24}$$

$$f''(x_i) = \frac{f(x_{i+1}) - 2f(x_i) + f(x_{i-1})}{h^2} - \frac{h^2}{24}(f^{(4)}(c_1) + f^{(4)}(c_2)) \tag{6.31}$$

The first fraction of Equation (6.31) is the centered divided difference approximation for the second derivative $\mathrm{CDD}_3''(x_i)$, using $T_3(x)$.

$$f''(x_i) \approx \frac{f(x_{i+1}) - 2f(x_i) + f(x_{i-1})}{h^2}$$

$$\mathrm{CDD}_3''(x_i) = \frac{f(x_{i+1}) - 2f(x_i) + f(x_{i-1})}{h^2} \tag{6.32}$$

The second fraction of Equation (6.31) is the truncation error of $O(h^2)$, an improvement over the truncation errors of the $\mathrm{FDD}_2''(x_i)$ and $\mathrm{BDD}_2''(x_i)$, both of $O(h)$.

Example 6.9. Use the $\mathrm{CDD}_3''(x_i)$ to approximate $f''(0.75)$ for the data of Example 6.1 and $h = 0.25$.
Solution. Using Equation (6.32) with $x_i = 0.75$, $x_{i-1} = 0.5$, and $x_{i+1} = 1$

$$f''(0.75) \approx \mathrm{CDD}_3''(0.75) = \frac{f(1) - 2f(0.75) + f(0.5)}{(0.25)^2}$$

$$= \frac{7.7 - 2(6.4) + 4.2}{0.0625} = -14.4 \, \mathrm{m/s}^2$$

Naturally, approximation formulae for $f''(x_i)$ which are more accurate (and complex) also exist, using more terms in the Taylor series and higher-degree Taylor polynomials. But the methods described in this section and the previous one are meant to provide guidelines for such extended derivations.

6.6 Richardson's extrapolation

Richardson's extrapolation combines two approximations, each over a different interval of equally spaced x values. In proportion, the better one is weighted more and positively, the worse one is weighted less and negatively, and both are used for an improved approximation. Generalized Richardson's extrapolation encompasses Romberg integration of Section 5.3.

For interval length h and $D_h(x_i)$, a derivative approximation of $f'(x_i)$ or $f''(x_i)$ at x_i for h, the weighted and improved Richardson's extrapolation derivative approximation $D_R(x_i)$ at x_i is defined

$$D_R(x_i) \approx \frac{4D_{h/2}(x_i) - D_h(x_i)}{3} \tag{6.33}$$

The more accurate approximation uses the smaller interval $h/2$ and is weighted four times positively, while the less accurate approximation uses the larger interval h and is weighted once negatively; the combination is therefore divided by a total of three. The derivative approximation $D_h(x_i)$ can employ *any* of the approximation formulae described in Sections 6.1 through 6.5. Examples using selected derivative approximations follow.

Example 6.10. Use Richardson's extrapolation with the $\text{FDD}'_1(x_i)$ to approximate $f'(1)$ for the data of Example 6.1 and $h = 0.5$.

Solution. Two derivative approximations using the $\text{FDD}'_1(x_i)$ of Equation (6.3) at $x_i = 1$ are calculated for interval lengths of h and $h/2$.

For $h = 0.5$, $x_{i+1} = x_i + h = 1.5$.

$$D_h(1) = D_{0.5}(1) = \text{FDD}'_1(1) = \frac{f(1.5) - f(1)}{0.5} = \frac{8.9 - 7.7}{0.5} = 2.4$$

For $h/2 = 0.5/2 = 0.25$, $x_{i+1} = x_i + h/2 = 1.25$.

$$D_{h/2}(1) = D_{0.25}(1) = \text{FDD}'_1(1) = \frac{f(1.25) - f(1)}{0.25} = \frac{9.3 - 7.7}{0.25} = 6.4$$

Weighting according to Equation (6.33) yields

$$D_R(1) = \frac{4D_{0.25}(1) - D_{0.5}(1)}{3} = \frac{4(6.4) - 2.4}{3} = 7.733 \text{ m/s}$$

Example 6.11. Use Richardson's extrapolation with the $\mathrm{CDD}_4'(x_i)$ to approximate $f'(1)$ for the data of Example 6.1 and $h = 0.5$.

Solution. The derivative approximations using the $\mathrm{CDD}_4'(x_i)$ of Equation (6.25) at $x_i = 1$ are solved for interval lengths h and $h/2$.

For $h = 0.5$, $x_{i-2} = 0$, $x_{i-1} = 0.5$, $x_{i+1} = 1.5$, and $x_{i+2} = 2$.

$$D_h(1) = D_{0.5}(1) = \mathrm{CDD}_4'(1) = \frac{-f(2) + 8f(1.5) - 8f(0.5) + f(0)}{12(0.5)}$$

$$= \frac{-7.9 + 8(8.9) - 8(4.2) + 0}{6} = 4.95$$

For $h/2 = 0.5/2 = 0.25$, $x_{i-2} = 0.5$, $x_{i-1} = 0.75$, $x_{i+1} = 1.25$, and $x_{i+2} = 1.5$.

$$D_{h/2}(1) = D_{0.25}(1) = \mathrm{CDD}_4'(1) = \frac{-f(1.5) + 8f(1.25) - 8f(0.75) + f(0.5)}{12(0.25)}$$

$$= \frac{-8.9 + 8(9.3) - 8(6.4) + 4.2}{3} = 6.167$$

Weighting according to Equation (6.33) produces

$$D_R(1) = \frac{4D_{0.25}(1) - D_{0.5}(1)}{3} = \frac{4(6.167) - 4.95}{3} = 6.573\,\mathrm{m/s}$$

Example 6.12. Use Richardson's extrapolation with the $\mathrm{BDD}_2''(x_i)$ to approximate $f''(2)$ for the data of Example 6.1 and $h = 1$.

Solution. The derivative approximations using the $\mathrm{BDD}_2''(x_i)$ of Equation (6.28) at $x_i = 2$ are calculated for interval lengths h and $h/2$.

For $h = 1$, $x_{i-2} = 0$ and $x_{i-1} = 1$.

$$D_h(2) = D_1(2) = \mathrm{BDD}_2''(2) = \frac{f(0) - 2f(1) + f(2)}{(1)^2} = 0 - 2(7.7) + 7.9 = -7.5$$

For $h/2 = 1/2 = 0.5$, $x_{i-2} = 1$ and $x_{i-1} = 1.5$.

$$D_{h/2}(2) = D_{0.5}(2) = \mathrm{BDD}_2''(2) = \frac{f(1) - 2f(1.5) + f(2)}{(0.5)^2} = \frac{7.7 - 2(8.9) + 7.9}{0.25} = -8.8$$

The result of weighting according to Equation (6.33) is

$$D_R(2) = \frac{4D_{0.5}(2) - D_1(2)}{3} = \frac{4(-8.8) - (-7.5)}{3} = -9.233\,\mathrm{m/s}^2$$

6.7 Approximation for unequally spaced points

The concise divided difference derivative (DDD) approximations of Sections 6.1 through 6.5 are possible because the points are equally spaced, i. e., $x_{i+1} - x_i = h$ for any valid index i. If $x_{i+1} - x_i$ is not constant for all i, one alternative is using the Lagrange polynomial function of Section 3.2, equivalent to a DD polynomial function but more general for the case of unequally spaced points.

The second-degree Lagrange polynomial function of Equation (3.33) through three given ordered points $(x_0, f(x_0))$, $(x_1, f(x_1))$, and $(x_2, f(x_2))$ is restated here for convenience.

$$\mathcal{L}_2(x) = \frac{(x - x_1)(x - x_2)}{(x_0 - x_1)(x_0 - x_2)} f(x_0) + \frac{(x - x_0)(x - x_2)}{(x_1 - x_0)(x_1 - x_2)} f(x_1) + \frac{(x - x_0)(x - x_1)}{(x_2 - x_0)(x_2 - x_1)} f(x_2)$$

$\mathcal{L}_2'(x)$ is solved analytically. For each fraction of $\mathcal{L}_2(x)$, simply differentiate the numerator, divide by the two constant factors, and multiply by the constant function value. For instance, the first numerator expands

$$(x - x_1)(x - x_2) = x^2 - x_2 x - x_1 x + x_2 x_1$$

The derivative of this expansion is

$$2x - x_2 - x_1.$$

Repeating this for the other two numerators and then dividing each by their corresponding factors and multiplying by their respective function values, the second-degree Lagrange polynomial function derivative $\mathcal{L}_2'(x)$ is

$$\mathcal{L}_2'(x) = \frac{2x - x_2 - x_1}{(x_0 - x_1)(x_0 - x_2)} f(x_0) + \frac{2x - x_2 - x_0}{(x_1 - x_0)(x_1 - x_2)} f(x_1) + \frac{2x - x_1 - x_0}{(x_2 - x_0)(x_2 - x_1)} f(x_2) \quad (6.34)$$

In addition to its application on unequally spaced points, there are two more significant advantages of using Lagrange polynomial function derivatives like Equation (6.34), even on equally spaced points. First, the more compact DDD approximations of Sections 6.1 through 6.5 are valid for derivatives *only* at the given points. In contrast, derivatives like Equation (6.34) on an associated interval, in this case $[x_0, x_2]$, can be used to approximate the derivative of *any* value within that interval.

For example, using the data of Example 6.1, it is not possible to approximate $f'(1.4)$ with one of the DDD approximations because the point $(1.4, f(1.4))$ is unknown. But Equation (6.34) is applicable using a set of three given consecutive points at $\{x_0, x_1, x_2\}$, in which the closed interval bounded by x_0 and x_2 includes $x = 1.4$, such as $x_0 = 1.25$, $x_1 = 1.5$, and $x_2 = 1.75$.

When approximating a derivative at an x_k which is not in the set $\{x_0, \ldots, x_n\}$, if $x_i < x_k < x_{i+1}$ for $i = 1, \ldots, n - 2$, the third x value employed could be either x_{i-1} or x_{i+2}. Since x values closer to the unknown x_k generally provide more relevant information, the potential third x value which has the smaller distance to x_k is used. Thus, x_{i-1} is used if

$$x_k - x_{i-1} < x_{i+2} - x_k$$

Conversely, x_{i+2} is used if

$$x_k - x_{i-1} > x_{i+2} - x_k.$$

Note that if $i = 0$ and $x_0 < x_k < x_1$, the third x value must be x_2. Similarly, if $i = n - 1$ and $x_{n-1} < x_k < x_n$, the third x value must be x_{n-2}.

Consider which set of three consecutive x values should be selected to approximate the derivative at $x_k = 3$ from among the following four x values.

$$x_0 = 1.8, \ x_1 = 2.9, \ x_2 = 3.8, \ x_3 = 4.5$$

Because $2.9 < 3 < 3.8$, $x_1 = 2.9$ and $x_2 = 3.8$ must be two of the values used, with a choice for the third value at either $x_0 = 1.8$ or $x_3 = 4.5$. The differences between $x_k = 3$ and each is

$$x_k - x_0 = 3 - 1.8 = 1.2$$
$$x_3 - x_k = 4.5 - 3 - 1.5$$

Since $1.2 < 1.5$, $x_k - x_0 < x_3 - x_k$, so $x_0 = 1.8$ is used. The set of three consecutive x values for the Lagrange polynomial function derivative $\mathcal{L}_2'(x)$ of Equation (6.34) is $\{1.8, 2.9, 3.8\}$.

If instead $x_k = 3.5$, which is also between $x_1 = 2.9$ and $x_2 = 3.8$, then

$$x_k - x_0 = 3.5 - 1.8 = 1.7$$
$$x_3 - x_k = 4.5 - 3.5 = 1$$

Here $x_k - x_0 > x_3 - x_k$ because $1.7 > 1$, so $x_3 = 4.5$ is used. In this instance the set of three consecutive x values for the Lagrange derivative $\mathcal{L}_2'(x)$ of Equation (6.34) is $\{2.9, 3.8, 4.5\}$.

A second advantage of Lagrange polynomial function derivatives like Equation (6.34) is not requiring a certain number of x values to the right or left of x_i like the forward, backward, and centered DDD approximations. For instance, using the data of Example 6.1, it is impossible to approximate $f'(1.75)$ with the $\mathrm{CDD}_4'(x_i)$ because two x values to the right of $x_i = 1.75$ are needed. Equation (6.34) handles this issue by using any set of three consecutive x values $\{x_0, x_1, x_2\}$ in which the closed interval bounded by x_0 and x_2 includes $x_i = 1.75$, such as $x_0 = 1.5$, $x_1 = 1.75$, and $x_2 = 2$.

Example 6.13. Use the Lagrange polynomial function derivative $\mathcal{L}_2'(x)$ to approximate $f'(1.7)$ for the data.

time x_i in s	1.1	1.3	1.7	2.3	2.9	3.1
distance $f(x_i)$ in m	23	27	39	57	60	77

Solution. There are three sets of consecutive x values $\{x_0, x_1, x_2\}$, in which the corresponding closed intervals, bounded by x_0 and x_2, include the point at $x = 1.7$: $\{1.1, 1.3, 1.7\}$, $\{1.3, 1.7, 2.3\}$, and $\{1.7, 2.3, 2.9\}$.

The centered one of $x_0 = 1.3$, $x_1 = 1.7$, and $x_2 = 2.3$ is chosen. Using Equation (6.34)

$$\mathcal{L}'_2(1.7) = \frac{2(1.7) - 2.3 - 1.7}{(1.3 - 1.7)(1.3 - 2.3)} f(1.3) + \frac{2(1.7) - 2.3 - 1.3}{(1.7 - 1.3)(1.7 - 2.3)} f(1.7) + \frac{2(1.7) - 1.7 - 1.3}{(2.3 - 1.3)(2.3 - 1.7)} f(2.3)$$

$$= \frac{-0.6}{0.4}(27) + \frac{-0.2}{-0.24}(39) + \frac{0.4}{0.6}(57) = -40.5 + 32.5 + 38 = 30 \text{ m/s}$$

Example 6.14. Use the Lagrange polynomial function derivative $\mathcal{L}'_2(x)$ to approximate $f'(2.5)$ for the data of Example 6.13.

Solution. Unlike Example 6.13, only two sets of consecutive x values $\{x_0, x_1, x_2\}$ exist in which the associated closed intervals bounded by x_0 and x_2 contain the unknown point at $x = 2.5$: $\{1.7, 2.3, 2.9\}$ and $\{2.3, 2.9, 3.1\}$ Since 3.1 is slightly closer than 1.7 is to 2.5, the set $x_0 = 2.3$, $x_1 = 2.9$, and $x_2 = 3.1$ is selected. By Equation (6.34)

$$\mathcal{L}'_2(2.5) = \frac{2(2.5) - 3.1 - 2.9}{(2.3 - 2.9)(2.3 - 3.1)} f(2.3) + \frac{2(2.5) - 3.1 - 2.3}{(2.9 - 2.3)(2.9 - 3.1)} f(2.9) + \frac{2(2.5) - 2.9 - 2.3}{(3.1 - 2.3)(3.1 - 2.9)} f(3.1)$$

$$= \frac{-1}{0.48}(57) + \frac{-0.4}{-0.12}(60) + \frac{-0.2}{0.16}(77) = -118.75 + 200 - 96.25 = -15 \text{ m/s}$$

Example 6.15. Use the Lagrange polynomial function derivative $\mathcal{L}'_2(x)$ to approximate $f'(1)$ for the equally spaced data of Example 6.1.

Solution. Three possible sets exist of consecutive x values $\{x_0, x_1, x_2\}$ where the corresponding closed intervals, bounded by x_0 and x_2, include the point at $x = 1$: $\{0.5, 0.75, 1\}$, $\{0.75, 1, 1.25\}$, and $\{1, 1.25, 1.5\}$. The centered one of $x_0 = 0.75$, $x_1 = 1$, and $x_2 = 1.25$ is chosen. According to Equation (6.34)

$$\mathcal{L}'_2(1) = \frac{2(1) - 1.25 - 1}{(0.75 - 1)(0.75 - 1.25)} f(0.75) + \frac{2(1) - 1.25 - 0.75}{(1 - 0.75)(1 - 1.25)} f(1) + \frac{2(1) - 1 - 0.75}{(1.25 - 0.75)(1.25 - 1)} f(1.25)$$

$$= \frac{-0.25}{0.125}(6.4) + \frac{0}{-0.0625}(7.7) + \frac{0.25}{0.125}(9.3) = -12.8 + 0 + 18.6 = 5.8 \text{ m/s}$$

The result of Example 6.15 is equivalent to result of Example 6.3, which used the $\text{CDD}'_2(x_i)$ and the same centered interval $[0.75, 1.25]$. This is not surprising; Section 3.2 established Lagrange polynomial functions $\mathcal{L}_n(x)$ and DD polynomial functions $f_n(x)$ of the same degree n are equivalent. The case for $n = 1$ was proved and Example 3.4 demonstrated an instance where $n = 2$.

Correspondingly, the derivatives of the Lagrange polynomial function $\mathcal{L}'_2(x)$ of Equation (6.34) and the $\text{CDD}'_2(x_i)$ of Equation (6.10) are both derived from the second-degree Taylor polynomial and are algebraically equivalent for three equally spaced points at x_0, x_1, and x_2, centered at x_1. Although not a proof, motivation for one lies in the following step from Example 6.15 using $\mathcal{L}'_2(x)$.

$$\mathcal{L}'_2(1) = \frac{-0.25}{0.125}(6.4) + \frac{0}{-0.0625}(7.7) + \frac{0.25}{0.125}(9.3)$$

Note that how the middle term is eliminated, analogous to the $\text{CDD}'_2(x_i)$ which does not use $x_i = x_1$ but $x_{i-1} = x_0$ and $x_{i+1} = x_2$. Rewriting the two remaining terms

$$\mathcal{L}_2'(1) = \frac{0.25}{0.125}(-6.4 + 9.3) = 2(9.3 - 6.4) = \frac{9.3 - 6.4}{0.5}$$

The fraction on the right is the last step of Example 6.3 using the $\text{CDD}_2'(x_i)$.

Differentiation concepts summary

1. Numerical derivatives are necessary for functions which are not closed-form expressions.
2. All divided difference derivative (DDD) approximations are defined by points an equally spaced distance d between x values.
3. DDD approximations at x_i use a Taylor polynomial and x values either to the right (forward, FDD), left (backward, BDD), or on both sides (centered, CDD) of x_i.
4. Using a higher-degree Taylor polynomial for a DDD increases the order of the truncation error and improves the DDD approximation, so that decreasing d decreases the truncation error by a power of d.
5. FDD and BDD approximations are defined entirely by, and may be overly reliant on, x values exclusively to the right or left of x_i, respectively, unlike CDD approximations.
6. The MVT for derivatives provides algebraic evidence supporting graphical intuition that CDD are better than FDD or BDD approximations, especially for functions that are nearly symmetric and do not change concavity for small d.
7. Richardson's extrapolation weights derivative approximations on h and $h/2$ spaced intervals for an improved approximation.
8. Richardson's extrapolation is usable with any FDD, BDD, or CDD approximation, constructed from any Taylor polynomial $T_n(x)$, and for any derivative $f^{(n)}(x)$.
9. For unequally spaced x values, the Lagrange polynomial function $\mathcal{L}_n(x)$ of Section 3.2, equivalent to a DD polynomial function, can be constructed and easily differentiated analytically to a closed form $\mathcal{L}_n'(x)$.
10. $\mathcal{L}_n'(x)$ can approximate a derivative at any x value in a given interval, not just at known x_i values like the DDD approximations.
11. $\mathcal{L}_n'(x_i)$ does not require a certain number of x values to the right, left, or on both sides of x_i like the DDD approximations.

Exercises

Section 6.1

Use the $\text{FDD}_1'(x_i)$ to approximate $f'(x_i)$.

x_i seconds	0.00	0.75	1.50	2.25	3.00	3.75	4.50	5.25	6.00
$f(x_i)$ feet	1.5	1.8	2.4	3.9	5.1	2.7	1.2	2.1	4.2

1. $f'(0.00)$
2. $f'(0.75)$
3. $f'(1.50)$
4. $f'(2.25)$
5. $f'(3.00)$
6. $f'(3.75)$
7. $f'(4.50)$
8. $f'(5.25)$

Use the $\text{BDD}_1'(x_i)$ to approximate $h'(x_i)$.

x_i minutes	1.0	1.4	1.8	2.2	2.6	3.0	3.4	3.8	4.2
$h(x_i)$ meters	27.9	32.1	34.4	30.9	26.5	24.9	25.2	25.8	30.6

9. $h'(1.4)$ 10. $h'(1.8)$ 11. $h'(2.2)$ 12. $h'(2.6)$
13. $h'(3.0)$ 14. $h'(3.4)$ 15. $h'(3.8)$ 16. $h'(4.2)$

Use either the $\text{FDD}_1'(x_i)$ or the $\text{BDD}_1'(x_i)$ to approximate $s'(x_i)$. Compare the results.

x_i hours	10.0	10.6	11.2	11.8	12.4	13.0	13.6	14.2
$s(x_i)$ miles	45.3	62.4	102.6	133.2	152.7	172.5	189.3	208.2

17. $\text{FDD}_1'(x_i)$ for $s'(10.0)$ and $\text{BDD}_1'(x_i)$ for $s'(10.6)$
18. $\text{FDD}_1'(x_i)$ for $s'(10.6)$ and $\text{BDD}_1'(x_i)$ for $s'(11.2)$
19. $\text{FDD}_1'(x_i)$ for $s'(11.2)$ and $\text{BDD}_1'(x_i)$ for $s'(11.8)$
20. $\text{FDD}_1'(x_i)$ for $s'(11.8)$ and $\text{BDD}_1'(x_i)$ for $s'(12.4)$
21. $\text{FDD}_1'(x_i)$ for $s'(12.4)$ and $\text{BDD}_1'(x_i)$ for $s'(13.0)$
22. $\text{FDD}_1'(x_i)$ for $s'(13.0)$ and $\text{BDD}_1'(x_i)$ for $s'(13.6)$

Section 6.2.1

Use the $\text{CDD}_2'(x_i)$ to approximate $f'(x_i)$ for the data of Exercises #1. through #8. above. Compare corresponding $f'(x_i)$ values.

23. $f'(0.75)$ 24. $f'(1.50)$ 25. $f'(2.25)$
26. $f'(3.00)$ 27. $f'(3.75)$ 28. $f'(4.50)$

Use the $\text{CDD}_2'(x_i)$ to approximate $h'(x_i)$ for the data of Exercises #9. through #16. above. Compare corresponding $h'(x_i)$ values.

29   $h'(1.4)$ 30. $h'(1.8)$ 31. $h'(2.2)$
32. $h'(2.6)$ 33. $h'(3.0)$ 34. $h'(3.4)$

Section 6.2.2

Use the $\text{FDD}_2'(x_i)$ to approximate $f'(x_i)$ for the data of Exercises #1. through #8. above. Compare corresponding $f'(x_i)$ values.

35. $f'(0.75)$ 36. $f'(1.50)$ 37. $f'(2.25)$
38. $f'(3.00)$ 39. $f'(3.75)$ 40. $f'(4.50)$

Use the $\text{BDD}_2'(x_i)$ to approximate $h'(x_i)$ for the data of Exercises #9. through #16. above. Compare corresponding $h'(x_i)$ values.

41. $h'(1.8)$ 42. $h'(2.2)$ 43. $h'(2.6)$
44. $h'(3.0)$ 45. $h'(3.4)$ 46. $h'(3.8)$

Section 6.3

Use the $CDD_4'(x_i)$ to approximate $s'(x_i)$ for the data of Exercises #17. through #22. above. Compare corresponding $s'(x_i)$ values.

47. $s'(11.2)$ 48. $s'(11.8)$ 49. $s'(12.4)$ 50. $s'(13.0)$

Section 6.4

Use the $FDD_2''(x_i)$ to approximate $h''(x_i)$.

x_i seconds	5.0	5.2	5.4	5.6	5.8	6.0	6.2	6.4	6.6
$h(x_i)$ meters	34.3	37.5	39.4	32.9	28.7	26.3	23.1	25.2	30.9

51. $h''(5.2)$ 52. $h''(5.4)$ 53. $h''(5.6)$
54. $h''(5.8)$ 55. $h''(6.0)$ 56. $h''(6.2)$

Use the $BDD_2''(x_i)$ to approximate $f''(x_i)$.

x_i minutes	1.00	1.8	2.6	3.4	4.2	5.0	5.8	6.6	7.4
$f(x_i)$ feet	12.6	16.3	15.2	14.2	13.9	13.2	12.1	13.4	13.8

57. $f''(2.6)$ 58. $f''(3.4)$ 59. $f''(4.2)$
60. $f''(5.0)$ 61. $f''(5.8)$ 62. $f''(6.6)$

Section 6.5

Use the $CDD_3''(x_i)$ to approximate $h''(x_i)$ for the data of Exercises #51. through #56. above. Compare corresponding $h''(x_i)$ values.

63. $h''(5.2)$ 64. $h''(5.4)$ 65. $h''(5.6)$
66. $h''(5.8)$ 67. $h''(6.0)$ 68. $h''(6.2)$

Use the $CDD_3''(x_i)$ to approximate $f''(x_i)$ for the data of Exercises #57. through #62. above. Compare corresponding $f''(x_i)$ values.

69. $f''(2.6)$ 70. $f''(3.4)$ 71. $f''(4.2)$
72. $f''(5.0)$ 73. $f''(5.8)$ 74. $f''(6.6)$

Sections 6.6 and 6.1

Use Richardson's extrapolation with the $FDD_1'(x_i)$ to approximate $f'(x_i)$ for the data of Exercises #1. through #8. above and $h = 1.5$. Compare corresponding $f'(x_i)$ values.

75. $f'(0.75)$ 76. $f'(1.50)$ 77. $f'(2.25)$
78. $f'(3.00)$ 79. $f'(3.75)$ 80. $f'(4.50)$

Use Richardson's extrapolation with the $BDD_1'(x_i)$ to approximate $h'(x_i)$ for the data of Exercises #9. through #16. above and $h = 0.8$. Compare corresponding $h'(x_i)$ values.

81. $h'(1.8)$ 82. $h'(2.2)$ 83. $h'(2.6)$
84. $h'(3.0)$ 85. $h'(3.4)$ 86. $h'(3.8)$

Sections 6.6 and 6.2.1

Use Richardson's extrapolation with the $CDD'_2(x_i)$ to approximate $f'(x_i)$ for the data of Exercises #1. through #8. above and $h = 1.5$. Compare corresponding $f'(x_i)$ values.

87. $f'(1.50)$ 88. $f'(2.25)$ 89. $f'(3.00)$ 90. $f'(3.75)$

Use Richardson's extrapolation with the $CDD'_2(x_i)$ to approximate $h'(x_i)$ for the data of Exercises #9. through #16. above and $h = 0.8$. Compare corresponding $h'(x_i)$ values.

91. $h'(2.2)$ 92. $h'(2.6)$ 93. $h'(3.0)$ 94. $h'(3.4)$

Sections 6.6 and 6.2.2

Use Richardson's extrapolation with the $FDD'_2(x_i)$ to approximate $f'(x_i)$ for the data of Exercises #1. through #8. above and $h = 1.5$. Compare corresponding $f'(x_i)$ values.

95. $f'(0.75)$ 96. $f'(1.50)$ 97. $f'(2.25)$ 98. $f'(3.00)$

Use Richardson's extrapolation with the $BDD'_2(x_i)$ to approximate $h'(x_i)$ for the data of Exercises #9. through #16. above and $h = 0.8$. Compare corresponding $h'(x_i)$ values.

99. $h'(2.6)$ 100. $h'(3.0)$ 101. $h'(3.4)$ 102. $h'(3.8)$

Sections 6.6 and 6.3

Use Richardson's extrapolation with the $CDD'_4(x_i)$ to approximate $s'(x_i)$ for the data of Exercises #17. through #22. above, the additional data below, and $h = 1.2$. Compare corresponding $s'(x_i)$ values.

x_i hours	14.8	15.4	16.0	16.6
$s(x_i)$ miles	207.8	196.5	185.1	180.4

103. $s'(12.4)$ 104. $s'(13.0)$ 105. $s'(13.6)$ 106. $s'(14.2)$

Sections 6.6 and 6.4

Use Richardson's extrapolation with the $FDD''_2(x_i)$ to approximate $h''(x_i)$ for the data of Exercises #51. through #56. above and $h = 0.4$. Compare corresponding $h''(x_i)$ values.

107. $h''(5.2)$ 108. $h''(5.4)$ 109. $h''(5.6)$ 110. $h''(5.8)$

Use Richardson's extrapolation with the $BDD''_2(x_i)$ to approximate $f''(x_i)$ for the data of Exercises #57. through #62. above and $h = 1.6$. Compare corresponding $f''(x_i)$ values.

111. $f''(4.2)$ 112. $f''(5.0)$ 113. $f''(5.8)$ 114. $f''(6.6)$

Sections 6.6 and 6.5

Use Richardson's extrapolation with the $\text{CDD}_3''(x_i)$ to approximate $h''(x_i)$ for the data of Exercises #51. through #56. above and $h = 0.4$. Compare corresponding $h''(x_i)$ values.

115. $h''(5.4)$ 116. $h''(5.6)$ 117. $h''(5.8)$ 118. $h''(6.0)$

Use Richardson's extrapolation with the $\text{CDD}_3''(x_i)$ to approximate $f''(x_i)$ for the data of Exercises #57. through #62. above and $h = 1.6$. Compare corresponding $f''(x_i)$ values.

119. $f''(3.4)$ 120. $f''(4.2)$ 121. $f''(5.0)$ 122. $f''(5.8)$

Section 6.7

Use $\mathcal{L}_2'(x)$ to approximate $f'(x)$ and let $x_1 = x$ where possible.

x_i seconds	1.2	3.5	5.8	7.2	8.9	11.4	12.3
$f(x_i)$ feet	104	121	139	128	132	137	142
$f'(x_i)$ ft/s	#123.	#124.	#125.	#126.	#127.	#128.	#129.

Use $\mathcal{L}_2'(x)$ to approximate $f'(x)$ and let $x_1 = x$ where possible.

x_i minutes	0	3	4	6	12	17	24
$f(x_i)$ meters	9.3	8.7	5.2	5.9	6.4	5.8	4.5
$f'(x_i)$ m/min	#130.	#131.	#132.	#133.	#134.	#135.	#136.

Use $\mathcal{L}_2'(x)$ to approximate $f'(x)$. Choose the set of three consecutive x values $\{x_0, x_1, x_2\}$ in which the approximated x value is closer to x_0 or x_2.

x_i seconds	10.0	11.7	13.1	14.2	15.5	16.3	17.4
$f(x_i)$ meters	45.1	50.3	39.8	40.7	41.5	43.2	39.6

137. $f'(11)$ 138. $f'(12)$ 139. $f'(14)$

140. $f'(16)$ 141. $f'(17)$

x_i minutes	1.12	1.45	1.78	2.26	2.81	3.22	3.54
$f(x_i)$ feet	1.012	0.941	0.823	0.659	0.574	0.968	1.037

142. $f'(1.35)$ 143. $f'(1.65)$ 144. $f'(2.52)$

145. $f'(2.95)$ 146. $f'(3.45)$

x_i hours	0	1.5	3.25	5.75	8.00	9.25	12.00
$f(x_i)$ kilometers	153.7	141.9	126.2	108.3	97.5	101.4	137.1

147. $f'(1.25)$ 148. $f'(3.00)$ 149. $f'(6.00)$
150. $f'(8.5)$ 151. $f'(9.5)$

x_i hours	20	28	32	35	40	48	57
$f(x_i)$ miles	15.4	19.2	20.3	21.4	26.1	19.8	8.5

152. $f'(26.8)$ 153. $f'(29.2)$ 154. $f'(38.4)$
155. $f'(45.7)$ 156. $f'(49.1)$

7 Differential equations

A mathematical model involving a rate or rates of change is a differential equation (DEQ) with one or more derivative terms. Solving a DEQ analytically can be difficult depending on its type, and like the integration problem, may not be possible.

The analytic solution to a DEQ is an equation before any differentiation has been applied. For example, the solution to a DEQ with a $\frac{dy}{dx}$ term is an equation in terms of just x and y. When the undifferentiated equation, i. e., the solution, along with all its relevant derivatives are substituted into the DEQ, the DEQ reduces to $0 = 0$. Even the simplest DEQs of the form $\frac{dy}{dx} = f(x)$ may not have a solution of the form $y = F(x) + C$ if no antiderivative $F(x)$ of $f(x)$ exists. Another example are DEQs of the form $\frac{dy}{dx} = f(x,y)$. These could be solved analytically if separable and rewritten in the form $p(y)dy = q(x)dx$, where antiderivatives for $p(y)$ and $q(x)$ exist. But that isn't always possible; the relatively simple DEQ $\frac{dy}{dx} = x + y$ is not separable.

The techniques in this chapter apply to DEQs of the form $\frac{dy}{dx} = f(x,y)$, in which a first derivative can be isolated to equal any combination of terms of x and y. The solutions are not closed-form equations but a set of points which describe the underlying undifferentiated equation. Each of the first four methods apply repeated linearizations using an approximated weighted slope defined by one or two points. The last method examines a common approximated slope comprised of four weighted slopes.

7.1 Euler's method

Approximating a function value using its linearization is a common calculus exercise. Presumably, over a sufficiently small interval, any function resembles a line. A tangent line to $y = F(x)$ at $x = x_0$ is

$$y - F(x_0) = F'(x_0)(x - x_0) \tag{7.1}$$

Using Leibnitz notation the derivative $F'(x_0)$ is expressed

$$\left.\frac{dy}{dx}\right|_{x=x_0}$$

Or, the derivative of $F(x)$ can be defined as its own function $f(x)$ where $F'(x) = f(x)$. Thus, the following three notations are all equivalent

$$F'(x_0) = \left.\frac{dy}{dx}\right|_{x=x_0} = f(x_0) \tag{7.2}$$

https://doi.org/10.1515/9783112221051-007

The tangent line of Equation (7.1) is the same line used now for the linearization of $y = F(x)$ at x_0. In this context, it is solved for y and renamed $L(x)$, and the notation for the derivative is substituted according to Equation (7.2), so that Equation (7.1) becomes

$$L(x) = F(x_0) + f(x_0)(x - x_0). \tag{7.3}$$

For x values near x_0, y values $L(x) \approx F(x)$, as shown in Figure 7.1. Naturally when $x = x_0$ exactly at the point of linearization, the y values of the approximated point and the true point are equal as well, $L(x_0) = F(x_0)$. For x_1 close to x_0, $L(x_1) \approx F(x_1)$, indicated by the short black vertical distance $|F(x_1) - L(x_1)|$, the absolute error between the true and approximated y values. For x_2 not close to x_0, the y values $L(x_2) \neq F(x_2)$ as depicted by the long black vertical distance $|F(x_2) - L(x_2)|$. This is not unusual, as an approximation $L(x_i)$ tends to get worse as x_i moves farther away from x_0 using a single linearization.

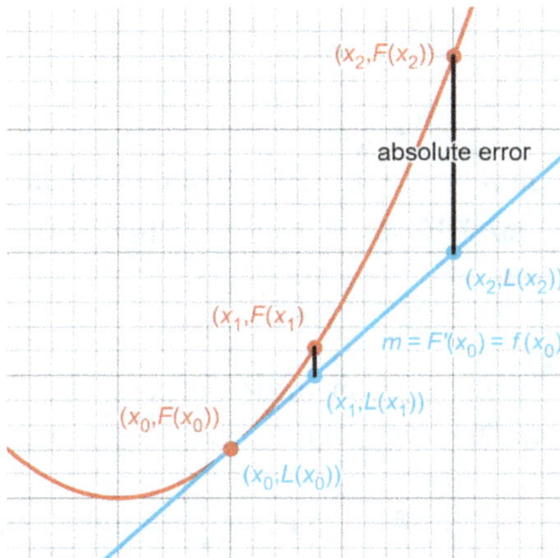

Figure 7.1: The black vertical distances between $F(x_i)$ and $L(x_i)$ represent the absolute errors between the true function values in red and the linearization approximations in blue.

Instead of using the single linearization of Equation (7.3) to approximate points any distance from an initial point $(x_0, F(x_0))$, Euler's method applies multiple distinct, smaller, and successive linearizations based on a previously approximated point.

7.1.1 The single variable case, $\frac{dy}{dx} = f(x)$

Before delving into the general case in which the derivative $\frac{dy}{dx} = f(x,y)$ is expressed in terms of both x and y, an example involving a derivative in terms of x only, $\frac{dy}{dx} = f(x)$, is dissected using simpler, intuitive graphs.

Along with Figure 7.1, Figures 7.2 and 7.3 illustrate the difference between Euler's method and a single linearization. In practice, the true red function $y = F(x)$ is unknown, while its derivative $F'(x)$, equal to the DEQ, is given and used to approximate points on

$y = F(x)$. The true red function $y = F(x)$ is displayed in Figures 7.1 through 7.3 only for comparison with the approximated blue points at $(x_i, L(x_i))$.

The general form of Equation (7.2) is used for the derivative of the unknown function $y = F(x)$ at all unknown $x = x_k$ values.

$$F'(x_k) = \left.\frac{dy}{dx}\right|_{x=x_k} = f(x_k) \tag{7.4}$$

Euler's method employs a sufficiently small constant distance h between x values for any integer $i > 0$ to determine where each of each equally spaced x values and approximated points occur, that is

$$h = x_{i+1} - x_i. \tag{7.5}$$

Solving for x_{i+1}, it follows the location of the next point is at

$$x_{i+1} = x_i + h. \tag{7.6}$$

In the first instance, the next approximated point is at $x_1 = x_0 + h$.

For the first associated approximated function value $L(x_1)$, the linearization of Equation (7.3) starting at the given initial point $(x_0, F(x_0))$ with $F'(x_0) = f(x_0)$ is applied.

$$L(x_1) = F(x_0) + f(x_0)(x_1 - x_0)$$

In Figure 7.1, this is indicated by the blue solid tangent line from the initial point at x_0 to the next x value at x_1. Note the first application of Euler's method mirrors using a single linearization.

But now, a new linearization is applied by evaluating the derivative of the unknown function $y = F(x)$ at the approximated point $(x_1, L(x_1))$ since the true point $(x_1, F(x_1))$ would be unknown. However, in this simple case the derivative, which is equal to the DEQ, $F'(x)$, $\frac{dy}{dx}$, and $f(x)$ according to Equation (7.4), is defined in terms of x only. Thus, the approximated y value at $L(x_1)$ does not affect the value of the derivative, so that

$$\left.\frac{dy}{dx}\right|_{(x_1, L(x_1))} = \left.\frac{dy}{dx}\right|_{(x_1, F(x_1))} = \left.\frac{dy}{dx}\right|_{x=x_1} = F'(x_1) = f(x_1).$$

Observe in Figure 7.2 the red dashed tangent line at the true point $(x_1, F(x_1))$ and the blue dashed tangent line at $(x_1, L(x_1))$ are parallel and have equal slopes of $F'(x_1)$, since the y values $F(x_1)$ and $L(x_1)$ of the points are irrelevant in calculating $F'(x_1)$, which is based on an x value only. However, the approximated y value $L(x_1)$ *is* relevant to the location of the next point at $(x_2, L(x_2))$, emanating from the position of the known approximated point $(x_1, L(x_1))$ and not the unknown true point at $(x_1, F(x_1))$.

One more iteration is shown in Figure 7.3, employing a linearization based on $F'(x_2)$ and the slope of the red dashed tangent line at the true point $(x_2, F(x_2))$. Since $F'(x_2)$ is

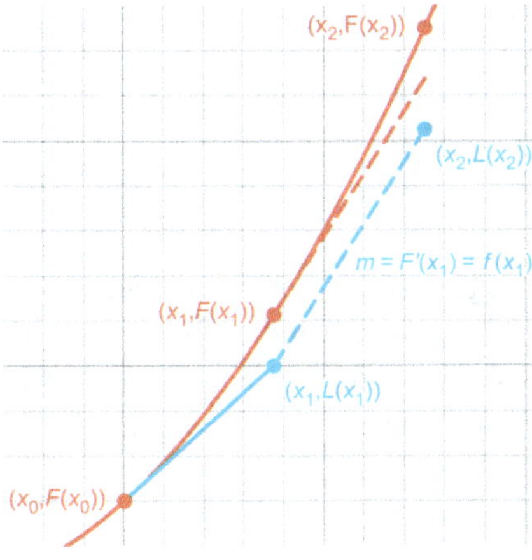

Figure 7.2: The red dashed line is tangent to $y = F(x)$ at the unknown true point $(x_1, F(x_1))$, but the parallel blue dashed line is applied to the known approximated point $(x_1, L(x_1))$.

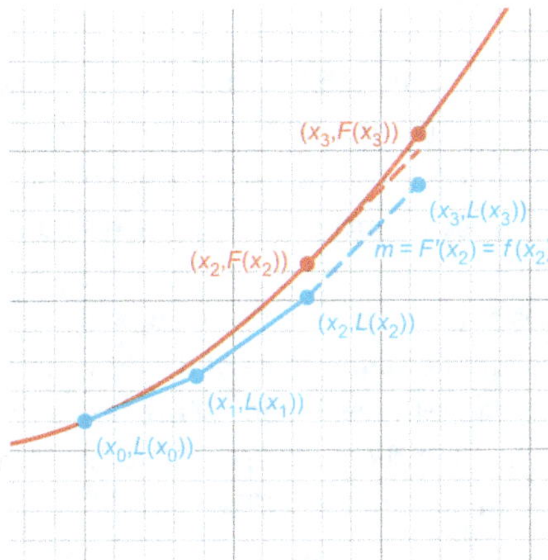

Figure 7.3: The red dashed line is tangent to $y = F(x)$ at the unknown true point $(x_2, F(x_2))$. The parallel blue dashed line is used instead at the previous known approximated point $(x_2, L(x_2))$ to find the next point at $(x_3, F(x_3))$.

based solely on x_2 regardless of the true and approximated y values $F(x_2)$ and $L(x_2)$, respectively, that same slope is used in conjunction with the previous approximated point $(x_2, L(x_2))$, indicated by the blue dashed line, which is parallel to the red dashed line.

Defining this process analytically consists of simply repeatedly applying the single linearization of Equation (7.3) to multiple, smaller, and consecutive intervals based on the previously approximated point. So instead of approximating every $L(x_{i+1})$ along the single linearization at x_0, $L(x_{i+1})$ is approximated by the linearization at x_i. Substituting x_{i+1} for x and x_i for x_0 in Equation (7.3)

$$L(x_{i+1}) = L(x_i) + f(x_i)(x_{i+1} - x_i)$$

Equation (7.5) defined the difference between consecutive x values as h, and, using the alternate notation for the derivative of Equation (7.4), these two substitutions produce the y-coordinate approximation iterative step of Euler's method for DEQs of the form $\frac{dy}{dx} = f(x)$.

$$L(x_{i+1}) = L(x_i) + \left(\frac{dy}{dx}\Big|_{x=x_i} \right) h \tag{7.7}$$

Finding the next $L(x_{i+1})$ value by Equation (7.7) is the crux of Euler's method because finding the next x value is trivial by Equation (7.6), $x_{i+1} = x_i + h$. Note the set of x values and the set of $L(x)$ values are found independently using Equations (7.6) and (7.7), respectively, and then paired to form the set of approximated points $(x_i, L(x_i))$ for any integer $i > 0$.

7.1.2 The multivariable case, $\frac{dy}{dx} = f(x, y)$

The derivatives and tangent line slopes in Figures 7.1 through 7.3 illustrated the case where the derivative is a function of x only, or $\frac{dy}{dx} = f(x)$. But Euler's method is applicable when the derivative is a function of both x and y as well, or $\frac{dy}{dx} = f(x, y)$. Implicit differentiation of an equation which is not necessarily a function commonly results in such a derivative.

The iterative step of Euler's method for DEQs of the form $\frac{dy}{dx} = f(x, y)$, in which the y value $L(x_i)$ of the approximated point also contributes to the value of the derivative, necessitates a slight modification to Equation (7.7). Since the evaluation of the derivative now requires both coordinate values x_i and $L(x_i)$, rather than just the x value at x_i, the following replaces Equation (7.4) in the linearization of Equation (7.7).

$$m_L = \frac{dy}{dx}\Big|_{(x_i, L(x_i))} \tag{7.8}$$

Note the tangent line slope m_L at the current point is also the left endpoint of the current interval $[x_i, x_{i+1}]$. All methods in this chapter use m_L in some capacity, lending to this standard definition and making methods easily comparable.

Substituting Equation (7.8) in Equation (7.7) produces the general iterative step for the y-coordinate $L(x_{i+1})$ using Euler's method for DEQs of the form $\frac{dy}{dx} = f(x, y)$.

$$L(x_{i+1}) = L(x_i) + m_L h \tag{7.9}$$

The specific DEQ example $\frac{dy}{dx} = 3x^2 y$, with analytic solution $y = 2e^{x^3}$ for comparison purposes, and Figures 7.4 through 7.6, demonstrate how a DEQ containing both x and y values further perturbs each approximated point from the true point. These deviations can propagate quickly depending on the solution to the DEQ and size of h.

With a starting point of $(0.1, 2)$ and distance $h = 0.5$, Equation (7.6) is used for the x value of the first approximated point.

$$x_1 = x_0 + h = 0.1 + 0.5 = 0.6$$

The slope at the current point, which in this first iteration is also the initial point, uses the given DEQ and Equation (7.8).

$$m_L = \left.\frac{dy}{dx}\right|_{(x_0, L(x_0))} = \left.\frac{dy}{dx}\right|_{(0.1, 2)} = 3(0.1)^2(2) = 0.06$$

Now, the first approximated y value can be found by Equation (7.9).

$$L(x_1) = L(x_0) + m_L h$$
$$= 2 + (0.06)0.5 = 2.03$$

Thus the first approximated point $(x_1, L(x_1)) = (0.6, 2.03)$. In the second and subsequent applications of Euler's method for DEQs of the form $\frac{dy}{dx} = f(x, y)$, the impact of using approximated y values at $L(x_i)$ is observable from both a numerical and graphical perspective.

The next x value by Equation (7.6) is at

$$x_2 = x_1 + h = 0.6 + 0.5 = 1.1$$

The next approximated y value $L(x_2)$ requires Equation (7.8) and the slope based on both the $x = x_1$ and $y = L(x_1)$ components of the previously approximated point, now the current point and left endpoint of the new current interval $[x_1, x_2]$.

$$m_L = \left.\frac{dy}{dx}\right|_{(x_1, L(x_1))} = \left.\frac{dy}{dx}\right|_{(0.6, 2.03)} = 3(0.6)^2(2.03) = 2.192$$

Completing the calculation of the approximated y value by substituting into Equation (7.9)

$$L(x_2) = L(x_1) + m_L h = 2.03 + (2.192)0.5 = 3.126$$

Here, the use of the approximated $y = L(x_1)$ component marks the difference between Equation (7.7) for DEQs of the form $\frac{dy}{dx} = f(x)$ and Equation (7.9) for DEQs of the form $\frac{dy}{dx} = f(x, y)$. The true y value at $x = 0.6$ is $F(0.6) = 2.482$, so the derivative of Equation (7.8) based on the true point is

$$m_L = \left.\frac{dy}{dx}\right|_{(x_1, F(x_1))} = \left.\frac{dy}{dx}\right|_{(0.6, 2.482)} = 3(0.6)^2(2.482) = 2.681$$

Applying this derivative to the previously approximated point using Equation (7.9) alters the next approximated y value from $L(x_2) = 3.126$ to $L_a(x_2)$.

$$L_a(x_2) = L(x_1) + m_L h = 2.03 + (2.681)0.5 = 3.371$$

Figure 7.4 highlights the derivative both at and using the approximated point by the slope of the orange dashed line. The derivative at the true point is represented by the slope of the red dashed tangent line. These are not parallel due to the use of different y values at $L(x_1)$ and $F(x_1)$ which produce different slopes by using $\frac{dy}{dx} = f(x,y)$. Consequently, the resulting next approximated point, even emanating from the same $(x_1, L(x_1)) = (0.6, 2.03)$, yields different results at the y values of $L(x_2) = 3.126$ and $L_a(x_2) = 3.371$.

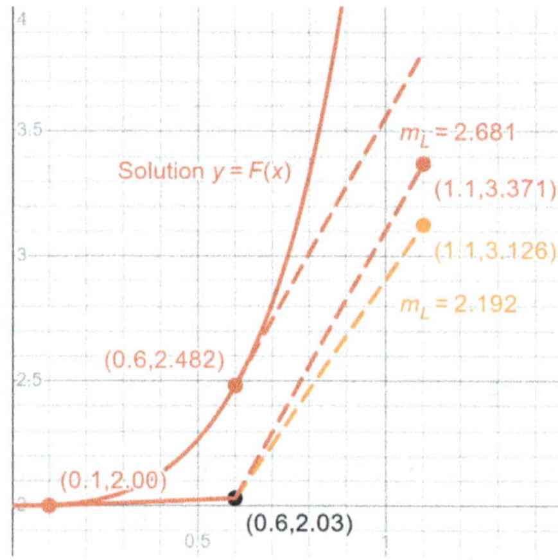

Figure 7.4: Since $\frac{dy}{dx} = f(x,y)$ uses a y value, the slopes at the approximated point $(0.6, 2.03)$ and the true point $(0.6, 2.482)$ are not equal, indicated by the nonparallel orange and red dashed lines. The termination of these lines impacts the location of the next approximated point at $x_2 = 1.1$.

In contrast, in Figure 7.2 the tangent lines beginning at the approximated point and true point *are* parallel because those slopes use $\frac{dy}{dx} = f(x)$ based on x values only, and both the approximated and true point use the same x value.

The disparity between points found using the slope at the approximated $L(x_i)$ and $F(x_i)$ values increases in the next two iterations. The next x value is

$$x_3 = x_2 + h = 1.1 + 0.5 = 1.6$$

The derivative using the approximated point at $(x_2, L(x_2))$ is

$$m_L = \left. \frac{dy}{dx} \right|_{(1.1,3.126)} = 3(1.1)^2(3.126) = 11.347$$

The corresponding approximated y value using Equation (7.9) and substituting the above is

$$L(x_3) = L(x_2) + m_L h = 3.126 + (11.347)0.5 = 8.800$$

The true y value at $x = 1.1$ is $F(1.1) = 7.570$, which alters the derivative and approximated y value, respectively,

$$m_L = \left.\frac{dy}{dx}\right|_{(1.1,7.570)} = 3(1.1)^2(7.570) = 27.479$$

$$L_a(x_3) = L(x_2) + m_L h = 3.126 + (27.479)0.5 = 16.866$$

In Figure 7.5, the difference between the slope of the green and red dashed lines is more pronounced than the difference between the slope of the orange and red dashed lines of Figure 7.4, as successive approximated points stray farther from the initial true point of $(0.1, 2)$. In turn, the vertical gap between the approximated green and red points is larger than the vertical gap between the orange and red points of Figure 7.4.

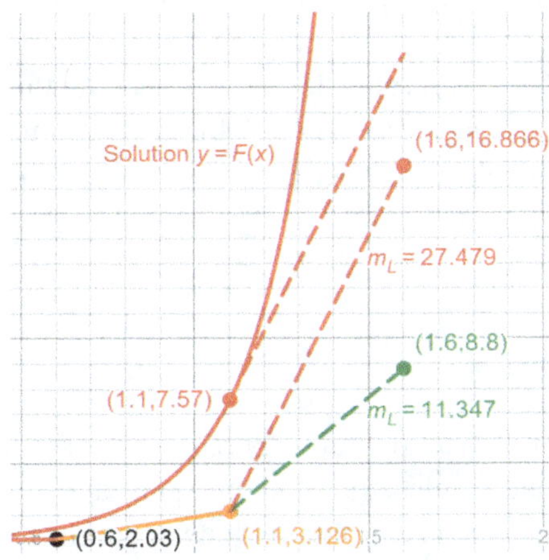

Figure 7.5: As the approximated y value $L(x_2) = 3.126$ strays farther from the true y value $F(x_2) = 7.570$, the disparity between their corresponding derivatives increases. In turn the deviation of the next approximated point at $x_3 = 1.6$ is greater than it was at $x_2 = 1.1$.

The exponential function solution to this particular DEQ and the relatively large distance of $h = 0.5$ directly contribute to the escalating absolute error between the approximated and true y values $L(x_i)$ and $F(x_i)$. Figure 7.6 implies how quickly by comparing the derivative and slope of the approximated and true point at $x_3 = 1.6$, which generates the next approximated point at $x_4 = 2.1$. At the approximated point $(1.6, 8.800)$ the slope is

$$\left.\frac{dy}{dx}\right|_{(1.6,8.800)} = 3(1.6)^2(8.800) = 67.584$$

But at the true point $(1.6, 120.199)$ the slope is

$$\left.\frac{dy}{dx}\right|_{(1.6,120.199)} = 3(1.6)^2(120.199) = 923.128$$

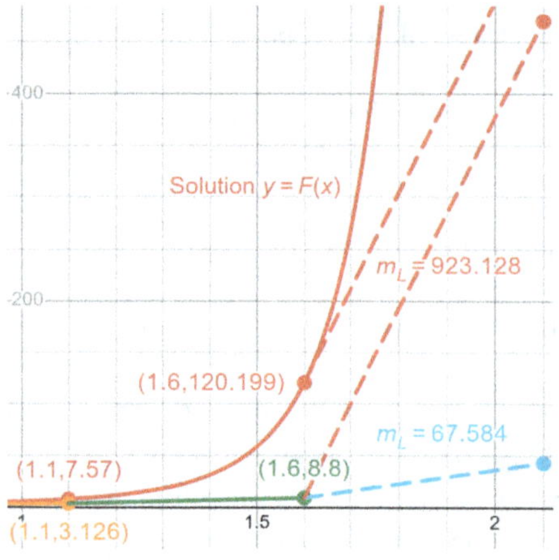

Figure 7.6: The approximations continue to worsen due to the exponential function solution $y = F(x)$ of the DEQ and relatively large $h = 0.5$. After only four approximations, not only are the blue and red points much farther apart than the green and red points in Figure 7.5, but the blue point will be nowhere near the true point on the solution curve $y = F(x)$.

Naturally, the impact of the derivative on the vertical distance between the blue and red points is noticeable.

Remember, in practice DEQs of the form $\frac{dy}{dx} = f(x,y)$ requiring a numerical solution lack the analytic solution $y = F(x)$ which is displayed in Figures 7.4 through 7.6 for comparison purposes. Therefore keeping a small distance h between x_i and x_{i+1} values is imperative.

7.1.3 Euler's method comprehensive template

All DEQ approximation methods in this chapter are computationally reliant on a set of current values which change upon each iteration. As such, they lend to organization by table templates in which each row is an iteration and each column is one of the changing values in the set. Such templates are easily portable to spreadsheet applications or implemented in computer algorithms.

A consistent template is provided for each DEQ approximation method in Sections 7.1 through 7.5, detailing three distinct row types convenient for implementation:
– One header row containing a column description of each changing value
– One row indicating the calculation of each changing value per iteration
– Multiple rows color-coded by the dependencies between columns

Table 7.1 is the template for generating approximation points for a DEQ using Euler's method. The first row describes all the changing values needed for each iteration. The second row describes how to calculate each of the changing values for every iteration. Each row containing three colored cells highlight how the first two are used in finding the third. Thus, the first two yellow cells, containing the current x_i and $L(x_i)$ values, are used for finding the current derivative m_L at the left endpoint of the current interval,

Table 7.1: A DEQ template for Euler's method. In each colored row, the first two colored cells are used in the calculation of the third colored cell.

x_i	$L(x_i)$	m_L	$L(x_{i+1})$
Equation (7.6)[*]	Previous $L(x_{i+1})$[*]	Given DEQ, Eq. (7.8)	Equation (7.9)
x_i	$L(x_i)$	m_L	
	$L(x_i)$	m_L	$L(x_{i+1})$

[*]The coordinates of the given point $(x_0, F(x_0))$ are used initially.

while the first two pink cells, containing the current $L(x_i)$ value and the current derivative m_L at the left endpoint of the current interval, are used to find the next approximated $L(x_{i+1})$ value.

Example 7.1. Use Euler's method to approximate three additional points for the DEQ $\frac{dy}{dx} = -\frac{4\sin x}{y}$ with given initial point $(1.9, 3.227)$ and $h = 0.4$.

Solution. The Table 7.1 template is used to organize each step. The given initial point and the $h = 0.4$ equally spaced x values are inserted easily.

i	x_i	$L(x_i)$	m_L	$L(x_{i+1})$
0	1.9	3.227		
1	2.3			
2	2.7			
3	3.1			

The slope column m_L is evaluated at the current (and in this case, initial) point using the values in the two previous columns. According to the group of yellow cells in Table 7.1, the given DEQ, and Equation (7.8)

$$m_L = \frac{dy}{dx}\bigg|_{(1.9, 3.227)} = -\frac{4\sin 1.9}{3.227} = -1.173$$

The last column $L(x_{i+1})$ uses Equation (7.9), the h value, and the values in the two previous columns, according to the group of pink cells in Table 7.1.

$$L(x_1) = 3.227 + (-1.173)0.4 = 2.758$$

This is the first approximated y value $L(x_1)$, which is carried to the corresponding column in the next row to complete the approximation of the first point.

i	x_i	$L(x_i)$	m_L	$L(x_{i+1})$
0	1.9	3.227	−1.173	2.758
1	2.3	2.758		
2	2.7			
3	3.1			

Repeating the process for row $i = 1$, the slope m_L is found using the current left endpoint $(2.3, 2.758)$.

$$m_L = \frac{dy}{dx}\bigg|_{(2.3, 2.758)} = -\frac{4\sin 2.3}{2.758} = -1.082$$

$L(x_2)$ is found with that derivative value m_L, the current y value $L(x_1)$, and the h value. $L(x_2)$ is then carried to the next row to complete the approximation of the second point.

$$L(x_2) = 2.758 + (-1.082)0.4 = 2.325$$

i	x_i	$L(x_i)$	m_L	$L(x_{i+1})$
0	1.9	3.227	−1.173	2.758
1	2.3	2.758	−1.082	2.325
2	2.7	2.325		
3	3.1			

For row $i = 2$, the current left endpoint is $(2.7, 2.325)$. The calculations for the associated derivative m_L and next approximated y value $L(x_3)$ are

$$m_L = \frac{dy}{dx}\bigg|_{(2.7, 2.325)} = -\frac{4\sin 2.7}{2.325} = -0.735$$

$$L(x_3) = 2.325 + (-0.735)0.4 = 2.031$$

i	x_i	$L(x_i)$	m_L	$L(x_{i+1})$
0	1.9	3.227	−1.173	2.758
1	2.3	2.758	−1.082	2.325
2	2.7	2.325	−0.735	2.031
3	3.1	2.031		

Given the initial point $(1.9, 3.227)$, the three additional approximated points are $(2.3, 2.758)$, $(2.7, 2.325)$, and $(3.1, 2.031)$.

Although Euler's method is used for DEQs that cannot be solved analytically, the DEQ of Example 7.1 is used for comparison purposes. It is separable and has the solution $y = \sqrt{13 + 8\cos x}$. Figure 7.7 displays this solution and the true points in red, and the approximated points in blue. If the goal is approximating a closed-form function solution to the DEQ rather than just approximations at selected points, any of the interpolation techniques of Chapter 3 could be employed, such as fitting a DD polynomial function, Lagrange polynomial function, a quadratic splines piecewise function, or cubic splines piecewise function through the approximated points, or any of the regression functions of Chapter 4 could be fit through the points. Solutions combining and comparing these techniques are explored in Chapter 8.

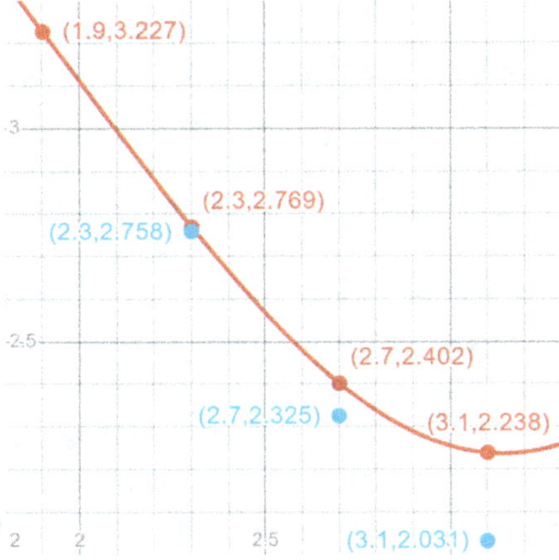

Figure 7.7: The DEQ $\frac{dy}{dx} = -\frac{4\sin x}{y}$ has the true solution $y = \sqrt{13 + 8\cos x}$ in red. The blue points are approximations using the DEQ and Euler's method; the red points are the exact values.

7.2 Midpoint method

One modified Euler's method, the midpoint method, does not use the linearization at $(x_i, L(x_i))$ and the slope at the current left endpoint m_L of Equation (7.8) to approximate $L(x_{i+1})$. Rather, the slope $\frac{dy}{dx}$ at the estimated midpoint of the interval $[x_i, x_{i+1}]$ is employed. An initial estimation of the true y-coordinate of the midpoint is necessary since the DEQ may be a function of both x and y and the only known true y-coordinate is the initial one of $F(x_0)$.

The x-coordinate of the midpoint of the current interval is notated

$$x_{i+\frac{1}{2}} = x_i + \frac{h}{2} \tag{7.10}$$

The y-coordinate of the estimated midpoint uses the derivative and slope at the current point $(x_i, L(x_i))$ like Equation (7.9). But since the location of the midpoint is only half the distance h away from the current point, $h/2$ is substituted in Equation (7.9) to estimate this y-coordinate.

$$L(x_{i+\frac{1}{2}}) = L(x_i) + m_L \frac{h}{2} \tag{7.11}$$

The estimated midpoint $(x_{i+\frac{1}{2}}, L(x_{i+\frac{1}{2}}))$ from Equations (7.10) and (7.11) now is used to estimate the derivative of the unknown true midpoint. The approximated tangent line slope for the midpoint method is

$$m_{\text{mid}} = \left.\frac{dy}{dx}\right|_{(x_{i+\frac{1}{2}}, L(x_{i+\frac{1}{2}}))} \tag{7.12}$$

Lastly, the tangent line slope of Equation (7.12) is applied to the starting point of the current interval $(x_i, L(x_i))$ and thus a full distance h is used to ultimately approximate the next y-coordinate $L(x_{i+1})$ using the midpoint method.

$$L(x_{i+1}) = L(x_i) + m_{\mathrm{mid}}h \tag{7.13}$$

Figures 7.8 and 7.9 illustrate the above process which defines the midpoint method. In Figure 7.8, the black dashed line is generated, using the derivative and slope at the current point $(x_i, L(x_i))$ to estimate the y-coordinate $L(x_{i+\frac{1}{2}})$ at the midpoint of the current interval, at a distance of half the interval length at $x_i + h/2$. This estimated midpoint in blue is used in lieu of the unknown true midpoint in red to approximate the derivative and slope of Equation (7.12), needed to locate the next approximated point. The slope of the red dashed tangent line at the true midpoint and the slope of the blue dashed line at the estimated midpoint are not equal and the lines are not quite parallel.

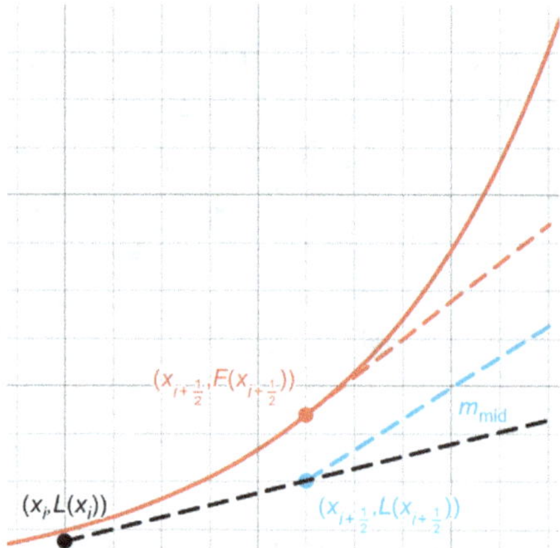

Figure 7.8: Using the midpoint method, the derivative and slope based on the current point generates the black dashed line, used to estimate the blue midpoint. The blue dashed line has a slope based on this midpoint, an estimation of the slope of the red dashed line at the true red midpoint.

In Figure 7.9, the slope at the estimated midpoint is used to connect the current point at $(x_i, L(x_i))$ to the next approximated point in blue at $(x_{i+1}, L(x_{i+1}))$ by Equation (7.13). Both blue lines are parallel. If the black dashed line, used to estimate the midpoint, is extended a full distance h from x_i, the resulting black point represents the next approximation using Euler's method. The true point at x_{i+1} is in red. In this example, the difference between the y value $L(x_{i+1})$ using Euler's method and the true y value $F(x_{i+1})$ is a little more than twice the difference between the y value $L(x_{i+1})$ using the midpoint method and the true y value $F(x_{i+1})$.

Roughly analogous to the derivative methods of Chapter 6, Euler's method is akin to a forward divided difference, while the midpoint method resembles a centered divided difference, further implying the midpoint method is an improvement.

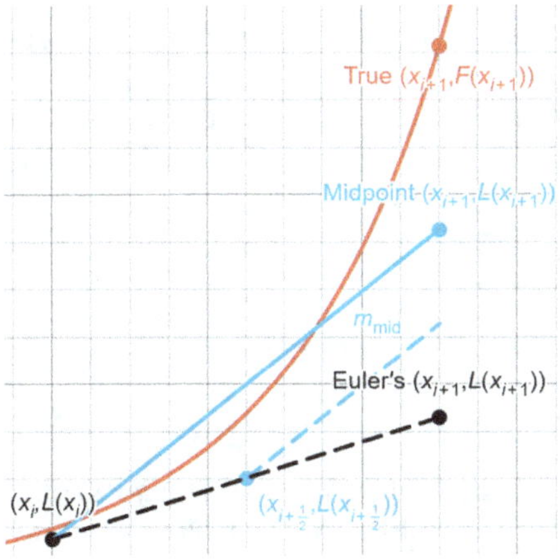

Figure 7.9: The slope of the blue dashed line is applied to the current point at x_i, forming the blue solid line that approximates the next blue point at x_{i+1} using the midpoint method. Using only the black dashed line shows the location of the next point in black using Euler's method. The red point is the true value on the true solution of the DEQ.

Equations (7.10) through (7.13), along with Equations (7.6) and (7.8), constitute the midpoint method. Implied is the reasonable assumption that for a sufficiently small interval distance h, the tangent line slope $\frac{dy}{dx}$ at the true midpoint is itself a close approximation of the secant line slope through the true endpoints of the current interval at $(x_i, F(x_i))$ and $(x_{i+1}, F(x_{i+1}))$.

Midpoint method comprehensive template

Like Table 7.1 for Euler's method, Table 7.2 is a template for generating approximation points for a DEQ using the midpoint method. Recall each row containing three colored cells indicates how the first two are used in finding the third. For instance, the first two green cells highlighting the current $L(x_i)$ value and derivative m_L at the current left endpoint are used to find the third green cell, highlighting the estimated y-coordinate of the midpoint $L(x_{i+\frac{1}{2}})$

Table 7.2: A DEQ template for the midpoint method. In each colored row, the first two colored cells are used in the calculation of the third colored cell.

x_i	$L(x_i)$	m_L	$x_{i+\frac{1}{2}}$	$L(x_{i+\frac{1}{2}})$	m_{mid}	$L(x_{i+1})$
Eq. (7.6)[*]	Previous $L(x_{i+1})$[*]	DEQ, Eq. (7.8)	Eq. (7.10)	Eq. (7.11)	DEQ, Eq. (7.12)	Eq. (7.13)
x_i	$L(x_i)$	m_L				
	$L(x_i)$	m_L		$L(x_{i+\frac{1}{2}})$		
			$x_{i+\frac{1}{2}}$	$L(x_{i+\frac{1}{2}})$	m_{mid}	
	$L(x_i)$				m_{mid}	$L(x_{i+1})$

[*]The coordinates of the given point $(x_0, F(x_0))$ are used initially.

Example 7.2. Use the midpoint method to approximate two additional points for the DEQ $\frac{dy}{dx} = \frac{3x^2}{y}$ with given initial point $(-0.7, 0.560)$ and $h = 0.3$.

Solution. The Table 7.2 template is used to organize each step. The initial point and the two additional points spaced $h = 0.3$ apart are inserted.

i	x_i	$L(x_i)$	m_L	$x_{i+\frac{1}{2}}$	$L(x_{i+\frac{1}{2}})$	m_{mid}	$L(x_{i+1})$
0	-0.7	0.560					
1	-0.4						
2	-0.1						

The first two yellow cells in Table 7.2 are used to find the value of the third yellow cell, the derivative at the current point, which the second row indicates is calculated with the given DEQ and Equation (7.8).

$$m_L = \left.\frac{dy}{dx}\right|_{(-0.7,0.56)} = \frac{3(-0.7)^2}{0.56} = 2.625$$

The first two three green cells in Table 7.2 are used to find the value of the third green cell, the estimated y-coordinate of the midpoint, and by the second row is found using Equation (7.11).

$$L(x_{0+\frac{1}{2}}) = L(-0.7) + m_L\frac{0.3}{2}$$

$$L(x_{\frac{1}{2}}) = 0.56 + (2.625)0.15 = 0.954$$

The first blue cell in Table 7.2 is the x value of the midpoint of the current interval. Using Equation (7.10)

$$x_{0+\frac{1}{2}} = x_{\frac{1}{2}} = x_0 + \frac{h}{2} = -0.7 + \frac{0.3}{2} = -0.55$$

Along with the second blue cell in Table 7.2, the value of the third blue cell, the derivative at the estimated midpoint, is calculated once again with the given DEQ and Equation (7.12), as noted in the second row.

$$m_{\mathrm{mid}} = \left.\frac{dy}{dx}\right|_{(x_{\frac{1}{2}},L(x_{\frac{1}{2}}))} = \left.\frac{dy}{dx}\right|_{(-0.55,0.954)} = \frac{3(-0.55)^2}{0.954} = 0.951$$

Lastly, the first two pink cells in Table 7.2 are used to find the value of the third pink cell, the y value of the next approximated point. In the second row, this is found using Equation (7.13).

$$L(x_{0+1}) = L(-0.7) + m_{\mathrm{mid}}h$$

$$L(x_1) = 0.56 + (0.951)0.3 = 0.845$$

This $L(x_1)$ value is carried to the next row and paired with the x_1 value, so that the first approximated point is $(x_1, L(x_1)) = (-0.4, 0.845)$. The current state of the table is

i	x_i	$L(x_i)$	m_L	$x_{i+\frac{1}{2}}$	$L(x_{i+\frac{1}{2}})$	m_{mid}	$L(x_{i+1})$
0	-0.7	0.560	2.625	-0.55	0.954	0.951	0.845
1	-0.4	0.845					
2	-0.1						

The process repeats for the second approximated point.

$$m_L = \left.\frac{dy}{dx}\right|_{(x_1, L(x_1))} = \left.\frac{dy}{dx}\right|_{(-0.4, 0.845)} = \frac{3(-0.4)^2}{0.845} = \boxed{0.568}$$

$$L(x_{\frac{3}{2}}) = L(x_1) + m_L\frac{h}{2} = 0.845 + (0.568)0.15 = \boxed{0.930}$$

$$x_{\frac{3}{2}} = x_1 + \frac{h}{2} = -0.4 + \frac{0.3}{2} = -0.25$$

$$m_{mid} = \left.\frac{dy}{dx}\right|_{(x_{\frac{3}{2}}, L(x_{\frac{3}{2}}))} = \left.\frac{dy}{dx}\right|_{(-0.25, 0.930)} = \frac{3(-0.25)^2}{0.930} = \boxed{0.202}$$

$$L(x_2) = L(x_1) + m_{mid}h = 0.845 + (0.202)0.3 = \boxed{0.906}$$

With $x_2 = -0.1$, the second approximated point is $(x_2, L(x_2)) = (-0.1, 0.906)$. Inserting the above values, the final state of the table is

i	x_i	$L(x_i)$	m_L	$x_{i+\frac{1}{2}}$	$L(x_{i+\frac{1}{2}})$	m_{mid}	$L(x_{i+1})$
0	-0.7	0.560	2.625	-0.55	0.954	0.952	0.845
1	-0.4	0.845	0.568	-0.25	0.930	0.202	0.906
2	-0.1	0.906					

Like Example 7.1, for comparison purposes the DEQ in Example 7.2 is separable and the analytic solution is $y = \sqrt{2x^3 + 1}$. Figure 7.10 shows this true solution in red. The two approximated points $(-0.4, 0.845)$ and $(-0.1, 0.906)$ in blue are found using the derivative or slope of the estimated midpoints at $x_{\frac{1}{2}}$ and $x_{\frac{3}{2}}$, the blue open circles, and linearizing by the blue dashed lines. The red dashed lines have the true tangent line slope at the true midpoints, the red open circles. The red and blue dashed lines at the same x values are not quite parallel because the y values also contribute to the value of the derivative or slope, and one uses the true value $y = F(x_i)$, while the other relies on an approximation $y = L(x_i)$.

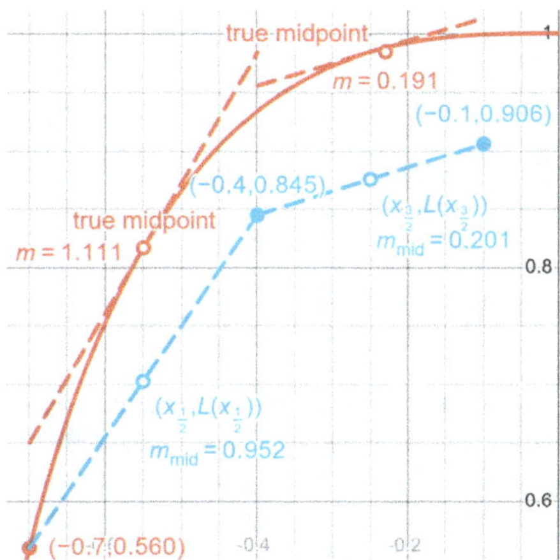

Figure 7.10: The solution to the DEQ $\frac{dy}{dx} = \frac{3x^2}{y}$ of Example 7.2 is $y = \sqrt{x^3 + 1}$ in red. The blue points are approximations using the midpoint method. The blue open circles are the estimated midpoints, where the derivative or slope is found and used to generate the blue dashed lines. The red dashed tangent lines at the true midpoints (red open circles) are close but not parallel to the blue dashed lines.

7.3 Heun's method

Another modified Euler's method similar to the midpoint method is Heun's method. Instead of estimating the y-coordinate of the midpoint of the interval $[x_i, x_{i+1}]$, the y-coordinate at the right endpoint x_{i+1} is estimated. The mean of the tangent line slopes at both endpoints of the interval is used in the linearization at $(x_i, L(x_i))$ to approximate the next point $(x_{i+1}, L(x_{i+1}))$.

The estimated y-coordinate of the right endpoint of the current interval uses the standard linearization of Equation (7.9). It is defined $L_p(x_{i+1})$, emphasizing it is only a temporary predicted y-coordinate at x_{i+1} and different from the final approximated y-coordinate $L(x_{i+1})$.

$$L_p(x_{i+1}) = L(x_i) + m_L h \tag{7.14}$$

The derivative and tangent line slope at the right endpoint of the current interval at x_{i+1} uses the predicted y-coordinate $L_p(x_{i+1})$ of Equation (7.14) and is defined

$$m_R = \frac{dy}{dx}\bigg|_{(x_{i+1}, L_p(x_{i+1}))} \tag{7.15}$$

Figure 7.11 displays the above beginning steps of Heun's method. The black dashed tangent line approximation is the standard Euler's method linearization from the current point $(x_i, L(x_i))$ to the next point, and provides the slope at the left endpoint m_L. The next point in blue has the relabeled y-coordinate $L_p(x_{i+1})$ indicating it is only a temporary predicted value. Using the DEQ, the slope of the dashed line at the predicted right endpoint $(x_{i+1}, L_p(x_{i+1}))$, both in blue, estimates the slope of the dashed line at the true right endpoint $(x_{i+1}, F(x_{i+1}))$, both in red.

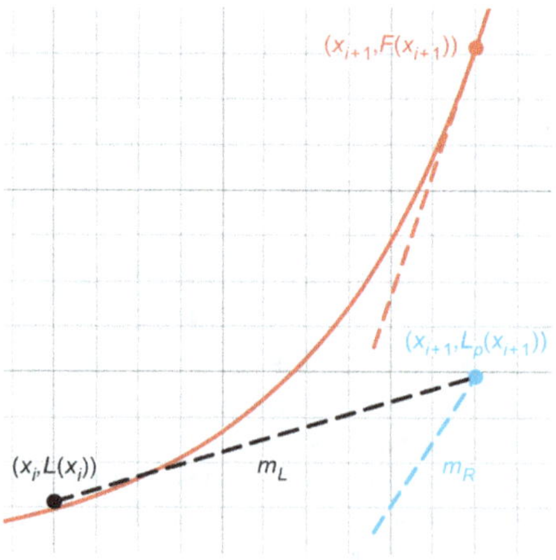

Figure 7.11: Heun's method uses the standard linearization of Euler's method, with the black dashed line applied to find the temporary next point in blue. The blue point and blue dashed line approximate the red point and red dashed line.

Next, the tangent line slope for Heun's method is the mean of the tangent line slope at each endpoint, found using Equations (7.8) and (7.15).

$$m_{\text{Heu}} = \overline{m} = \frac{m_L + m_R}{2} \tag{7.16}$$

The next approximated y-coordinate $L(x_{i+1})$ using Heun's method is found with the mean slope of Equation (7.16), linearizing at the current point at x_i to the next equally spaced point at x_{i+1}, using the full interval length h.

$$L(x_{i+1}) = L(x_i) + m_{\text{Heu}}h \tag{7.17}$$

The remaining steps of Heun's method are illustrated in Figure 7.12. From Figure 7.11, the black and blue dashed lines use the tangent line slope approximations at the left and right endpoints, respectively. The mean of these slopes is slope of the parallel green solid lines, a compromise graphed between the black and blue dashed lines. The linearization at the current point at x_i applies the green dashed line to approximate the next point at x_{i+1} in green.

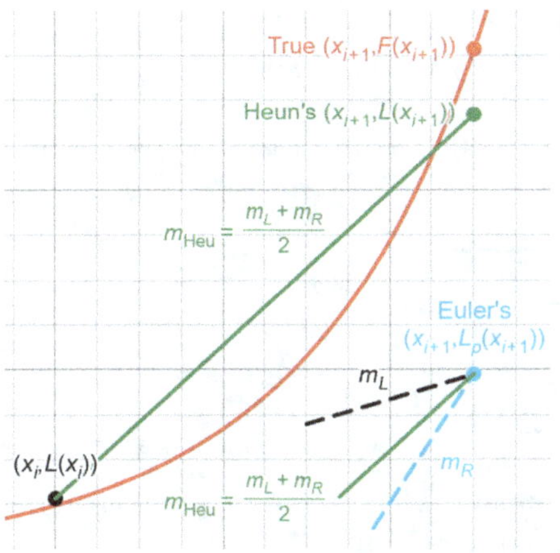

Figure 7.12: The slope of Heun's method is the mean of the slope at the left and right endpoint, the slope of the green solid lines. Graphed between the black and blue dashed lines depicting the compromise, the slope of the green solid line is applied to the current point to approximate the next point in green. The temporary point of Heun's method in blue is equivalent to the actual next point of Euler's method.

Note the temporary point $(x_{i+1}, L_p(x_{i+1}))$ in blue is equivalent to the next approximated point of Euler's method. In this case, the point produced by Euler's method is noticeably farther from the true next point in red than the next point generated by Heun's method in green is to the red point.

Heun's method comprehensive template

Equations (7.6), (7.8), and (7.14) through (7.17) define Heun's method algebraically. Table 7.3 is a template for generating the approximation points of the solution to a DEQ

Table 7.3: A DEQ template for Heun's method. In each colored row, the first two colored cells are used in the calculation of the third colored cell.

x_i	$L(x_i)$	m_L	x_{i+1}	$L_p(x_{i+1})$	m_R	m_{Heu}	$L(x_{i+1})$
Eq. (7.6)*	Previous $L(x_{i+1})^*$	DEQ, Eq. (7.8)	Eq. (7.6)	Eq. (7.14)	DEQ, Eq. (7.15)	Eq. (7.16)	Eq. (7.17)
x_i	$L(x_i)$	m_L					
	$L(x_i)$	m_L		$L_p(x_{i+1})$			
			x_{i+1}	$L_p(x_{i+1})$	m_R		
		m_L			m_R	m_{Heu}	
	$L(x_i)$					m_{Heu}	$L(x_{i+1})$

*The coordinates of the given point $(x_0, F(x_0))$ are used initially.

using Heun's method. Like the previous templates of Tables 7.1 and 7.2, each row containing three colored cells indicate how the first two are used in finding the third. For instance, the first two blue cells highlighting the next x-coordinate x_{i+1} and its associated temporary predicted y-coordinate $L_p(x_{i+1})$ are used to find the third blue cell, highlighting the derivative or slope m_R at the right endpoint of the current interval.

Example 7.3. Use Heun's method to approximate two additional points for the DEQ $\frac{dy}{dx} = 2xy$ with given initial point $(1.0, 1.087)$ and $h = 0.4$.

Solution. Using Table 7.3 to organize each step, the initial point and two additional points spaced $h = 0.4$ apart are inserted.

i	x_i	$L(x_i)$	m_L	x_{i+1}	$L_p(x_{i+1})$	m_R	m_{Heu}	$L(x_{i+1})$
0	1.0	1.087						
1	1.4							
2	1.8							

The first two yellow cells in Table 7.3 are used to find the value of the third yellow cell, the derivative at the current left endpoint, which Table 7.3 states is calculated with the given DEQ and Equation (7.8).

$$m_L = \left.\frac{dy}{dx}\right|_{(1.0,1.087)} = 2(\,1.0\,)(\,1.087\,) = 2.174$$

The first two three green cells in Table 7.3 are used to find the value of the third green cell, the temporary predicted y-coordinate at the right endpoint of the current interval. By the second row of Table 7.3, using Equation (7.14)

$$L_p(x_{0+1}) = L(x_0) + m_L h$$

$$L_p(x_1) = \boxed{1.087} + (\,\boxed{2.174}\,)0.4 = \boxed{1.957}$$

The first blue cell in Table 7.3 is the x value of the right endpoint of the current interval. Using Equation (7.6)

$$x_{0+1} = x_1 = x_0 + h = 1.0 + 0.4 = 1.4$$

Along with the second blue cell in Table 7.3, the value of the third blue cell, the derivative at the predicted right endpoint, is found. Calculated with the given DEQ and Equation (7.15) by the second row of Table 7.3

$$m_R = \left.\frac{dy}{dx}\right|_{(1.4,1.957)} = 2(1.4)(1.957) = 5.480$$

The first two magenta cells in Table 7.3 are used to find the slope of the third magenta cell, the mean tangent line slope used in Heun's method. Using Equation (7.16)

$$m_{Heu} = \frac{2.174 + 5.480}{2} = 3.827$$

The final y-coordinate approximation of the third pink cell in Table 7.3 uses the first two pink cells and Equation (7.17).

$$L(x_1) = L(x_0) + m_{Heu}h = L(1.0) + (3.827)0.4$$

$$L(1.4) = 1.087 + (3.827)0.4 = 2.618$$

This value is carried to the next row in the $L(x_i)$ column, and with the x_i value for $i = 1$ the first approximated point is $(x_1, L(x_1)) = (1.4, 2.618)$. All above calculated values are inserted in the table.

i	x_i	$L(x_i)$	m_L	x_{i+1}	$L_p(x_{i+1})$	m_R	m_{Heu}	$L(x_{i+1})$
0	1.0	1.087	2.174	1.4	1.957	5.480	3.827	2.618
1	1.4	2.618						
2	1.8							

All above steps repeat for the second approximated point.

$$m_L = \left.\frac{dy}{dx}\right|_{(1.4,2.618)} = 2(1.4)(2.618) = 7.330$$

$$L_p(x_{1+1}) = L(x_1) + m_L h$$

$$L_p(x_2) = L(1.4) + (7.330)0.4 = 2.618 + (7.330)0.4 = 5.550$$

$$x_2 = x_1 + h = 1.4 + 0.4 = 1.8$$

$$m_R = \left.\frac{dy}{dx}\right|_{(1.8,5.550)} = 2(1.8)(5.550) = 19.980$$

$$m_{Heu} = \frac{7.330 + 19.980}{2} = 13.655$$

$$L(x_2) = L(x_1) + m_{Heu}h = L(1.4) + (13.655)0.4$$

$$L(1.8) = 2.618 + (13.655)0.4 = 8.080$$

With $x_2 = 1.8$ the second approximated point is $(x_2, L(x_2)) = (1.8, 8.080)$. The final table is

i	x_i	$L(x_i)$	m_L	x_{i+1}	$L_p(x_{i+1})$	m_R	m_{Heu}	$L(x_{i+1})$
0	1.0	1.087	2.174	1.4	1.957	5.480	3.827	2.618
1	1.4	2.618	7.330	1.8	5.550	19.980	13.655	8.080
2	1.8	8.080						

7.4 Ralston's method

For interval $[x_i, x_{i+1}]$, the midpoint method uses the tangent line slope m_{mid}, Equation (7.12), at the estimated midpoint to generate the next approximated point. Heun's method uses an evenly weighted average slope m_{Heu}, Equation (7.16), of the mean of the left endpoint and the predicted right endpoint of the interval, to approximate the next point. Ralston's method combines both ideas.

Instead of employing the midpoint and its x-coordinate located half an interval length from x_i of the left endpoint of the current interval, the point located at the x-coordinate three-quarters an interval length from x_i is used (the seemingly arbitrary three-quarter point is justified after Theorem 7.3 in Section 7.5.1). Analogous to Equation (7.10), this x-coordinate is defined

$$x_{i+\frac{3}{4}} = x_i + \frac{3h}{4} \tag{7.18}$$

The estimated y-coordinate of the three-quarter point uses the linearization at the current left endpoint and the value of the derivative there, similar to Equation (7.9). But just as the estimation of the y-coordinate of the midpoint of Equation (7.11) replaced the full interval length of h with $\frac{h}{2}$, in this case $\frac{3h}{4}$ replaces h. The corresponding altered linearization of Equations (7.9) and (7.11) is

$$L(x_{i+\frac{3}{4}}) = L(x_i) + m_L \frac{3h}{4} \tag{7.19}$$

The derivative or tangent line slope at the three-quarter point is denoted

$$m_{3/4} = \left.\frac{dy}{dx}\right|_{(x_{i+\frac{3}{4}},\, L(x_{i+\frac{3}{4}}))} \tag{7.20}$$

Figure 7.13 illustrates how the black dashed line, using Equation (7.19), estimates the y-coordinate of the blue point, located three-quarters of the length of the interval from the x-coordinate of the left endpoint. The slope at this estimated three-quarter point estimates the slope at the true three-quarter point on the curve, represented by the slope of the red dashed line. The blue and red dashed lines are almost parallel.

Rather than constructing an equally weighted mean slope using the slope at the left and right endpoint, as for m_{Heu} by Equation (7.16), an unequally weighted average slope defines the tangent line slope for Ralston's method. The slope at the left endpoint is weighted once and the slope at the three-quarter point is weighted twice; justification is provided after Theorem 7.3 in Section 7.5.1.

$$m_{\mathrm{Ral}} = \frac{m_L + 2m_{3/4}}{3} \tag{7.21}$$

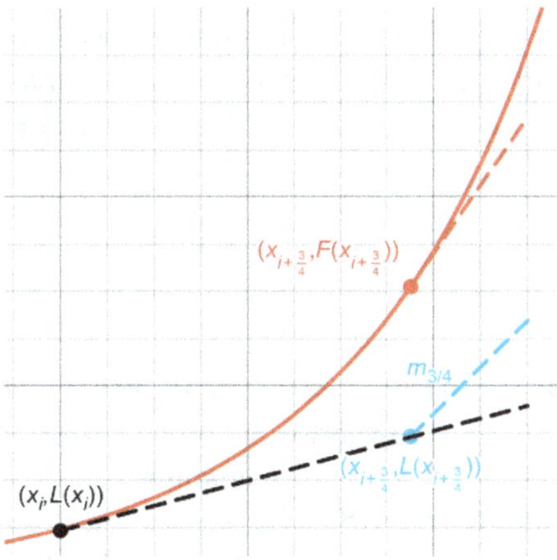

Figure 7.13: Using Ralston's method, the point at three-quarters of the interval length from the current black point is estimated and a slope is generated, represented by the blue dashed line. The red dashed line is the slope at the true point three-quarters of the interval length from the current point.

Now, the next y-coordinate $L(x_{i+1})$ using Ralston's method can be found by the slope defined by Equation (7.21), applied to a full interval length h from the current point at x_i.

$$L(x_{i+1}) = L(x_i) + m_{\mathrm{Ral}}h \tag{7.22}$$

Figure 7.14 depicts the final steps of Ralston's method. The blue dashed line has the slope at the estimated three-quarter point, duplicated from Figure 7.13. The orange dashed line has a slope two times the slope of the blue dashed line, and the black dashed line has the slope at the current left endpoint. Per Equation (7.21), the green solid line between the black and orange dashed lines has the corresponding weighted slope of Ralston's method, and is applied to the current black point to determine the next approximated point in green.

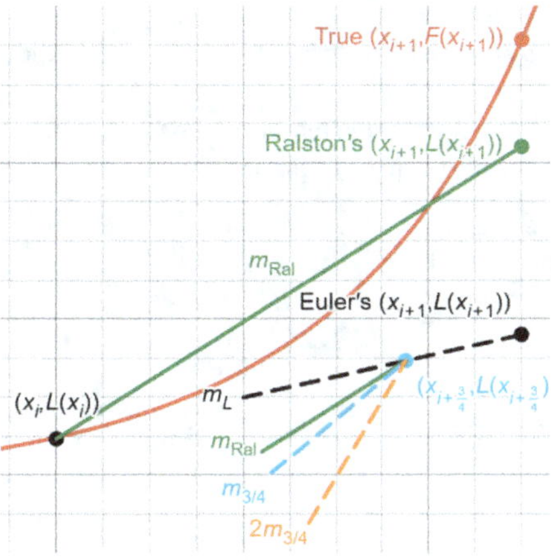

Figure 7.14: The slope of the black dashed line of the current point, and the slope of the orange dashed line, twice the slope of the blue dashed line of the three-quarter point, are the weighted components for the slope of the green solid line used in Ralston's method. The green point at x_{i+1} is from Ralston's method while the black point at x_{i+1} is from Euler's method; the true point at x_{i+1} is in red.

Note that if the black dashed line is extended to the x_{i+1} value, it terminates at the next approximated point in black using Euler's method. In this instance, the approximation of Ralston's method is noticeably closer to the true point in red than the approximation of Euler's method.

Ralston's method comprehensive template

Equations (7.6), (7.8), and (7.18) through (7.22) algebraically constitute Ralston's method. The companion Table 7.4 is a template for generating the approximation points of the solution, and like the templates of Tables 7.1 through 7.3, each row containing three colored cells indicate how the first two are used in finding the third. For instance, the first two magenta cells containing the slope m_L at the current left endpoint and the slope $m_{3/4}$ at the three-quarter point are used to calculate the weighted slope m_{Ral} of Ralston's method, the third magenta cell.

Table 7.4: A DEQ template for Ralston's method. In each colored row, the first two colored cells are used in the calculation of the third colored cell.

x_i	$L(x_i)$	m_L	$x_{i+\frac{3}{4}}$	$L(x_{i+\frac{3}{4}})$	$m_{3/4}$	m_{Ral}	$L(x_{i+1})$
Eq. (7.6)[*]	Previous $L(x_{i+1})$[*]	DEQ, Eq. (7.8)	Eq. (7.18)	Eq. (7.19)	DEQ, Eq. (7.20)	Eq. (7.21)	Eq. (7.22)
x_i	$L(x_i)$	m_L					
	$L(x_i)$	m_L		$L(x_{i+\frac{3}{4}})$			
			$x_{i+\frac{3}{4}}$	$L(x_{i+\frac{3}{4}})$	$m_{3/4}$		
		m_L			$m_{3/4}$	m_{Ral}	
	$L(x_i)$					m_{Ral}	$L(x_{i+1})$

[*]The coordinates of the given point $(x_0, F(x_0))$ are used initially.

Example 7.4. Use Ralston's method to approximate two additional points for the DEQ $\frac{dy}{dx} = -\frac{2x}{e^y}$ with given initial point $(2.0, 3.332)$ and $h = 1.5$.

Solution. The colored cells and equations of the Table 7.4 template are referenced throughout. Inserting the initial point and two $h = 1.5$ equally spaced x values into the template begins the process.

i	x_i	$L(x_i)$	m_L	$x_{i+\frac{3}{4}}$	$L(x_{i+\frac{3}{4}})$	$m_{3/4}$	m_{Ral}	$L(x_{i+1})$
0	2.0	3.332						
1	3.5							
2	5.0							

Common to the previous methods, the first two yellow cells, the given DEQ, and Equation (7.8) determine the slope at the current left endpoint of the third yellow cell.

$$m_L = \left.\frac{dy}{dx}\right|_{(2.0,3.332)} = -\frac{2(\,2.0\,)}{e^{\,3.332}} = \boxed{-0.143}$$

The first two green cells containing the current left endpoint and slope at that endpoint are used to find the third green cell, the y-coordinate at the estimated three-quarter point. By Equation (7.19)

$$L(x_{0+\frac{3}{4}}) = L(x_0) + m_L\frac{3h}{4} = L(2.0) + (-0.143)\frac{3(1.5)}{4}$$

$$L(x_{\frac{3}{4}}) = \boxed{3.332} + (\,\boxed{-0.143}\,)1.125 = \boxed{3.171}$$

The first blue call is the corresponding x value of the three-quarter point, and by Equation (7.18)

$$x_{0+\frac{3}{4}} = x_{\frac{3}{4}} = x_0 + \frac{3h}{4} = 2.0 + \frac{3(1.5)}{4} = 3.125$$

Along with the second blue cell, the slope at the estimated three-quarter point of the third blue cell is calculated using the given DEQ with Equation (7.20).

$$m_{3/4} = \left.\frac{dy}{dx}\right|_{(3.125,3.171)} = -\frac{2(\,3.125\,)}{e^{\,3.171}} = \boxed{-0.262}$$

The weighted slope of Ralston's method in the third magenta cell is found with the slope values at the left endpoint and three-quarter point from the first two magenta cells. Using Equation (7.21)

$$m_{\text{Ral}} = \frac{m_L + 2m_{3/4}}{3} = \frac{\boxed{-0.143} + 2(\,\boxed{-0.262}\,)}{3} = \boxed{-0.222}$$

Lastly, using the first two pink cells and Equation (7.22) solves the y-coordinate of the approximated point of the third pink cell.

$$L(x_1) = L(x_0) + m_{\text{Ral}}h = L(2.0) + (-0.222)1.5$$

$$L(3.5) = \boxed{3.332} + (\,\boxed{-0.222}\,)1.5 = \boxed{2.999}$$

Carrying this y value to the next $i = 1$ row, the first approximated point is $(x_1, L(x_1)) = (3.5, 2.999)$. The table template after the first iteration follows.

i	x_i	$L(x_i)$	m_L	$x_{i+\frac{3}{4}}$	$L(x_{i+\frac{3}{4}})$	$m_{3/4}$	m_{Ral}	$L(x_{i+1})$
0	2.0	3.332	−0.143	3.125	3.171	−0.262	−0.222	2.999
1	3.5	2.999						
2	5.0							

All above steps repeat for the second approximated point.

$$m_L = \left.\frac{dy}{dx}\right|_{(3.5,2.999)} = -\frac{2(3.5)}{e^{2.999}} = \boxed{-0.349}$$

$$L(x_{1+\frac{3}{4}}) = L(x_1) + m_L\frac{3h}{4}$$

$$L(x_{\frac{3}{4}}) = L(3.5) + (-0.349)\frac{3(1.5)}{4} = 2.999 + (-0.349)1.125 = \boxed{2.606}$$

$$x_{\frac{7}{4}} = x_1 + \frac{3h}{4} = 3.5 + \frac{3(1.5)}{4} = 4.625$$

$$m_{3/4} = \frac{dy}{dx}\bigg|_{(4.625,2.606)} = -\frac{2(4.625)}{e^{2.606}} = \boxed{-0.683}$$

$$m_{\text{Ral}} = \frac{m_L + 2m_{3/4}}{3} = \frac{-0.349 + 2(-0.683)}{3} = \boxed{-0.572}$$

$$L(x_2) = L(x_1) + m_{\text{Ral}}h = L(3.5) + (-0.572)1.5$$

$$L(5.0) = 2.999 + (-0.572)1.5 = \boxed{2.141}$$

Thus the second approximated point is $(x_2, L(x_2)) = (5.0, 2.141)$ for the final table.

i	x_i	$L(x_i)$	m_L	$x_{i+\frac{3}{4}}$	$L(x_{i+\frac{3}{4}})$	$m_{3/4}$	m_{Ral}	$L(x_{i+1})$
0	2.0	3.332	−0.143	3.125	3.171	−0.262	−0.222	2.999
1	3.5	2.999	−0.349	4.625	2.606	−0.683	−0.572	2.141
2	5.0	2.141						

Applying the tangent line slope at the estimated midpoint (Section 7.2) or the mean of the tangent line slopes at the estimated endpoints (Section 7.3) to approximate the next point using linearization intuitively makes sense. However, as mentioned at the outset, using the tangent line slope at the estimated three-quarter point and counting it twice the value of the tangent line slope at the current left endpoint in a weighted average seems arbitrary. This is addressed in the following section.

7.5 Runge–Kutta methods

The midpoint method, Heun's method, and Ralston's method of Sections 7.2 through 7.4 are all particular cases of a general type called Runge–Kutta (RK) methods. Specifically, all three are second-order methods because they are derived from a second-degree Taylor polynomial. The general calculation of the next y-coordinate using a second-order RK method is

$$L(x_{i+1}) = L(x_i) + m_{2\text{RK}}h \tag{7.23}$$

This resembles Euler's method of Equation (7.9).

$$L(x_{i+1}) = L(x_i) + m_L h$$

However, instead of simply using the slope m_L at the current left endpoint, m_L is weighted with a second slope value at an additional estimated point in the current interval. Let p equal the fraction of the interval length h. The x-coordinate of the additional point is then $x_i + ph$ and its corresponding estimated y-coordinate is $L(x_i) + m_L ph$. The second slope value is calculated at this point.

$$m_2 = \left.\frac{dy}{dx}\right|_{(x_i + ph, L(x_i) + m_L ph)} \tag{7.24}$$

For constant weighting factors w_1 and w_2 where $w_1 + w_2 = 1$, $0 < w_2 \leq 1$, and the condition $p = \frac{1}{2w_2}$, any valid tangent line slope for a second-order RK method of Equation (7.23) is defined

$$m_{2RK} = w_1 m_L + w_2 m_2 \tag{7.25}$$

The comprehensive theory behind RK methods, their derivations and error analysis, comprise entire texts themselves and are beyond the scope of this textbook. However, Equations (7.23) and (7.25) are offered for the motivations behind the general nth order RK method and to contextualize the methods described in Sections 7.2 through 7.4.

The slope m_{2RK} of Equation (7.25) is recursive. By Equation (7.24), the derivative factor m_2 of the w_2 term is calculated at a point using the derivative factor m_L of the previous term w_1. By extension a third-order RK method has a w_3 term that calculates its derivative factor m_3 using the derivative factor m_2 of the previous term w_2. In general, an nth order RK method has a w_n term that calculates its derivative factor m_n using the derivative factor m_{n-1} of the previous term w_{n-1}. Some of the $w_1, \ldots, w_{n-1}$ terms may equal zero, but $w_1 + w_2 + \cdots + w_n = 1$, which has infinite solutions so that there are infinite RK methods of any order. The challenge is finding a set of $w_1, \ldots, w_n$ values that optimizes the approximation of $L(x_{i+1})$.

The salient feature of RK methods is the nth-degree Taylor polynomial used to generate them includes terms with derivatives up to and including $\frac{d^n y}{dx^n}$. But the corresponding RK methods only ever use first derivatives $\frac{dy}{dx}$, and therein lies their value.

7.5.1 Particular second-order RK methods

Consider the approximation of the next y-coordinate using the midpoint method, Heun's method, and Ralston's method. According to Equations (7.13), (7.17), and (7.22), respectively, each is found by

$$L(x_{i+1}) = L(x_i) + m_{\text{mid}} h$$
$$L(x_{i+1}) = L(x_i) + m_{\text{Heu}} h$$
$$L(x_{i+1}) = L(x_i) + m_{\text{Ral}} h$$

Note how each definition matches the general form of a second-order RK method of Equation (7.23).

$$L(x_{i+1}) = L(x_i) + m_{2RK} h$$

The following three theorems respectively prove m_{mid}, m_{Heu}, and m_{Ral} satisfy the definition of m_{2RK} for a specific value of w_2 with the requirements $w_1 + w_2 = 1$ and $p = \frac{1}{2w_2}$. Therefore, each is a particular instance of a second-order RK method.

Theorem 7.1. *The midpoint method is a second-order RK method since m_{mid} satisfies the definition of m_{2RK}.*

Proof. Suppose $w_2 = 1$ so that

$$w_1 + w_2 = 1 \implies w_1 = 0$$

Satisfying m_{2RK} of Equation (7.25) means

$$m_{\text{2RK}} = w_1 m_L + w_2 m_2$$
$$= 0 m_L + 1 m_2$$
$$= m_2$$

Also, for $w_2 = 1$,

$$p = \frac{1}{2w_2} = \frac{1}{2(1)} = \frac{1}{2}$$

Substituting into Equation (7.24)

$$m_2 = \left. \frac{dy}{dx} \right|_{(x_i+ph,L(x_i)+m_L ph)} = \left. \frac{dy}{dx} \right|_{(x_i+\frac{h}{2},L(x_i)+m_L \frac{h}{2})}$$

The coordinates of the point where the derivative is evaluated can be rewritten according to Equations (7.10) and (7.11).

$$x + \frac{h}{2} = x_{i+\frac{1}{2}}$$
$$L(x_i) + m_L \frac{h}{2} = L(x_{i+\frac{1}{2}})$$

Substituting yields the derivative

$$m_2 = \left. \frac{dy}{dx} \right|_{(x_{i+\frac{1}{2}},L(x_{i+\frac{1}{2}}))}$$

This is precisely the slope defined by Equation (7.12) as m_{mid}, used in the midpoint method of Equation (7.13), which completes the proof.

$$m_{\text{2RK}} = m_2 = \left. \frac{dy}{dx} \right|_{(x_{i+\frac{1}{2}},L(x_{i+\frac{1}{2}}))} = m_{\text{mid}} \qquad \square$$

Theorem 7.2. *Heun's method is a second-order RK method since m_{Heu} satisfies the definition of m_{2RK}.*

Proof. Let $w_2 = \frac{1}{2}$ so that

$$p = \frac{1}{2w_2} = \frac{1}{2(\frac{1}{2})} = 1$$

Substituting into Equation (7.24),

$$m_2 = \left.\frac{dy}{dx}\right|_{(x_i+ph,\,L(x_i)+m_L ph)} = \left.\frac{dy}{dx}\right|_{(x_i+h,\,L(x_i)+m_L h)}$$

The coordinates of the point where the derivative is evaluated can be rewritten according to Equations (7.6) and (7.14).

$$x_i + h = x_{i+1}$$
$$L(x_i) + m_L h = L_p(x_{i+1})$$

Substituting notation, the derivative is now expressed as

$$m_2 = \left.\frac{dy}{dx}\right|_{(x_{i+1},\,L_p(x_{i+1}))}$$

This is exactly the slope at the predicted right endpoint of the current interval by Equation (7.15), thus,

$$m_2 = m_R$$

Since $w_2 = \frac{1}{2}$,

$$w_1 + w_2 = 1 \implies w_1 = \frac{1}{2}$$

Satisfying m_{2RK} of Equation (7.25) requires

$$m_{2RK} = w_1 m_L + w_2 m_2 = \frac{1}{2}m_L + \frac{1}{2}m_2$$

Because $m_2 = m_R$, as shown above, substituting gives

$$m_{2RK} = \frac{1}{2}m_L + \frac{1}{2}m_R = \frac{m_L + m_R}{2}$$

The right side is equivalent to m_{Heu} by Equation (7.16), completing the proof.

$$m_{2RK} = m_{Heu} \qquad \square$$

> ⚡ **Theorem 7.3.** *Ralston's method is a second-order RK method since m_{Ral} satisfies the definition of m_{2RK}.*

Proof. Let $w_2 = \frac{2}{3}$, so that

$$p = \frac{1}{2w_2} = \frac{1}{2(\frac{2}{3})} = \frac{3}{4}$$

Substituting into Equation (7.24),

$$m_2 = \left.\frac{dy}{dx}\right|_{(x_i+ph,\,L(x_i)+m_Lph)} = \left.\frac{dy}{dx}\right|_{(x_i+\frac{3h}{4},\,L(x_i)+m_L\frac{3h}{4})}$$

The coordinates of the point where the derivative is calculated can be expressed according to Equations (7.18) and (7.19) as

$$x_i + \frac{3h}{4} = x_{i+\frac{3}{4}}$$

$$L(x_i) + m_L\frac{3h}{4} = L(x_{i+\frac{3}{4}})$$

Rewriting the derivative using this notation yields

$$m_2 = \left.\frac{dy}{dx}\right|_{(x_{i+\frac{3}{4}},\,L(x_{i+\frac{3}{4}}))}$$

By Equation (7.20), this is equivalent to the slope at the estimated three-quarter point of the interval, or

$$m_2 = m_{3/4}$$

For $w_2 = \frac{2}{3}$,

$$w_1 + w_2 = 1 \implies w_1 = \frac{1}{3}$$

Satisfying m_{2RK} of Equation (7.25) means

$$m_{2RK} = w_1 m_L + w_2 m_2 = \frac{1}{3}m_L + \frac{2}{3}m_2$$

As demonstrated above, $m_2 = m_{3/4}$, so substituting produces

$$m_{2RK} = \frac{1}{3}m_L + \frac{2}{3}m_{3/4} = \frac{m_L + 2m_{3/4}}{3}$$

By Equation (7.21), the right side is precisely m_{Ral} which completes the proof.

$$m_{2RK} = m_{Ral} \qquad \qquad \square$$

Because a second-order RK method must satisfy $w_1 + w_2 = 1$ and $p = \frac{1}{2w_2}$, the connection between the weights and the derivative m_2 at the additional point used in Ralston's method is now justified. By letting $w_2 = 2/3$, w_1 *must* equal 1/3 and thus w_2 is twice the weight of w_1. Furthermore, if $w_2 = 2/3$ then p must equal $\frac{1}{2w_2} = \frac{1}{4/3} = \frac{3}{4}$, and in turn $ph = \frac{3h}{4}$, meaning the additional point is found at the three-quarter point of the current interval. The significance of letting $w_2 = 2/3$ was proved in Ralston and Rabinowitz [1]; it produces a minimum bound on the truncation error for all second-order RK methods defined by Equation (7.23).

7.5.2 A fourth-order RK method

There is often a trade-off between the complexity of the RK method implemented and the precision required. This section concludes by presenting a fourth-order RK method commonly used in practice which strikes a good balance between simplicity and accuracy. Observe both the increased complexity in the slope calculation, yet the same simple, underlying structure of approximating the next y-coordinate using linearization.

Recall the initial linearization of Euler's method using Equation (7.9).

$$L(x_{i+1}) = L(x_i) + m_L h$$

The second-order RK method replaced the single slope at the current left endpoint with a weighted slope using an additional second point in the current interval according to Equations (7.23) and (7.25).

$$L(x_{i+1}) = L(x_i) + m_{2RK} h$$
$$m_{2RK} = w_1 m_L + w_2 m_2$$

The fourth-order RK method uses an additional two points and m_{4RK} is weighted by four slopes, but still within the context of linearization.

$$L(x_{i+1}) = L(x_i) + m_{4RK} h \tag{7.26}$$
$$m_{4RK} = w_1 m_L + w_2 m_2 + w_3 m_3 + w_4 m_4 \tag{7.27}$$

The slope m_{Ral} of Equation (7.21) used in Ralston's method was proved in Theorem 7.3 to be a particular instance of a second-order RK slope m_{2RK}, weighting the slope at the point in the interval twice as much as the slope at the left endpoint. Similarly, the tangent line slope for a fourth-order RK method employed here is defined by the particular weighting where the slope at each of the two points in the interval are weighted twice as much as the slope at the left and right endpoint. Thus, $w_1 = w_4 = \frac{1}{6}$ and $w_2 = w_3 = \frac{2}{6}$, giving

$$m_{4RK} = \frac{m_L + 2m_2 + 2m_3 + m_4}{6} \tag{7.28}$$

The weighted slopes are defined recursively as follows. The second slope is the same as the midpoint slope of Equation (7.12), which uses the first slope at the left endpoint m_L, so that

$$m_2 = m_{\text{mid}} = \left.\frac{dy}{dx}\right|_{(x_{i+\frac{1}{2}},L(x_{i+\frac{1}{2}}))} \tag{7.29}$$

The third slope uses the same x-coordinate of the estimated midpoint, but the y-coordinate is now predicated on the second slope m_2 rather than the first slope m_L. Define

$$L_3(x_{i+\frac{1}{2}}) = L(x_i) + m_2\frac{h}{2} \tag{7.30}$$

The third slope is therefore

$$m_3 = \left.\frac{dy}{dx}\right|_{(x_{i+\frac{1}{2}},L_3(x_{i+\frac{1}{2}}))} \tag{7.31}$$

The fourth slope uses the x-coordinate of the estimated right endpoint and the y-coordinate uses the previous third slope m_3. Let

$$L_4(x_{i+1}) = L(x_i) + m_3 h \tag{7.32}$$

The fourth and final slope is

$$m_4 = \left.\frac{dy}{dx}\right|_{(x_{i+1},L_4(x_{i+1}))} \tag{7.33}$$

7.5.3 Fourth-order RK comprehensive template

Equations (7.6), (7.8), (7.10), (7.11), (7.26), and (7.28) through (7.33) define a particular fourth-order RK method. The template for the fourth-order RK method for approximating the next y-coordinate is Table 7.5. The interpretation of each row with three colored cells remains unchanged; the first two cells determine the third. For example, in the sixth row of Table 7.5, the first two brown cells $L(x_i)$ and m_3 are used to find the third brown cell $L_4(x_{i+1})$. Note the eighth row of Table 7.5 contains five salmon cells; the first four are the slopes used for the final weighted slope in the fifth salmon cell $m_{4\text{MK}}$.

The entire process using the Table 7.5 template is detailed in Example 7.5 and illustrated step-by-step in Figures 7.15 through 7.20.

Table 7.5: The DEQ template for a fourth-order RK method. In each colored row, the first two colored cells are used in the calculation of the third colored cell. In the eighth row, the first four salmon cells are used to find the fifth salmon cell.

x_i	$L(x_i)$	m_L	$x_{i+\frac{1}{2}}$	$L(x_{i+\frac{1}{2}})$	m_2	$L_3(x_{i+\frac{1}{2}})$	m_3	x_{i+1}	$L_4(x_{i+1})$	m_4	m_{4RK}	$L(x_{i+1})$
Eq. (7.6)*	**Previous** $L(x_{i+1})$*	**DEQ,** **Eq.** (7.8)	**Eq.** (7.10)	**Eq.** (7.11)	**DEQ,** **Eq.** (7.29)	**Eq.** (7.30)	**Eq.** (7.31)	**Eq.** (7.6)	**Eq.** (7.32)	**DEQ,** **Eq.** (7.33)	**Eq.** (7.28)	**Eq.** (7.26)
x_i	$L(x_i)$	m_L										
	$L(x_i)$	m_L		$L(x_{i+\frac{1}{2}})$								
			$x_{i+\frac{1}{2}}$	$L(x_{i+\frac{1}{2}})$	m_2							
	$L(x_i)$				m_2	$L_3(x_{i+\frac{1}{2}})$						
			$x_{i+\frac{1}{2}}$			$L_3(x_{i+\frac{1}{2}})$	m_3					
	$L(x_i)$						m_3		$L_4(x_{i+1})$			
								x_{i+1}	$L_4(x_{i+1})$	m_4		
		m_L			m_2		m_3			m_4	m_{4RK}	
	$L(x_i)$										m_{4RK}	$L(x_{i+1})$

*The coordinates of the given point $(x_0, F(x_0))$ are used initially.

Example 7.5. Use the fourth-order RK method of Equations (7.26) and (7.28) to approximate the next point for the DEQ $\frac{dy}{dx} = -\frac{2x}{e^y}$ with given initial point $(2, 3.332)$ and $h = 3$.

Solution. Tables according to the template of Tables 7.5 track each step. The initial point is placed in the Table 7.5 template.

i	x_i	$L(x_i)$	m_L	$x_{i+\frac{1}{2}}$	$L(x_{i+\frac{1}{2}})$	m_2	$L_3(x_{i+\frac{1}{2}})$	m_3
0	2	3.332						

The initial slope at the left endpoint m_L is found by the three yellow cells of the first row of Table 7.5 and Equation (7.8), as illustrated by the black dashed line in Figure 7.15.

$$m_L = \frac{dy}{dx}\Big|_{(2,3.332)} = -\frac{2(\,2\,)}{e^{\,3.332}} = -0.143$$

This first slope m_L is used for the y-coordinate $L(x_{i+\frac{1}{2}})$ of the estimated midpoint in the interval. By the three green cells of the second row of Table 7.5 and Equation (7.11)

$$L(x_{i+\frac{1}{2}}) = L(x_{0+\frac{1}{2}}) = L(x_0) + m_L \frac{3}{2}$$

$$L(x_{\frac{1}{2}}) = 3.332 + (\,-0.143\,)1.5 = 3.118$$

By Equation (7.10) the x-coordinate of the midpoint for $i = 0$ is

$$x_{i+\frac{1}{2}} = x_i + \frac{h}{2}$$

$$x_{\frac{1}{2}} = x_0 + \frac{3}{2} = 2 + 1.5 = 3.5$$

Figure 7.15 notes this estimated midpoint $(x_{\frac{1}{2}}, L(x_{\frac{1}{2}})) = (3.5, 3.118)$ in black. The updated Table 7.5 template follows.

i	x_i	$L(x_i)$	m_L	$x_{i+\frac{1}{2}}$	$L(x_{i+\frac{1}{2}})$	m_2	$L_3(x_{i+\frac{1}{2}})$	m_3
0	2	3.332	−0.143	3.5	3.118			

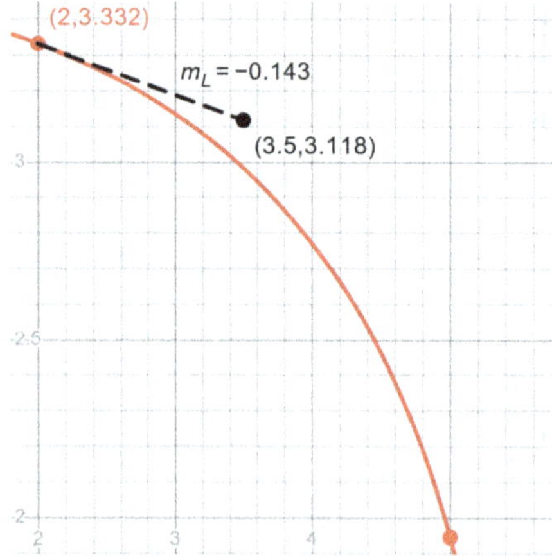

Figure 7.15: The first of four slopes used in the fourth-order RK method is m_L, represented by the black dashed line. Linearized from the left endpoint of the current interval, which in this case is also the given initial point in red, it is used for the black estimated midpoint of the current interval needed for the next, second slope. The true DEQ solution is in red.

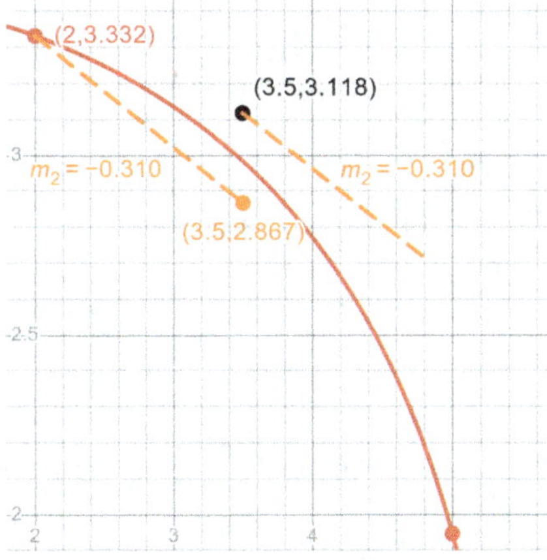

Figure 7.16: The second slope m_2 is represented by the parallel orange dashed lines. Found at the previously estimated midpoint in black, it is applied at the start of the interval and used to re-estimate the midpoint in orange.

The second slope m_2 is calculated at the previously estimated midpoint according to the three blue cells of the third row of Table 7.5 and Equation (7.29). Thus, in Figure 7.16, starting at the black midpoint, the orange dashed line has slope m_2.

$$m_2 = \left.\frac{dy}{dx}\right|_{(3.5,3.118)} = -\frac{2(\;3.5\;)}{e^{\;3.118}} = -0.310$$

Applying this slope to the start of the interval, the third y value $L_3(x_{i+\frac{1}{2}})$ is found, a re-estimation of the y-coordinate of the midpoint in the interval. The three magenta cells of the fourth row of Table 7.5 and Equation (7.30) produce this new estimation of the y-coordinate of the midpoint.

$$L_3(x_{i+\frac{1}{2}}) = L_3(x_{0+\frac{1}{2}}) = L(x_0) + m_2\frac{3}{2}$$

$$L_3(x_{\frac{1}{2}}) = \boxed{3.332} + (\;\boxed{-0.310}\;)1.5 = \boxed{2.867}$$

The x-coordinate of the midpoint does not change and is still $x_{\frac{1}{2}} = 3.5$. Figure 7.16 shows the new estimated midpoint $(x_{\frac{1}{2}}, L_3(x_{\frac{1}{2}})) = (3.5, 2.867)$ in orange. The current Table 7.5 template follows.

i	x_i	$L(x_i)$	m_L	$x_{i+\frac{1}{2}}$	$L(x_{i+\frac{1}{2}})$	m_2	$L_3(x_{i+\frac{1}{2}})$	m_3
0	2	3.332	−0.143	3.5	3.118	−0.310	2.867	

The third slope m_3 is calculated at the previous and second estimation of the midpoint using the values in the three violet cells of the fifth row of Table 7.5 and Equation (7.31). Figure 7.17 depicts m_3 by the green dashed line emanating from the orange midpoint.

$$m_3 = \left.\frac{dy}{dx}\right|_{(3.5,2.867)} = -\frac{2(\;3.5\;)}{e^{\;2.867}} = -0.398$$

Applying this slope as always to the start of the interval, the fourth y value $L_4(x_{i+1})$ at the estimated right endpoint is ultimately calculated by Equation (7.32) according to the three brown cells of the sixth row of Table 7.5.

$$L_4(x_{i+1}) = L_4(x_{0+1}) = L(x_0) + m_3 h$$

$$L_4(x_1) = \boxed{3.332} + (\boxed{-0.398})3 = \boxed{2.138}$$

The x-coordinate of the right endpoint for $i = 0$ is

$$x_{i+1} = x_i + h$$

$$x_1 = x_0 + 3 = 2 + 3 = 5$$

Figure 7.17 labels the estimated right endpoint $(x_1, L_4(x_1)) = (5, 2.138)$ in green. The updated Table 7.5 template follows, split and displayed as two rows because of its length.

i	x_i	$L(x_i)$	m_L	$x_{i+\frac{1}{2}}$	$L(x_{i+\frac{1}{2}})$	m_2	$L_3(x_{i+\frac{1}{2}})$	m_3
0	2	3.332	−0.143	3.5	3.118	−0.310	2.867	−0.398

i	x_{i+1}	$L_4(x_{i+1})$	m_4	m_{4RK}	$L(x_{i+1})$
0	5	2.138			

The fourth and final slope m_4 is found at the previously estimated right endpoint by the three gray cells of the seventh row of Table 7.5 and Equation (7.33). In Figure 7.18, the violet dashed line beginning from the green right endpoint has slope m_4.

$$m_4 = \left.\frac{dy}{dx}\right|_{(5,2.138)} = -\frac{2(\boxed{5})}{e^{\boxed{2.138}}} = \boxed{-1.179}$$

Whereas the three previous slopes m_L, m_2, and m_3 determined the estimated y-coordinates of the midpoint, the revised midpoint, and the right endpoint, respectively, the fourth slope m_4 is not used for estimating a y-coordinate. Nevertheless, in Figure 7.18, a parallel violet dashed line is drawn from the initial left endpoint, like the three previous dashed lines, for comparison purposes and to illustrate the weighted slope later in Figure 7.19.

The four slopes, one at the left endpoint m_L, one at the right endpoint m_4, and one at each estimated midpoint m_2 and m_3, which are both twice the weight of the endpoint slopes, are now combined to produce m_{4RK}, the definitive slope in the linearization that determines the next y-coordinate approximation $L(x_{i+1})$. According to Equation (7.28) and the five salmon cells of the eighth row of Table 7.5

$$m_{4RK} = \frac{m_L + 2m_2 + 2m_3 + m_4}{6}$$

$$= \frac{\boxed{-0.143} + 2(\boxed{-0.310}) + 2(\boxed{-0.398}) + (\boxed{-1.179})}{6} = \boxed{-0.456}$$

All four slopes are represented in Figure 7.19 by short dashed lines starting at the left endpoint. The two middle slopes m_2 and m_3 are shown multiplied by a factor of two indicating their double weighting.

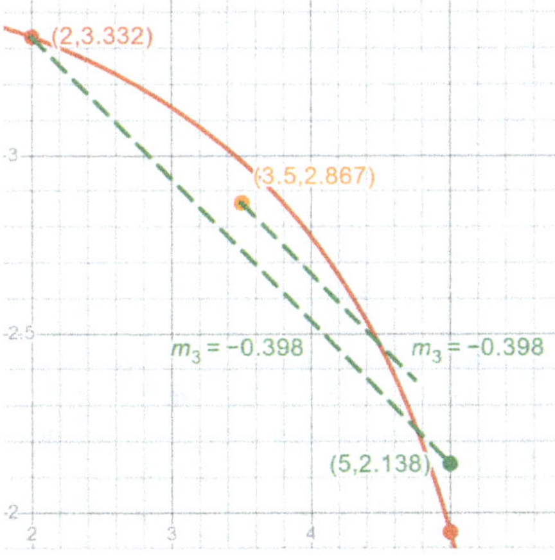

Figure 7.17: The third slope m_3 is represented by the parallel green dashed lines. Calculated at the previous and second estimated midpoint in orange, it is applied at the start of the interval and used to estimate the right endpoint in green.

The longer blue solid line uses the resultant slope m_{4RK} to approximate the next y-coordinate, found by Equation (7.26) and the three pink cells of the last row of Table 7.5.

$$L(x_{i+1}) = L(x_i) + m_{4MK}h$$

$$L(x_{0+1}) = L(x_0) + m_{4MK}h$$

$$L(x_1) = \boxed{3.332} + (\boxed{-0.456})3 = \boxed{1.964}$$

The x-coordinate of this point $x_1 = 5$ was found earlier and placed in the first column of the Table 7.5 template, so the next approximated point is $(x_1, L(x_1)) = (5, 1.964)$. The completed Table 7.5 template with the remaining intermediate values inserted follows, split again because of its length.

i	x_i	$L(x_i)$	m_L	$x_{i+\frac{1}{2}}$	$L(x_{i+\frac{1}{2}})$	m_2	$L_3(x_{i+\frac{1}{2}})$	m_3
0	2	3.332	−0.143	3.5	3.118	−0.310	2.867	−0.398
1	5	1.964						

i	x_{i+1}	$L_4(x_{i+1})$	m_4	m_{4RK}	$L(x_{i+1})$
0	5	2.138	−1.179	−0.456	1.964

The DEQ of Example 7.5 is the same used in Example 7.4. Separable with the analytic solution $y = \ln(32 - x^2)$, the exact value of the y-coordinate at $x = 5$ is

$$y = \ln(32 - 5^2) = \ln 7 = 1.946$$

In Figure 7.19, the true point $(5, 1.946)$ on the analytic solution, both in red, nearly coincides with the approximated point $(5, 1.964)$ in blue. In practice, the exact value is unknown because the analytic solution is unavailable, hence the use of an approximation method. The purpose here is for comparison only.

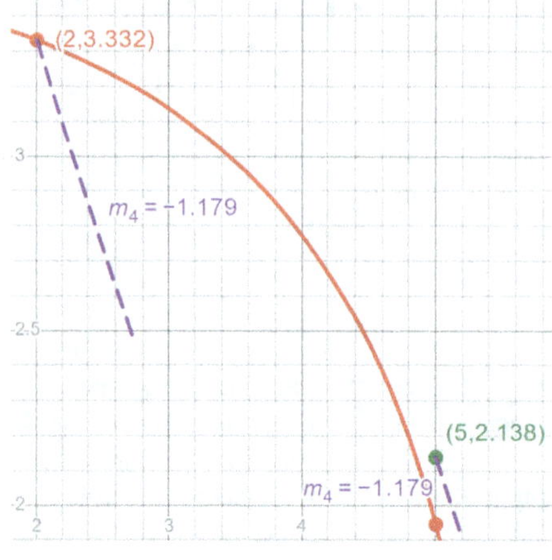

Figure 7.18: The fourth slope m_4 is represented by the violet dashed line, beginning from and determined by the previously estimated right endpoint in green. A parallel violet dashed line is drawn from the start of the interval like the three previous dashed lines for comparison only.

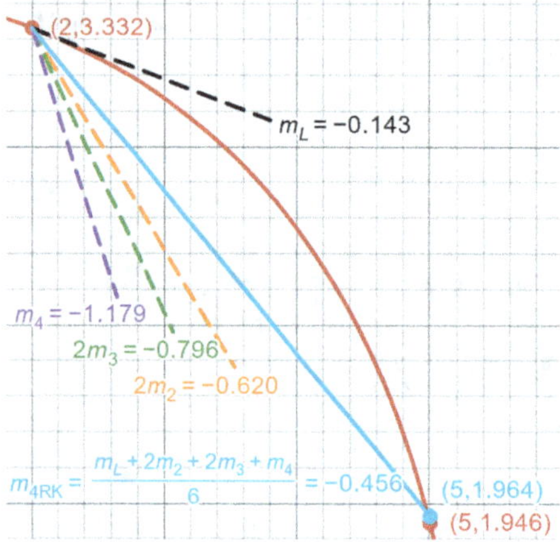

Figure 7.19: The slopes of the four short dashed lines comprise the weighted slope of the longer blue solid line used in the linearization to approximate the next point (5, 1.964) in blue. This approximation almost lies on the exact value of the red point (5, 1.946).

The same DEQ and initial value of Example 7.4 employed Ralston's method and in the explanation following that example it was noted it takes six approximated points with $h = 0.5$ to reach $x_6 = 5$. The y-coordinate approximation is then $L(x_6) = L(5) = 1.985$ with an absolute error of $|1.985 - 1.946| = 0.039$. In comparison, the fourth-order RK method of Example 7.5 needs just one approximated point with $h = 3$ to reach $x_1 = 5$ and $L(x_1) = L(5) = 1.964$. But despite the much larger h value used, the fourth-order RK method produces a smaller absolute error of $|1.964 - 1.946| = 0.018$ versus 0.039.

7.5.4 Observations on the composition of slope m_{4RK}

In Figure 7.20, the orange and green dashed lines highlight the respective middle slopes m_2 and m_3 used in the composite slope m_{4MK} of Equation (7.28) for the fourth-order

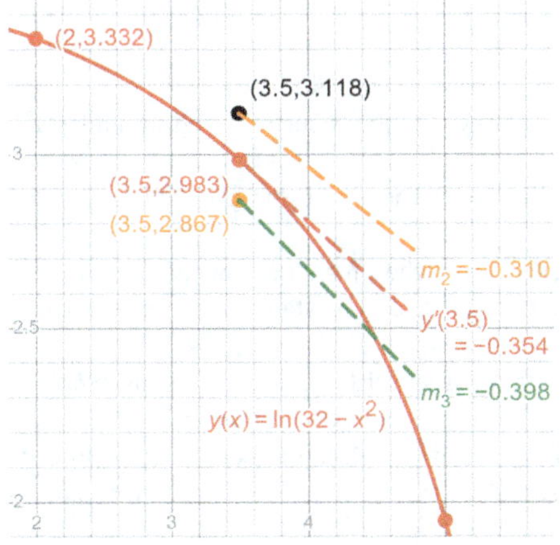

Figure 7.20: The red dashed line at the true midpoint with slope y' is not quite parallel to the orange and green dashed lines with slopes m_2 and m_3, both originating from estimated midpoints. The mean of m_2 and m_3 is a better approximation of y' than m_2 or m_3 alone, because the red dashed line lies between the orange and green dashed lines.

RK method. Each emanates from midpoints at $x = 3.5$ with y-coordinates of $y = 3.118$ $y = 2.867$, both of which are estimations from a DEQ that is a function of both x and y.

At the true midpoint $(3.5, 2.983)$, the tangent line slope is $y'(3.5) = -0.354$, a value between $m_2 = -0.310$ and $m_3 = -0.398$, and correspondingly represented by the red dashed line between the orange and green dashed lines. The mean of m_2 and m_3 therefore generates a better estimation of the true tangent line slope y' and, coincidentally in this case, the exact slope since

$$\frac{m_2 + m_3}{2} = \frac{-0.310 + (-0.398)}{2} = \frac{-0.708}{2} = -0.354 = y'(3.5)$$

Nevertheless, the ultimate goal is not the best estimation of the slope at the midpoint, but the slope that produces the next point at $x = 5$ where the y-coordinate is closest to the true y value.

The exact point on the true solution is $(5, 1.946)$ as previously found. Working backwards from the given initial point $(2, 3.332)$ and $h = 3$, the true slope m_t between these two points, and the target value of m_{4MK}, is

$$m_t = \frac{1.946 - 3.332}{5 - 2} = \frac{-1.386}{3} = -0.462$$

Thus, the mean -0.354 of the two middle slopes m_2 and m_3 is not negative enough, the impetus for including the modifying slopes m_L and m_4 at the endpoints. However, because they are at the extremes of the interval, m_L and m_4 are weighted once versus a double weighting of the two middle, inside slopes m_2 and m_3. The result of $m_{4RK} = -0.456$ confirms the corrective nature of the endpoint slopes in estimating $m_t = -0.462$.

Differential equations concepts summary

1. Not every differentiable equation (DEQ) is solvable by analytic methods, such as using an antiderivative or a separation of variables.
2. The numerical solution of a DEQ is a set of points; the solution can be fit to any closed form interpolation or regression function (see Chapter 8).
3. Euler's method DEQ solutions use successive linearizations at the left endpoint of each interval.
4. Midpoint method DEQ solutions use successive linearizations at the estimated midpoint of each interval.
5. Heun's method DEQ solutions use successive linearizations in which the slope is the mean of the slope at the left endpoint and predicted right endpoint of each interval.
6. Ralston's method DEQ solutions use successive linearizations in which the slope is the weighted average of the slope at the left endpoint and twice the slope at the estimated three-quarter point of each interval.
7. Runge–Kutta (RK) DEQ solutions recursively use previous slopes at various points, weighted.
8. The midpoint, Heun's, and Ralston's methods are all second-order RK methods derived from the second-degree Taylor polynomial $T_2(x)$.
9. RK methods incorporate derivatives up to $f^{(n)}(x)$ inclusive using the Taylor polynomial $T_n(x)$, but only first derivatives are needed for the composite slope calculation.
10. A fourth-order RK method is a good compromise between precision and derivation complexity.

Exercises

Section 7.1

Use Euler's method to approximate three additional points for the DEQ dy/dx, initial point $(x_0, L(x_0))$ and h value.

1. $dy/dx = \sqrt{2x + 5y}$ $(3, 2)$ $h = 0.25$
2. $dy/dx = (3x + 2y)^{1.5}$ $(0.7, 0.25)$ $h = 0.1$
3. $dy/dx = \ln(x + y)$ $(0.6, 0.5)$ $h = 0.3$
4. $dy/dx = \log(4x - y)$ $(1.5, 0.7)$ $h = 0.5$
5. $dy/dx = x^2 - 2xy - 3y^2$ $(9, 0.75)$ $h = 0.05$
6. $dy/dx = 0.2x^3 - 0.1y^3$ $(5, 6)$ $h = 0.35$
7. $dy/dx = 0.4e^x + 0.2e^{-y}$ $(5.7, 3.3)$ $h = 0.1$
8. $dy/dx = 6^x - y^4$ $(0.75, 1.9)$ $h = 0.25$
9. $dy/dx = \cos(x - 10y)$ $(4.15, -1.65)$ $h = 0.15$
10. $dy/dx = \sin(7x + 3y)$ $(3, 2.1)$ $h = 0.75$
11. $dy/dx = \dfrac{x+y^2}{x^2+y}$ $(1, 3.2)$ $h = 1.5$
12. $dy/dx = \dfrac{8xy}{x^2+y^2}$ $(2, 5.5)$ $h = 0.4$
13. $dy/dx = \tan^{-1}(1.5xy)$ $(6, 0.25)$ $h = 0.8$
14. $dy/dx = \tan^{-1}(3x^2 - 6.4y)$ $(6, 1.3)$ $h = 1.1$
15. $dy/dx = 2^x + \log_2 y$ $(0, 8)$ $h = 0.2$
16. $dy/dx = \log x - 10^y$ $(1, -2)$ $h = 1.4$

Use Euler's method to approximate points up to $(x_i, L(x_i))$ for the DEQ dy/dx, initial point $(x_0, L(x_0))$ and h value. Solve the DEQ analytically by separating variables and find the exact value $y(x_i)$.

17.	$dy/dx = 0.5y/x$	$(0.25, 0.65)$	$x_5 = 0.5$	$h = 0.05$
18.	$dy/dx = -y^2 e^x$	$(0, 0.8)$	$x_4 = 0.3$	$h = 0.075$
19.	$dy/dx = x^2 e^{3y}$	$(-5, -1.612)$	$x_6 = -4.25$	$h = 0.125$
20.	$dy/dx = x^3 e^{-2y}$	$(2.4, 1.5)$	$x_6 = 4.5$	$h = 0.35$
21.	$dy/dx = \cos(2.5x)y^{-4}$	$(3.65, 0.9)$	$x_6 = 3.5$	$h = -0.025$
22.	$dy/dx = y^2/(x+1)$	$(5.6, 2.11)$	$x_7 = 5.25$	$h = -0.05$

Section 7.2

Use the midpoint method to approximate three additional points for the DEQ dy/dx, initial point $(x_0, L(x_0))$ and h value.

23.	$dy/dx = 3.8x^2 - 1.3y$	$(2.4, 3.5)$	$h = 0.2$
24.	$dy/dx = 7.9x + 0.017y^2$	$(0, 0.8)$	$h = 0.5$
25.	$dy/dx = (4.5x + 1.9y)^{2.5}$	$(0.15, 0.7)$	$h = 0.02$
26.	$dy/dx = \sqrt{x - y}$	$(1, -2)$	$h = 0.25$
27.	$dy/dx = x^2 y^3 - 4x + y$	$(-2.05, -1.1)$	$h = 0.05$
28.	$dy/dx = 0.3y^2 - 2.2x + 1.15y$	$(3.3, 1.7)$	$h = 0.4$
29.	$dy/dx = \log_3(3.15x - 0.45y)$	$(4, 1.75)$	$h = 0.5$
30.	$dy/dx = \ln(1.84x + 2.95y)$	$(10.2, 4)$	$h = 0.1$
31.	$dy/dx = \sin(4x + 2y^2)$	$(2.9, -1.3)$	$h = 0.3$
32.	$dy/dx = \cos(x^2 - y^2)$	$(1.57, 0.8)$	$h = 0.15$
33.	$dy/dx = 4.8^x - 1.4^y$	$(-0.5, 6.5)$	$h = 0.25$
34.	$dy/dx = 1.35e^x + 1.8e^y$	$(-1, -4)$	$h = 0.55$
35.	$dy/dx = \dfrac{24y}{x^5 + 6}$	$(3.2, -21.1)$	$h = -0.4$
36.	$dy/dx = \dfrac{3x + y}{y^2 + 2}$	$(9.7, 14.5)$	$h = -0.6$
37.	$dy/dx = \ln y - e^{-x}$	$(3.25, 6.4)$	$h = 0.35$
38.	$dy/dx = \ln(x + 1) + e^y$	$(1.25, -1.5)$	$h = 0.45$

Use the midpoint method to approximate points up to $(x_i, L(x_i))$ for the DEQ dy/dx, initial point $(x_0, L(x_0))$ and h value. Solve the DEQ analytically by separating variables and find the exact value $y(x_i)$.

39.	$dy/dx = 3y/x^2$	$(2, 1.785)$	$x_5 = 2.75$	$h = 0.15$
40.	$dy/dx = y^2 e^{0.8x}$	$(-6, 1.35)$	$x_4 = -4.8$	$h = 0.3$
41.	$dy/dx = 0.2x(1 + y^2)$	$(2.9, -0.31)$	$x_7 = 4.02$	$h = 0.16$
42.	$dy/dx = y \sin 4x$	$(3.6, 7.47)$	$x_6 = 3.9$	$h = 0.05$
43.	$dy/dx = \sqrt{y} \sec^2 x$	$(3, 3.72)$	$x_6 = 3.6$	$h = 0.1$
44.	$dy/dx = e^y/\sqrt{x}$	$(9, 0)$	$x_6 = 10.68$	$h = 0.28$

Section 7.3

Use Heun's method to approximate three additional points for the DEQ dy/dx, initial point $(x_0, L(x_0))$ and h value.

45. $dy/dx = 4.1\sqrt{x} + 0.6y^{1.5}$ $\quad\quad$ $(4, 0.1)$ $\quad\quad$ $h = 0.35$
46. $dy/dx = x^3 - 9y$ $\quad\quad$ $(2.15, 5)$ $\quad\quad$ $h = 0.15$
47. $dy/dx = 1.02xy^2 + x^2 - 4.92y^4$ $\quad$ $(-4.15, 1.28)$ $\quad$ $h = 0.1$
48. $dy/dx = 0.617x^3y^2 - 0.991xy^4$ $\quad$ $(-2, 1.5)$ $\quad\quad$ $h = 0.45$
49. $dy/dx = (8x - 1.24y)^{1.5}$ $\quad\quad$ $(2.4, 6)$ $\quad\quad$ $h = 0.2$
50. $dy/dx = (0.64y - 0.56x)^{0.75}$ $\quad\quad$ $(1.25, 3.15)$ $\quad$ $h = 0.25$
51. $dy/dx = \ln(x^2 + 1.5y)$ $\quad\quad$ $(2, 4.5)$ $\quad\quad$ $h = 0.5$
52. $dy/dx = \log_5(10x - y^2)$ $\quad\quad$ $(11.1, 7.85)$ $\quad$ $h = 0.3$
53. $dy/dx = \tan^{-1}(1.62x) - 4y$ $\quad\quad$ $(5, 8)$ $\quad\quad$ $h = 0.15$
54. $dy/dx = 2.3x + \tan^{-1}y$ $\quad\quad$ $(-2.2, 7)$ $\quad\quad$ $h = 0.4$
55. $dy/dx = 12x^{-1} + 8y^{-1}$ $\quad\quad$ $(-5.75, -6.5)$ $\quad$ $h = 0.25$
56. $dy/dx = x^{-2} - 10y^{-2}$ $\quad\quad$ $(5.55, 1)$ $\quad\quad$ $h = 0.35$
57. $dy/dx = x\cos(2x + y)$ $\quad\quad$ $(1, 0.75)$ $\quad\quad$ $h = 0.6$
58. $dy/dx = \sin(y - 5x) + y$ $\quad\quad$ $(7, 8)$ $\quad\quad$ $h = 0.2$
59. $dy/dx = (0.5)^y - \log_{0.5}x$ $\quad\quad$ $(3.85, 5.85)$ $\quad$ $h = 0.05$
60. $dy/dx = 3^x + \log_3(2y + 5)$ $\quad\quad$ $(-3.5, 4.45)$ $\quad$ $h = 0.5$

Use Heun's method to approximate points up to $(x_i, L(x_i))$ for the DEQ dy/dx, initial point $(x_0, L(x_0))$ and h value. Solve the DEQ analytically by separating variables and find the exact value $y(x_i)$.

61. $dy/dx = 0.1ye^{0.5x}$ $\quad$ $(2.8, 2.25)$ $\quad$ $x_6 = 1$ $\quad$ $h = -0.3$
62. $dy/dx = -0.2x^4y^3$ $\quad$ $(1.4, 1.221)$ $\quad$ $x_6 = 0.68$ $\quad$ $h = -0.12$
63. $dy/dx = e^{-y}\sin x$ $\quad$ $(1.3, 0.55)$ $\quad$ $x_7 = 4.1$ $\quad$ $h = 0.4$
64. $dy/dx = 0.25xe^y$ $\quad$ $(1.48, 1.12)$ $\quad$ $x_6 = 1.96$ $\quad$ $h = 0.08$
65. $dy/dx = 12(xy)^{-1}$ $\quad$ $(6.4, 3.09)$ $\quad$ $x_6 = 10.9$ $\quad$ $h = 0.75$
66. $dy/dx = 0.7xy^2$ $\quad$ $(2.86, -0.62)$ $\quad$ $x_7 = 3.56$ $\quad$ $h = 0.1$

Section 7.4

Use Ralston's method to approximate three additional points for the DEQ dy/dx, initial point $(x_0, L(x_0))$ and h value.

67. $dy/dx = y - 0.2x^3$ $\quad\quad$ $(1.75, 1.25)$ $\quad$ $h = 0.5$
68. $dy/dx = 1.3y^3 - 1.6x^2$ $\quad\quad$ $(2.7, 2)$ $\quad\quad$ $h = 0.15$
69. $dy/dx = \log_2(6x + 1.5y^2)$ $\quad\quad$ $(-3, 3.75)$ $\quad\quad$ $h = 0.4$
70. $dy/dx = \ln(2.45y - 0.32x^2)$ $\quad$ $(-1.5, 5)$ $\quad\quad$ $h = 0.35$

71. $dy/dx = \sqrt{3.7x^2 + 4.8y^2}$ $\quad\quad$ $(3, -5)$ $\quad\quad$ $h = 0.25$
72. $dy/dx = (0.9x^7 + 3y)^{0.4}$ $\quad\quad$ $(5.2, 10)$ $\quad\quad$ $h = 0.1$
73. $dy/dx = \log y + 3x$ $\quad\quad$ $(10, 7.1)$ $\quad\quad$ $h = 0.3$
74. $dy/dx = 2.4y - \ln x$ $\quad\quad$ $(1.25, 1.6)$ $\quad\quad$ $h = 0.05$

75. $dy/dx = 12y - 7xy - x$ $\qquad$ $(1.4, -1)$ $\qquad$ $h = 0.2$
76. $dy/dx = 0.5x^3 + 1.85x^2 - 3.15y$ $\qquad$ $(0.5, 2.3)$ $\qquad$ $h = 0.4$
77. $dy/dx = e^{-x^2} - y^2$ $\qquad$ $(-1.75, 0.75)$ $\quad$ $h = -0.35$
78. $dy/dx = 2^{-x} + 5^{-y}$ $\qquad$ $(0.5, 4)$ $\qquad$ $h = -0.25$
79. $dy/dx = -0.8x + \sin 1.6y$ $\qquad$ $(6, 7)$ $\qquad$ $h = 0.75$
80. $dy/dx = y^2 \cos(1.2y - 0.64x)$ $\qquad$ $(4.1, -2.5)$ $\qquad$ $h = 0.3$
81. $dy/dx = (3y - 7x)/(13x + 17y)$ $\qquad$ $(7.1, 1)$ $\qquad$ $h = 0.05$
82. $dy/dx = (0.75x^2 + 4)/(2.85x + 1.15y)$ $\quad$ $(1.5, 1.25)$ $\qquad$ $h = 0.5$

Use Ralston's method to approximate points up to $(x_i, L(x_i))$ for the DEQ dy/dx, initial point $(x_0, L(x_0))$ and h value. Solve the DEQ analytically by separating variables and find the exact value $y(x_i)$.

83. $dy/dx = 4(y/x)^3$ $\qquad$ $(3.4, 2.016)$ $\qquad$ $x_6 = 4.3$ $\qquad$ $h = 0.15$
84. $dy/dx = ye^{0.25x}$ $\qquad$ $(-7.5, 5.54)$ $\qquad$ $x_7 = -5.4$ $\quad$ $h = 0.3$
85. $dy/dx = 10/(y + x^2 y)$ $\qquad$ $(-1.95, 1.75)$ $\quad$ $x_7 = 1.2$ $\qquad$ $h = 0.45$
86. $dy/dx = e^{0.4x}y^{-3}$ $\qquad$ $(2.5, 2.29)$ $\qquad$ $x_6 = 6.1$ $\qquad$ $h = 0.6$
87. $dy/dx = y \ln x$ $\qquad$ $(2.35, 1.06)$ $\qquad$ $x_6 = 2.95$ $\quad$ $h = 0.1$
88. $dy/dx = (x - 5)/y$ $\qquad$ $(8.4, 1.6)$ $\qquad$ $x_6 = 9.6$ $\qquad$ $h = 0.2$

Section 7.5

Use a fourth-order Runge–Kutta method to approximate three additional points for the DEQ dy/dx, initial point $(x_0, L(x_0))$ and h value.

89. $dy/dx = 5.5 - 1.6x - \sqrt{y}$ $\qquad$ $(9.75, 10.89)$ $\quad$ $h = 0.25$
90. $dy/dx = 32 - 0.9xy$ $\qquad$ $(2.8, 3.5)$ $\qquad$ $h = 0.4$
91. $dy/dx = (1.1y^2 - 6.2x^2)^{0.2}$ $\qquad$ $(-2, 0.125)$ $\qquad$ $h = 0.5$
92. $dy/dx = (3x - 0.56y^2)^{1.6}$ $\qquad$ $(2, -2.5)$ $\qquad$ $h = 0.15$
93. $dy/dx = \ln(x + 2) - 0.28y^3$ $\qquad$ $(4, 2.65)$ $\qquad$ $h = 0.35$
94. $dy/dx = 2x^2 + 7\log_2 y$ $\qquad$ $(3.2, 15)$ $\qquad$ $h = 0.2$
95. $dy/dx = 0.415e^{y^2} + 3\sqrt{x}$ $\qquad$ $(2.45, 0)$ $\qquad$ $h = 0.1$
96. $dy/dx = e^{-xy} - xy$ $\qquad$ $(1, 5.5)$ $\qquad$ $h = 0.45$
97. $dy/dx = 4.7\cos x - 3.3\sin y$ $\qquad$ $(3.95, 4)$ $\qquad$ $h = 0.3$
98. $dy/dx = 2.25\sin x + 1.5\cos y$ $\qquad$ $(11, 3)$ $\qquad$ $h = 0.5$
99. $dy/dx = 6\tan^{-1}(1.44y - 0.72x)$ $\qquad$ $(5.8, 3.25)$ $\qquad$ $h = 0.4$
100. $dy/dx = -2\tan^{-1}(0.16x^2 + 0.9y^2)$ $\qquad$ $(1.5, -1.25)$ $\quad$ $h = 0.35$
101. $dy/dx = -1 + x + 0.2y + 0.35y^2 - 0.6x^3$ $\quad$ $(4.1, 6.25)$ $\qquad$ $h = -0.2$
102. $dy/dx = x^2 + 24\sqrt{y} - 50$ $\qquad$ $(1.5, 9)$ $\qquad$ $h = -0.1$
103. $dy/dx = (4.2x + 3.9y)/(y^3 + y + 3)$ $\qquad$ $(1, -2)$ $\qquad$ $h = 0.75$
104. $dy/dx = 2.7y^2/(0.3x^3 - 4xy + 12.5)$ $\qquad$ $(7, 8)$ $\qquad$ $h = 0.3$

Use a fourth-order Runge–Kutta method to approximate points up to $(x_i, L(x_i))$ for the DEQ dy/dx, initial point $(x_0, L(x_0))$ and h value. Solve the DEQ analytically by separating variables and find the exact value $y(x_i)$.

105.	$dy/dx = (4x^3 + 1)/y$	$(-0.7, 2.02)$	$x_4 = 1.3$	$h = 0.5$
106.	$dy/dx = 20x/y^4$	$(0, 1.07)$	$x_5 = 1.25$	$h = 0.25$
107.	$dy/dx = 16 \ln x/(xy^{1.5})$	$(1.1, 1.26)$	$x_5 = 1.85$	$h = 0.15$
108.	$dy/dx = \frac{y}{5} \sec(\frac{x}{5}) \tan(\frac{x}{5})$	$(5, 6.365)$	$x_4 = 5.8$	$h = 0.2$
109.	$dy/dx = x + xy^2$	$(-0.25, -0.818)$	$x_5 = 1.75$	$h = 0.4$
110.	$dy/dx = e^y/(x + 3)$	$(6.6, 1.522)$	$x_5 = 8.35$	$h = 0.35$

Sections 7.2–7.5

Use the midpoint method, Heun's method, Ralston's method, and a fourth-order Runge–Kutta method to approximate points up to $(x_i, L(x_i))$ for the DEQ dy/dx, initial point $(x_0, L(x_0))$ and h value. Solve the DEQ analytically by separating variables and find the exact value $y(x_i)$. Round all answers to the thousandth place and note the accuracy of each method based on the absolute error, i. e., the magnitude of the difference between the approximated value generated by the method and the exact value.

111.	$dy/dx = y^2 \sin x$	$(2, 1.276)$	$x_8 = 5.6$	$h = 0.45$
112.	$dy/dx = y/(2x - 3)$	$(2, 1.46)$	$x_4 = 2.6$	$h = 0.15$
113.	$dy/dx = x \cos^2 y$	$(0.5, 0.14)$	$x_5 = 1.75$	$h = 0.25$
114.	$dy/dx = 5e^{4x}y^{-1}$	$(-0.5, 1.26)$	$x_4 = -0.1$	$h = 0.1$
115.	$dy/dx = \ln x/(xy)$	$(0.8, 0.7)$	$x_6 = 2$	$h = 0.2$
116.	$dy/dx = e^{0.1x-y}$	$(2.75, -1.8)$	$x_8 = 5.15$	$h = 0.3$
117.	$dy/dx = 2x \sqrt{y} \cos x$	$(3, 1.751)$	$x_8 = 0.2$	$h = -0.35$
118.	$dy/dx = \frac{3x^2 e^y}{1+x^3}$	$(0.7, -1.342)$	$x_4 = -0.9$	$h = -0.4$

8 Multiple methods

Each of the previous chapters offered various options for solving one specific numerical question. In this culminating chapter, previous techniques from multiple chapters are combined to solve more complex problems.

Previously, Example 3.10 in Section 3.4 featured multiple methods. The question of inverse interpolation was answered using standard interpolating techniques in Chapter 3 to fit a function, equating that function to a known value, and recasting the equation as a zero-finding problem, solved by the techniques in Chapter 1.

As mentioned in the last chapter, numerical DEQ solutions are sets of points. But a closed-form function could be gleaned using either interpolation or regression techniques. Each section in this chapter is structured around the solution from a specific DEQ method in Chapter 7, additionally fit to various interpolation and regression functions as described in Chapters 3 and 4, respectively. Although algebraically possible to construct any interpolation or regression function for most of the DEQ solutions referenced, justifications are provided in situations where a certain function was or was not considered.

8.1 Euler's method solutions

Functions in this section are fit to the solution of the DEQ $\frac{dy}{dx} = -\frac{4\sin x}{y}$ of Example 7.1, reproduced for convenience in Table 8.1.

Table 8.1: Euler's method DEQ solution to $\frac{dy}{dx} = -\frac{4\sin x}{y}$ of Example 7.1.

i	x_i	$L(x_i)$
0	1.9	3.227
1	2.3	2.758
2	2.7	2.325
3	3.1	2.031

8.1.1 A DD polynomial interpolation function

A third-degree DD polynomial function is fit exactly through the points of Table 8.1 according to the interpolation technique outlined in Section 3.1.3. First, the DD values are calculated and organized according to Table 3.1.

	1st DD	2nd DD	3rd DD
(1.9, 3.227)			
(2.3, 2.758)	−1.173		
(2.7, 2.325)	−1.083	0.113	
(3.1, 2.031)	−0.735	0.434	0.268

https://doi.org/10.1515/9783112221051-008

Using Equation (3.18), and the red and blue values in the above table, the third-degree DD polynomial function $f_3(x)$ is

$$f_3(x) = 3.227 - 1.173(x - 1.9) + 0.113(x - 1.9)(x - 2.3)$$
$$+ 0.268(x - 1.9)(x - 2.3)(x - 2.7)$$

Figure 8.1 compares this interpolated DD polynomial function $f_3(x)$ to the true solution $y = \sqrt{13 + 8\cos x}$ over the domain of the x values $[1.9, 3.1]$.

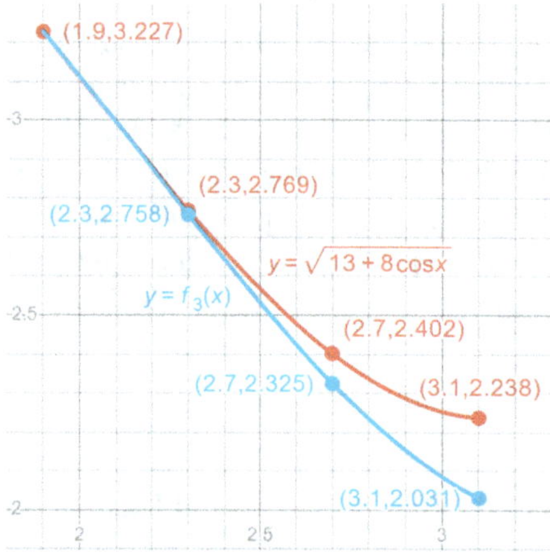

Figure 8.1: The blue third-degree DD polynomial function $y = f_3(x)$ fit through the solution points generated by Euler's method of Table 8.1. The true solution $y = \sqrt{13 + 8\cos x}$ is red.

Two regression functions that minimize the least squares accumulation error for the points of Table 8.1 are now examined in contrast to interpolating a function precisely through those points.

8.1.2 A logarithmic regression function

The points and third-degree DD polynomial interpolation function of Figure 8.1 suggest an underlying function which is decreasing and concave up, suitable for a logarithmic regression function according to Section 4.3.2. Section 4.3.1 established any base can be used without altering the logarithmic regression function fit; the natural log regression function $r_{\ln}(x)$ is used for simplicity.

The Table 4.2 template is employed for all necessary sums, relabeling $y_i = L(x_i)$.

i	x_i	y_i	$\ln x_i$	$(\ln x_i)^2$	$(\ln x_i)y_i$
0	1.9	3.227	0.642	0.412	2.072
1	2.3	2.758	0.833	0.694	2.297
2	2.7	2.325	0.993	0.986	2.309
3	3.1	2.031	1.131	1.279	2.297
		10.341	3.599	3.371	8.975

Parameter m is found by Equation (4.17) with sums from the table and $n = 4$.

$$m = \frac{n \sum_{i=1}^{n}(\ln x_i)y_i - \sum_{i=1}^{n}\ln x_i \sum_{i=1}^{n}y_i}{n \sum_{i=1}^{n}(\ln x_i)^2 - (\sum_{i=1}^{n}\ln x_i)^2} = \frac{4(8.975) - (3.599)(10.341)}{4(3.371) - (3.599)^2} = -2.480$$

This m, the sums from the table, $n = 4$, and Equation (4.18) solve parameter b.

$$b = \frac{\sum_{i=1}^{n}y_i - m \sum_{i=1}^{n}\ln x_i}{n} = \frac{10.341 - (-2.48)(3.599)}{4} = 4.817$$

The logarithmic regression function $r_{\ln}(x)$ according to Equation (4.15) is thus

$$r_{\ln}(x) = -2.48 \ln x + 4.817$$

Figure 8.2 depicts this logarithmic regression function $r_{\ln}(x)$ with the true solution $y = \sqrt{13 + 8\cos x}$ over the domain of the x values $[1.9, 3.1]$.

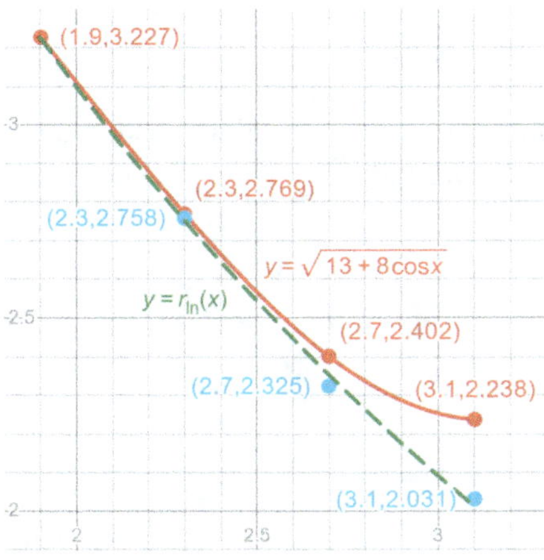

Figure 8.2: The green dashed logarithmic regression function $r_{\ln}(x) = -2.48 \ln x + 4.817$ fit through the solution points generated by Euler's method of Table 8.1. The true solution $y = \sqrt{13 + 8\cos x}$ is red.

8.1.3 A quadratic regression function

The Table 4.4 template is used for all needed sums, relabeling $y_i = L(x_i)$.

x_i	y_i	x_i^2	x_i^3	x_i^4	$x_i^2 y_i$	$x_i y_i$
1.9	3.227	3.61	6.859	13.032	11.649	6.131
2.3	2.758	5.29	12.167	27.984	14.590	6.343
2.7	2.325	7.29	19.683	53.144	16.949	6.278
3.1	2.031	9.61	29.791	92.352	19.518	6.296
10.0	10.341	25.80	68.500	186.512	62.706	25.048

Each sum and $n = 4$ is substituted in the augmented matrix of Equation (4.30).

$$\begin{bmatrix} 186.512 & 68.5 & 25.8 & \vdots & 62.706 \\ 68.5 & 25.8 & 10 & \vdots & 25.048 \\ 25.8 & 10 & 4 & \vdots & 10.341 \end{bmatrix}$$

Transforming to RREF by Gauss–Jordan elimination produces

$$\begin{bmatrix} 1 & 0 & 0 & \vdots & 0.285 \\ 0 & 1 & 0 & \vdots & -2.430 \\ 0 & 0 & 1 & \vdots & 6.822 \end{bmatrix}$$

By Equation (4.29), the three parameters are, in order, $a = 0.285$, $b = -2.43$, and $c = 6.822$. According to Equation (4.25), the quadratic regression function is

$$r_2(x) = 0.285x^2 - 2.43x + 6.822$$

In Figure 8.3, this quadratic regression function $r_2(x)$ is graphed with the true solution $y = \sqrt{13 + 8\cos x}$ over the domain of the x values $[1.9, 3.1]$.

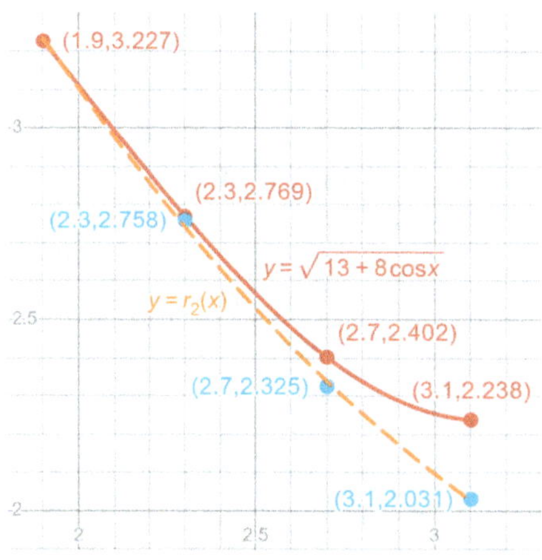

Figure 8.3: The orange dashed quadratic regression function $r_2(x) = 0.285x^2 - 2.43x + 6.822$ fit through the solution points generated by Euler's method of Table 8.1. The true solution $y = \sqrt{13 + 8\cos x}$ is red.

8.1.4 Regression function comparison

The logarithmic and quadratic regression functions of Figures 8.2 and 8.3, respectively, appear extremely comparable, supported by Figure 8.4 displaying both.

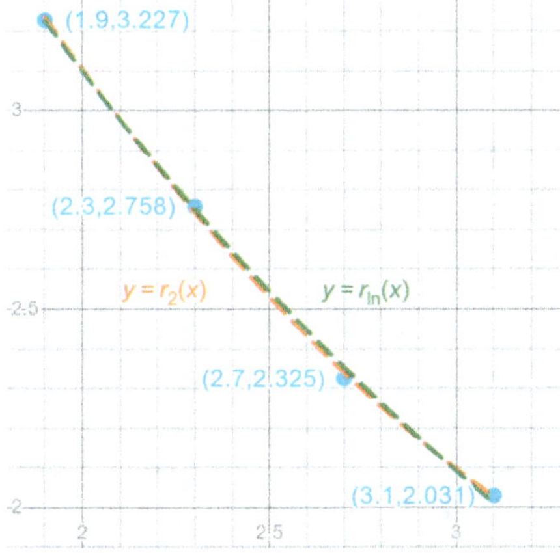

Figure 8.4: The green dashed logarithmic regression function $r_{\ln}(x) = -2.48 \ln x + 4.817$ of Figure 8.2 and the orange dashed quadratic regression function $r_2(x) = 0.285x^2 - 2.43x + 6.822$ of Figure 8.3, fit through the data of Table 8.1.

Table 8.2: Rounded to thousandths place, the logarithmic and quadratic regression functions produce near-true fits to the solution points of Table 8.1, evaluated by the least squares accumulation error E_a.

i	x_i	y_i	$r_{\ln}(x_i)$	$(y_i - r_{\ln}(x_i))^2$	$r_2(x_i)$	$(y_i - r_2(x_i))^2$
0	1.9	3.227	3.225	0.000	3.234	0.000
1	2.3	2.758	2.751	0.000	2.741	0.000
2	2.7	2.325	2.354	0.001	2.339	0.000
3	3.1	2.031	2.012	0.000	2.028	0.000
			$E_a =$	0.001	$E_a =$	0.000

Table 8.2 lists the approximated $r_{\ln}(x_i)$ and $r_2(x_i)$ values rounded to the thousandth decimal place and the least squares accumulation error E_a for each regression function, algebraically verifying both are nearly true fits.

8.1.5 Interpolation and regression comparison

In the absence of algebraic metrics comparing the efficacy of interpolation and regression approaches to fitting an appropriate function to the Euler's method DEQ solution points of Table 8.1, one function representing each technique is overlaid and observed in a single graph.

As discussed in Section 8.1.4, the logarithmic and quadratic regression functions are virtually identical graphically and algebraically. Therefore, only the quadratic regression function $r_2(x)$, a polynomial function, is compared to the third-degree DD polynomial interpolation function $f_3(x)$ in Figure 8.5. Note since $n = 4$ is small, the functions are quite similar.

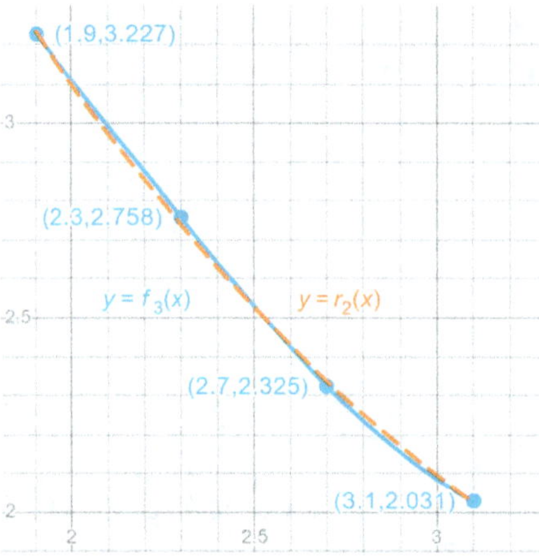

Figure 8.5: The blue solid third-degree DD polynomial function $y = f_3(x)$ and the orange dashed quadratic regression function $y = r_2(x)$ fit through the data of Table 8.1.

8.2 Midpoint method solutions

Functions in this section are fit to the solution of the DEQ $\frac{dy}{dx} = \frac{3x^2}{y}$ of Example 7.2, reproduced for convenience in Table 8.3.

Table 8.3: The midpoint method DEQ solution to $\frac{dy}{dx} = \frac{3x^2}{y}$ of Example 7.2.

i	x_i	$L(x_i)$
0	−0.7	0.560
1	−0.4	0.845
2	−0.1	0.906

8.2.1 A linear splines interpolation function

According to Equation (3.34) of Section 3.3.1, a linear spline between consecutive points is defined

$$s_i(x) = f(x_i) + \frac{f(x_{i+1}) - f(x_i)}{x_{i+1} - x_i}(x - x_i)$$

Using the current notation $f(x_i) = L(x_i)$ is relabeled for every function value, so in this example

$$s_0(x) = L(x_0) + \frac{L(x_1) - L(x_0)}{x_1 - x_0}(x - x_0)$$

$$s_1(x) = L(x_1) + \frac{L(x_2) - L(x_1)}{x_2 - x_1}(x - x_1)$$

Using the data of Table 8.3,

$$s_0(x) = 0.560 + \frac{0.845 - 0.560}{-0.4 - (-0.7)}(x - (-0.7))$$

$$= 0.560 + 0.950(x + 0.7) = 0.950x + 1.225$$

$$s_1(x) = 0.845 + \frac{0.906 - 0.845}{-0.1 - (-0.4)}(x - (-0.4))$$

$$= 0.845 + 0.203(x + 0.4) = 0.203x + 0.926$$

By Equation (3.35) the linear splines piecewise function is therefore

$$p_1(x) = \begin{cases} 0.950x + 1.225, & -0.7 \le x < -0.4 \\ 0.203x + 0.926, & -0.4 \le x \le -0.1 \end{cases}$$

Figure 8.6 illustrates this interpolated linear splines function $p_1(x)$ in blue, fit to points generated by the midpoint method, and an interpolated linear splines function $p_{1^*}(x)$ in black, fit to points generated by Euler's method. Clearly, over the domain of the x values $[-0.7, -0.1]$ the blue $p_1(x)$ better approximates the true solution in red than the black $p_{1^*}(x)$.

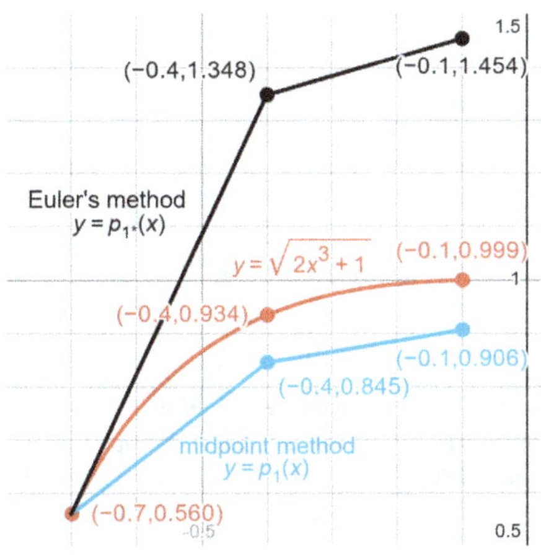

Figure 8.6: The red true solution $y = \sqrt{2x^3 + 1}$ to the DEQ $\frac{dy}{dx} = \frac{3x^2}{y}$, and linear splines functions $p_1(x)$ in blue and $p_{1^*}(x)$ in black using solution points generated by the midpoint method in Table 8.3 and Euler's method, respectively.

8.2.2 A linear regression function

The Table 4.1 template is populated for all needed sums, relabeling $y_i = L(x_i)$.

i	x_i	y_i	x_i^2	x_iy_i
1	−0.7	0.560	0.49	−0.392
2	−0.4	0.845	0.16	−0.338
3	−0.1	0.906	0.01	−0.091
	−1.2	2.311	0.66	−0.821

Parameter m is found by Equation (4.8) using the sums from the table and $n = 3$.

$$m = \frac{n \sum_{i=1}^{n} x_i y_i - \sum_{i=1}^{n} x_i \sum_{i=1}^{n} y_i}{n \sum_{i=1}^{n} x_i^2 - (\sum_{i=1}^{n} x_i)^2} = \frac{3(-0.821) - (-1.2)(2.311)}{3(0.66) - (-1.2)^2} = 0.574$$

This m, the sums from the table, $n = 3$, and Equation (4.9) solve parameter b.

$$b = \frac{\sum_{i=1}^{n} y_i - m \sum_{i=1}^{n} x_i}{n} = \frac{2.311 - (0.574)(-1.2)}{3} = 1.000$$

The linear regression function $r_1(x)$ according to Equation (4.2) is therefore

$$r_1(x) = 0.574x + 1.000$$

Figure 8.7 compares this linear regression function $r_1(x)$ with the true solution $y = \sqrt{2x^3 + 1}$ over the domain of the x values $[-0.7, -0.1]$.

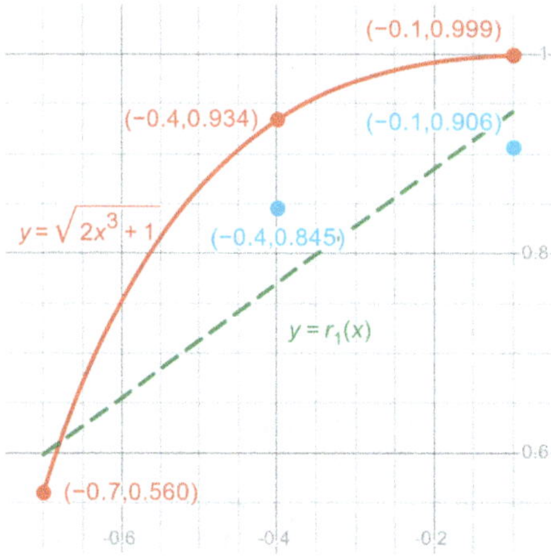

Figure 8.7: The green dashed linear regression function $r_1(x) = 0.574x + 1.000$ fit through the solution points generated by the midpoint method of Table 8.3. The true solution $y = \sqrt{2x^3 + 1}$ is red.

8.2.3 Interpolation and regression comparison

Figure 8.8 contrasts two solution approaches using linear functions and n points. With interpolation, $n - 1$ linear spline functions are found between each pair of consecutive points and define a piecewise function. Using regression, a single linear function is found which minimizes the accumulation error between the given values and those approximated by the regression function. In this example, the graph suggests interpolation is better, but in general, the data determines whether interpolation or regression, using linear functions, is superior.

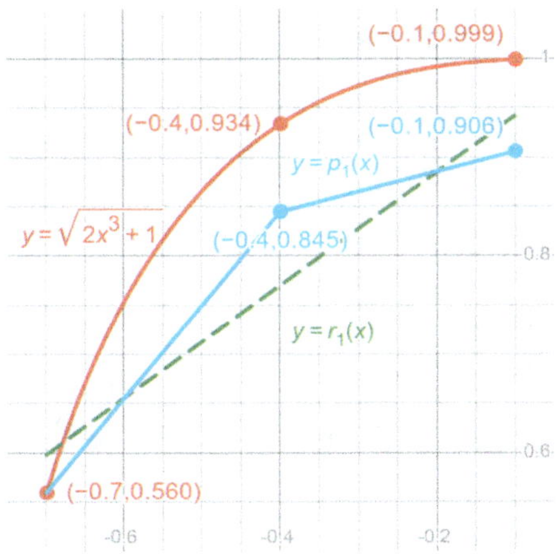

Figure 8.8: The blue solid linear splines interpolation function $y = p_1(x)$ and the green dashed linear regression function $y = r_1(x)$ fit through the data of Table 8.3, with the red true solution $y = \sqrt{2x^3 + 1}$.

8.3 Heun's method solutions

Functions in this section are fit to the solution of the DEQ $\frac{dy}{dx} = 2xy$ of Example 7.3, reproduced for convenience in Table 8.4.

Table 8.4: Heun's method DEQ solution to $\frac{dy}{dx} = 2xy$ of Example 7.3.

i	x_i	$L(x_i)$
0	1.0	1.087
1	1.4	2.618
2	1.8	8.080

8.3.1 A quadratic splines interpolation function

Referring to Section 3.3.2, three points require two quadratic splines defined by Equation (3.36).

$$s_0(x) = a_0 + b_0(x - x_0) + c_0(x - x_0)^2$$
$$s_1(x) = a_1 + b_1(x - x_1) + c_1(x - x_1)^2$$

Determining parameters $a_i, b_i,$ and c_i for $i = 0, 1$ for the points of Table 8.4 follows Steps 1 through 4 of Table 3.5. Note subscripts start at $i = 0$ rather than $i = 1$ as in Section 3.3.2, because these solution points distinguish the one given point at x_0 and the points generated by Heun's method at x_1 and x_2. In Section 3.3.2 *all* points were given.

Similarly, Section 3.3.2 used the notation $f(x_i)$ for the given y-coordinates of all points, whereas here notation for the approximated y-coordinates $L(x_i)$ replaces $f(x_i)$. Note, however, $L(x_0) = F(x_0)$, the one given y-coordinate.

Step 1 and Equation (3.37) defines the two a_i parameters.

$$a_0 = f(x_0) = L(x_0) = 1.087$$
$$a_1 = f(x_1) = L(x_1) = 2.618$$

Equation (3.43) designates the first b_i as b_1, and by Step 2

$$b_1 = \frac{f(x_2) - f(x_1)}{x_2 - x_1}$$

Here, the first b_i is b_0, since all subscripts are decreased by one. Substituting $L(x_i)$ for $f(x_i)$ as described above

$$b_0 = \frac{L(x_1) - L(x_0)}{x_1 - x_0} = \frac{2.618 - 1.087}{1.4 - 1.0} = 3.828$$

Step 2 further states the first two b_i parameters are equal, so that $b_1 = b_0 = 3.828$. Because only two b_i parameters are needed, Step 3 is not referenced.

Step 4 first notes because the first quadratic spline is actually linear, $c_1 = 0$, here notated as $c_0 = 0$. Second, subsequent c_i parameters starting at c_2 use Equation (3.39) so that

$$c_2 = \frac{f(x_3) - f(x_2) - b_2(x_3 - x_2)}{(x_3 - x_2)^2}$$

Decreasing all subscripts by one, the next c_i is c_1. Using the approximated y-coordinates with the adjusted subscripts and substituting in the above equation yields

$$c_1 = \frac{L(x_2) - L(x_1) - b_1(x_2 - x_1)}{(x_2 - x_1)^2} = \frac{8.080 - 2.618 - 3.828(1.8 - 1.4)}{(1.8 - 1.4)^2} = 24.568$$

The linear and quadratic splines are as follows, subsequently written as the piecewise function according to Equation (3.44).

$$s_0(x) = 1.087 + 3.828(x - 1)$$

$$s_1(x) = 2.618 + 3.828(x - 1.4) + 24.568(x - 1.4)^2$$

$$p_2(x) = \begin{cases} 1.087 + 3.828(x - 1), & 1 \leq x < 1.4 \\ 2.618 + 3.828(x - 1.4) + 24.568(x - 1.4)^2, & 1.4 \leq x \leq 1.8 \end{cases}$$

Figure 8.9 displays this interpolated quadratic splines function $p_2(x)$ with the true solution $y = 0.4e^{x^2}$ over the domain of the x values $[1.0, 1.8]$.

Note the difference in the vertical disparity between each approximated blue point and its corresponding true red point. At $x_1 = 1.4$

$$\left| L(x_1) - y(x_1) \right| = \left| L(1.4) - y(1.4) \right| = \left| 2.618 - 2.84 \right| = 0.222$$

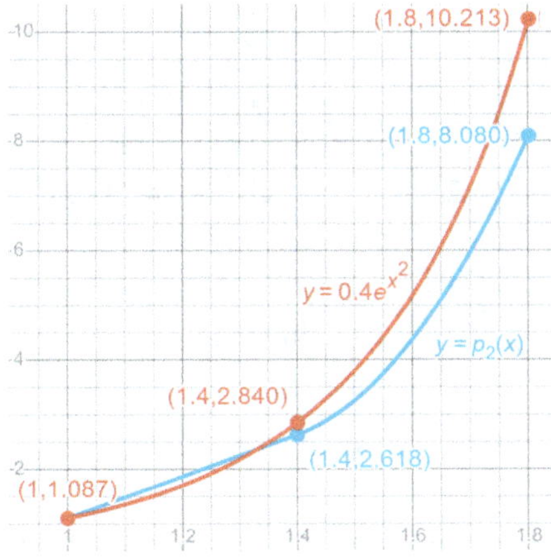

Figure 8.9: The blue quadratic splines function $y = p_2(x)$ fit through the solution points generated by Heun's method of Table 8.4. The true solution $y = 0.4e^{x^2}$ is red.

And at $x_2 = 1.8$

$$\left|L(x_2) - y(x_2)\right| = \left|L(1.8) - y(1.8)\right| = |8.08 - 10.213| = 2.133$$

The quality of any approximation not only hinges on the method used but also the nature of the true solution. The exponentially increasing analytic solution to the DEQ of Example 7.3 contributes to the increasing absolute error between the approximated and true y-coordinates as x values increase at the same constant interval length, here $h = 0.4$. Additionally, keeping h as sufficiently small as is reasonably possible is always a sound approach no matter which method is applied and regardless of the underlying solution.

Revisiting Example 7.3, suppose the only condition changed is setting $h = 0.2$. Now $x = 1.8$ is at x_4 instead of x_2 because the interval length h has been halved, doubling the number of x values. The re-approximated y-coordinate for $h = 0.2$ is $L(x_4) = 9.389$ and the absolute error drops from 2.133 to

$$\left|L(x_4) - y(x_4)\right| = \left|L(1.8) - y(1.8)\right| = |9.389 - 10.213| = 0.824$$

Or, if the interval is one-fourth the original length, $h = 0.1$, then there are four times as many x values and $x_2 = 1.8$ is now $x_8 = 1.8$. In turn the approximated y-coordinate adjusts to $L(x_8) = 9.952$ for a further shrinking absolute error.

$$\left|L(x_8) - y(x_8)\right| = \left|L(1.8) - y(1.8)\right| = |9.952 - 10.213| = 0.261$$

8.3.2 A Lagrange polynomial interpolation function

Instead of using two separate quadratic spline functions between both pairs of points, a single quadratic function is now fit through all three points using a second-degree Lagrange polynomial function. According to Equation (3.33) and the data of Table 8.4

$$\mathcal{L}_2(x) = \frac{(x - x_1)(x - x_2)}{(x_0 - x_1)(x_0 - x_2)}f(x_0) + \frac{(x - x_0)(x - x_2)}{(x_1 - x_0)(x_1 - x_2)}f(x_1) + \frac{(x - x_0)(x - x_1)}{(x_2 - x_0)(x_2 - x_1)}f(x_2)$$

$$= \frac{(x - 1.4)(x - 1.8)(1.087)}{(1 - 1.4)(1 - 1.8)} + \frac{(x - 1)(x - 1.8)(2.618)}{(1.4 - 1)(1.4 - 1.8)} + \frac{(x - 1)(x - 1.4)(8.08)}{(1.8 - 1)(1.8 - 1.4)}$$

$$= \frac{(x^2 - 3.2x + 2.52)(1.087)}{0.32} + \frac{(x^2 - 2.8x + 1.8)(2.618)}{-0.16} + \frac{(x^2 - 2.4x + 1.4)(8.08)}{0.32}$$

$$= (x^2 - 3.2x + 2.52)(3.396875) - 16.3625(x^2 - 2.8x + 1.8) + 25.25(x^2 - 2.4x + 1.4)$$

$$= 3.396875x^2 - 10.87x + 8.560125 - 16.3625x^2 + 45.815x - 29.4525$$

$$+ 25.25x^2 - 60.6x + 35.35$$

$$\mathcal{L}_2(x) = 12.284375x^2 - 25.655x + 14.457625 \tag{8.1}$$

The decimal values terminate and all places were retained to demonstrate a significant result discussed in Section 8.3.4.

Figure 8.10 compares this Lagrange polynomial interpolation function $\mathcal{L}_2(x)$ to the true solution $y = 0.4e^{x^2}$ over the domain of x values $[1.0, 1.8]$.

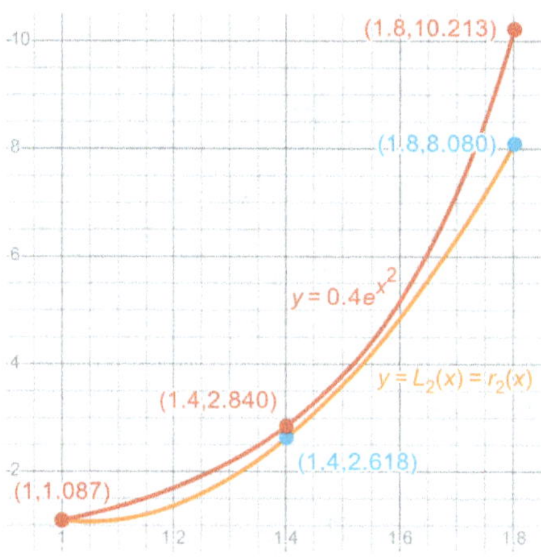

Figure 8.10: The orange second-degree Lagrange polynomial function $y = \mathcal{L}_2(x)$ of Equation (8.1) fit through the solution points generated by Heun's method of Table 8.4. The true solution $y = 0.4e^{x^2}$ is red. As discussed in Section 8.3.4, the quadratic regression function $y = r_2(x)$ of Equation (8.2) is equivalent to $y = \mathcal{L}_2(x)$.

The quadratic splines and second-degree Lagrange polynomial interpolation functions in Figures 8.9 and 8.10, respectively, are shown together in Figure 8.11.

8.3.3 A quadratic regression function

Analogous to finding $\mathcal{L}_2(x)$ of Equation (8.1) in Section 8.3.2, all decimal values in the following example terminate and all places are retained in solving $r_2(x)$; the significance is addressed in Section 8.3.4.

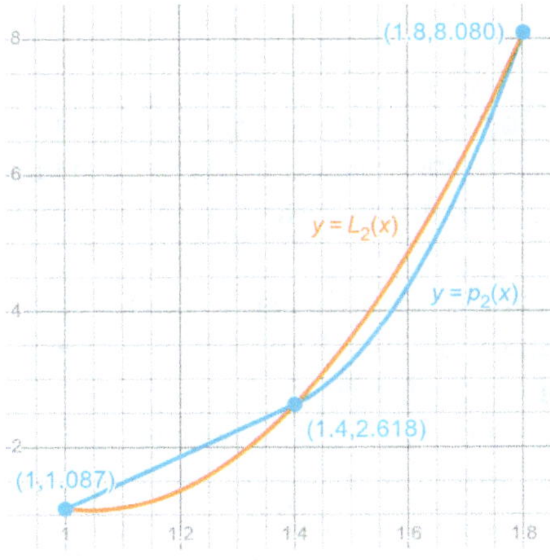

Figure 8.11: The blue quadratic splines function $y = p_2(x)$ of Figure 8.9 and the orange second-degree Lagrange polynomial function $y = \mathcal{L}_2(x)$ of Figure 8.10. Albeit differently, both use interpolated quadratic polynomial functions fit through the data of Table 8.4.

The Table 4.4 template is filled with all needed sums, relabeling $y_i = L(x_i)$.

x_i	y_i	x_i^2	x_i^3	x_i^4	$x_i^2 y_i$	$x_i y_i$
1.0	1.087	1.00	1.000	1.0000	1.08700	1.0870
1.4	2.618	1.96	2.744	3.8416	5.13128	3.6652
1.8	8.080	3.24	5.832	10.4976	26.17920	14.5440
4.2	11.785	6.20	9.576	15.3392	32.39748	19.2962

Each sum and $n = 3$ is substituted in the augmented matrix of Equation (4.30).

$$\left[\begin{array}{ccc:c} 15.3392 & 9.576 & 6.2 & 32.39748 \\ 9.576 & 6.2 & 4.2 & 19.2962 \\ 6.2 & 4.2 & 3 & 11.785 \end{array} \right]$$

Using Gauss–Jordan elimination for RREF produces

$$\left[\begin{array}{ccc:c} 1 & 0 & 0 & 12.284375 \\ 0 & 1 & 0 & -25.655 \\ 0 & 0 & 1 & 14.457625 \end{array} \right]$$

According to Equation (4.29), the parameters are, in order, $a = 12.284375$, $b = -25.655$, and $c = 14.457625$. By Equation (4.25), the quadratic regression function is

$$r_2(x) = 12.284375x^2 - 25.655x + 14.457625 \tag{8.2}$$

8.3.4 Interpolation and regression comparison

Retaining all decimal places in the calculation of both the second-degree Lagrange polynomial interpolation function $\mathcal{L}_2(x)$ of Equation (8.1) and the quadratic regression function $r_2(x)$ of Equation (8.2) is now justified; they are identical. The orange function of Figure 8.10 thus represents both $y = \mathcal{L}_2(x)$ and $y = r_2(x)$.

Although interpolation and regression techniques are motivated differently, the equality of $\mathcal{L}_2(x)$ and $r_2(x)$ in this example stems from both the particular regression function applied and the specific number of points. Recall the goal of regression is minimizing some accumulated difference between the given values and those approximated by a regression function, with an optimal difference of zero. If the regression function is an nth-degree polynomial and the data contains $n + 1$ points, such a regression function fits exactly through those given points and achieves the optimal difference of zero.

And precisely, the goal of one interpolation technique is fitting an nth degree polynomial exactly through $n + 1$ points. Since that nth degree polynomial is unique, it must be identical to the nth degree polynomial found through the same $n + 1$ points using regression in which the accumulated difference between the given and approximated values is zero.

For the data of Table 8.4, a quadratic regression function $r_2(x)$ is a second-degree polynomial function that minimizes the accumulated difference between each of the three given $L(x_i)$ values and the corresponding approximated $r_2(x_i)$ values. That difference is zero because the fit is exact for an nth degree polynomial function through $n + 1$ points for $n = 2$. And by definition, $\mathcal{L}_2(x)$ is the second degree polynomial function through the three points exactly. Since there exists only one second-degree polynomial function through the same three points, $r_2(x) = \mathcal{L}_2(x)$, which is why Equations (8.1) and (8.2) are equal. Generalizing, an nth degree DD or Lagrange polynomial interpolation function always equals an nth degree regression function fit through the same $n + 1$ points.

8.4 Ralston's method solutions

Functions in this section are fit to the solution of the DEQ $\frac{dy}{dx} = -\frac{2x}{e^y}$ of Example 7.4, reproduced for convenience in Table 8.5.

Table 8.5: Ralston's method DEQ solution to $\frac{dy}{dx} = -\frac{2x}{e^y}$ of Example 7.4.

i	x_i	$L(x_i)$
0	2.0	3.332
1	3.5	2.999
2	5.0	2.141

8.4.1 A cubic splines interpolation function

Revisiting Section 3.3.3, two cubic splines fit to three points are defined by Equation (3.45).

$$s_0(x) = a_0 + b_0(x - x_0) + c_0(x - x_0)^2 + d_0(x - x_0)^3$$
$$s_1(x) = a_1 + b_1(x - x_1) + c_1(x - x_1)^2 + d_1(x - x_1)^3$$

The parameters a_i, b_i, c_i, and d_i for $i = 0,1$ for the points of Table 8.5 are solved following Steps 1 through 5 of Table 3.6 with two notation modifications. All subscripts are decreased by one, and all $f(x_i)$ are replaced by $L(x_i)$; identical notation changes were applied and explained in Section 8.3.1.

Step 1 and Equation (3.37) defines the two a_i parameters.

$$a_0 = f(x_0) = L(x_0) = 3.332$$
$$a_1 = f(x_1) = L(x_1) = 2.999$$

In Step 2, an augmented matrix is formed using Equation (3.57).

$$\begin{bmatrix} 1 & 0 & 0 \\ x_2 - x_1 & 2(x_3 - x_1) & x_3 - x_2 \\ 0 & 0 & 1 \end{bmatrix} \times \begin{bmatrix} c_1 \\ c_2 \\ c_3 \end{bmatrix} = \begin{bmatrix} 0 \\ 3(\frac{f(x_3)-f(x_2)}{x_3-x_2}) - 3(\frac{f(x_2)-f(x_1)}{x_2-x_1}) \\ 0 \end{bmatrix}$$

According to the altered notation, this is equivalent to

$$\begin{bmatrix} 1 & 0 & 0 \\ x_1 - x_0 & 2(x_2 - x_0) & x_2 - x_1 \\ 0 & 0 & 1 \end{bmatrix} \times \begin{bmatrix} c_0 \\ c_1 \\ c_2 \end{bmatrix} = \begin{bmatrix} 0 \\ 3(\frac{L(x_2)-L(x_1)}{x_2-x_1}) - 3(\frac{L(x_1)-L(x_0)}{x_1-x_0}) \\ 0 \end{bmatrix}$$

$$\begin{bmatrix} 1 & 0 & 0 \\ 3.5 - 2 & 2(5 - 2) & 5 - 3.5 \\ 0 & 0 & 1 \end{bmatrix} \times \begin{bmatrix} c_0 \\ c_1 \\ c_2 \end{bmatrix} = \begin{bmatrix} 0 \\ 3(\frac{2.141-2.999}{5.0-3.5}) - 3(\frac{2.999-3.332}{3.5-2.0}) \\ 0 \end{bmatrix}$$

$$\begin{bmatrix} 1 & 0 & 0 & \vdots & 0 \\ 1.5 & 6 & 1.5 & \vdots & -1.05 \\ 0 & 0 & 1 & \vdots & 0 \end{bmatrix}$$

Using Gauss–Jordan elimination to put in RREF yields

$$\begin{bmatrix} 1 & 0 & 0 & \vdots & 0 \\ 0 & 1 & 0 & \vdots & -0.175 \\ 0 & 0 & 1 & \vdots & 0 \end{bmatrix}$$

The rightmost column indicates $c_0 = 0$ and $c_1 = -0.175$.

Step 3 uses Equation (3.52) for b_1.

$$b_1 = \frac{f(x_2) - f(x_1)}{x_2 - x_1} - \frac{(x_2 - x_1)(c_2 + 2c_1)}{3}$$

Using the reduced subscripts and $L(x_i)$ values, the first b_i here is

$$b_0 = \frac{L(x_1) - L(x_0)}{x_1 - x_0} - \frac{(x_1 - x_0)(c_1 + 2c_0)}{3}$$

$$= \frac{2.999 - 3.332}{3.5 - 2} - \frac{(3.5 - 2)(-0.175 + 2 \cdot 0)}{3} = -0.135$$

Although subsequent b_i parameters could be solved with Equation (3.52) again, the more efficient alternate Equation (3.51) is used per Step 4.

$$b_1 = b_0 + (x_1 - x_0)(c_1 + c_0) = -0.135 + (3.5 - 2)(-0.175 + 0) = -0.398$$

Lastly, in Step 5 d_0 and d_1 are found by Equation (3.50).

$$d_0 = \frac{c_1 - c_0}{3(x_1 - x_0)} = \frac{-0.175 - 0}{3(3.5 - 2)} = -0.039$$

$$d_1 = \frac{c_2 - c_1}{3(x_2 - x_1)} = \frac{0 - (-0.175)}{3(5 - 3.5)} = 0.039$$

The two cubic splines follow.

$$s_0(x) = 3.332 - 0.135(x - 2) - 0.039(x - 2)^3$$

$$s_1(x) = 2.999 - 0.398(x - 3.5) - 0.175(x - 3.5)^2 + 0.039(x - 3.5)^3$$

Using Equation (3.58) the cubic splines piecewise function is

$$p_3(x) = \begin{cases} 3.332 - 0.135(x - 2) - 0.039(x - 2)^3, & 2.0 \le x < 3.5 \\ 2.999 - 0.398(x - 3.5) - 0.175(x - 3.5)^2 + 0.039(x - 3.5)^3, & 3.5 \le x \le 5.0 \end{cases}$$

Figure 8.12 compares this interpolated cubic splines function $p_3(x)$ to the true solution $y = \ln(32 - x^2)$ over the domain of the x values [2, 5].

As mentioned in Section 8.3.1, the underlying true solution, here a natural log function, contributes to the fit of the approximated y-coordinates as the x-coordinates increase by a constant interval length h. The absolute error of the y-coordinates at $x_1 = 3.5$ is

$$|L(x_1) - y(x_1)| = |L(3.5) - y(3.5)| = |2.999 - 2.983| = 0.016$$

While at $x_2 = 5$ the absolute error of the y-coordinates is

$$|L(x_2) - y(x_2)| = |L(5) - y(5)| = |2.141 - 1.946| = 0.195$$

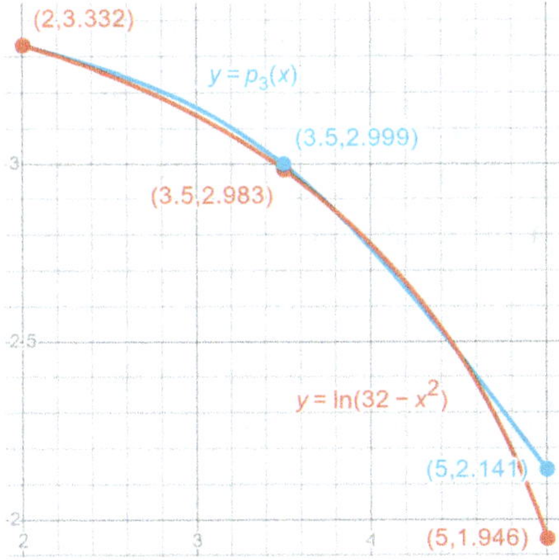

Figure 8.12: The blue cubic splines piecewise function $y = p_3(x)$ fit through the solution points generated by Ralston's method of Table 8.5. The true solution $y = \ln(32 - x^2)$ is red.

Also, as observed in Section 8.3.1, decreasing the size of h improved the approximated points. Suppose the interval length of Example 7.4 is divided by three, $h = 0.5$ replacing $h = 1.5$. With three times the number of points, $x_2 = 5$ becomes $x_6 = 5$. The re-approximated y-coordinate with $h = 0.5$ is $L(x_6) = 1.985$. In turn, the vertical absolute error is reduced from 0.195 to

$$\left|L(x_6) - y(x_6)\right| = \left|L(5) - y(5)\right| = |1.985 - 1.946| = 0.039$$

8.4.2 A linear regression function

The points and cubic splines piecewise interpolation function of Figure 8.12 indicate an underlying function which is decreasing and concave down, not a good candidate for a logarithmic regression function as analyzed in Section 4.3.2. And as explained in Section 8.3.4, a quadratic regression function using three points is equivalent to a second-degree DD or Lagrange polynomial interpolation function. Hence a linear regression function is applied to the data of Table 8.5, the only regression function both applicable to the data and not duplicated by an interpolation technique.

The Table 4.1 template is used for all necessary sums, relabeling $y_i = L(x_i)$.

i	x_i	y_i	x_i^2	$x_i y_i$
0	2.0	3.332	4.00	6.664
1	3.5	2.999	12.25	10.497
2	5.0	2.141	25.00	10.705
	10.5	8.472	41.25	27.866

Parameter m is found by Equation (4.8) with the sums in the table and $n = 3$.

$$m = \frac{n \sum_{i=1}^{n} x_i y_i - \sum_{i=1}^{n} x_i \sum_{i=1}^{n} y_i}{n \sum_{i=1}^{n} x_i^2 - (\sum_{i=1}^{n} x_i)^2} = \frac{3(27.866) - (10.5)(8.472)}{3(41.25) - (10.5)^2} = -0.397$$

This m, the sums from the table, $n = 3$, and Equation (4.9) solve parameter b.

$$b = \frac{\sum_{i=1}^{n} y_i - m \sum_{i=1}^{n} x_i}{n} = \frac{8.472 - (-0.397)(10.5)}{3} = 4.214$$

The linear regression function $r_1(x)$ according to Equation (4.2) is thus

$$r_1(x) = -0.397x + 4.214$$

Figure 8.13 displays this linear regression function $r_1(x)$ with the true solution $y = \ln(32 - x^2)$ over the domain of the x values $[2.0, 5.0]$.

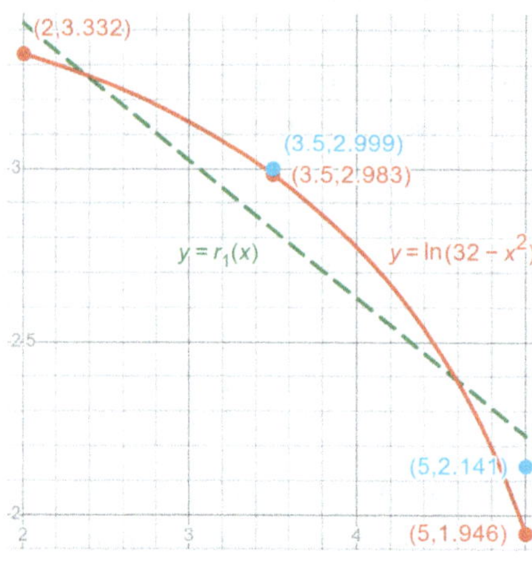

Figure 8.13: The green dashed linear regression function $r_1(x) = -0.397x + 4.214$ fit through the solution points generated by Ralston's method of Table 8.5. The true solution $y = \ln(32 - x^2)$ is red.

Exercises

Sections 8.1–8.4

Use any DEQ method from Chapter 7 to approximate six additional points for the DEQ, initial point $(x_0, L(x_0))$, and h value. Use an interpolation function from Chapter 3 or an appropriate regression function from Chapter 4 to model all seven points.

The numerical solutions of the points and the modeling function for each DEQ are open-ended; finding an optimal fit using multiple combinations of DEQ methods and interpolation or regression functions is encouraged. Each DEQ is separable and the analytic solution is given in the Appendix so that every numerical solution can be compared and evaluated against the exact solution.

1. $dy/dx = -e^x y^3$ $(0, 2.5)$ $h = 0.275$
2. $dy/dx = 72y/x^4$ $(2.15, 0.42)$ $h = 0.7$
3. $dy/dx = 5xe^{-y}$ $(1.5, 1.75)$ $h = 1.25$
4. $dy/dx = 9y^2/x$ $(3.5, 0.2)$ $h = 0.4$
5. $dy/dx = 0.7x^3/y$ $(1, 3.5)$ $h = 0.5$
6. $dy/dx = -3y^2$ $(0.3, 2)$ $h = 0.1$
7. $dy/dx = \dfrac{-4.78\sqrt{y}}{1+(x-10)^2}$ $(2.1, 11.925)$ $h = 1.1$
8. $dy/dx = 3.1x/(y + 2)$ $(0, 1)$ $h = 2$
9. $dy/dx = y^3/(4x + 1)$ $(4.3, 1.5)$ $h = 0.8$
10. $dy/dx = 0.87(y - 5.7)/x^2$ $(0.5, 6.3)$ $h = 0.6$
11. $dy/dx = (0.6x^2 + 3.46)/y^4$ $(0.4, 1)$ $h = 0.15$
12. $dy/dx = -2.5 \sin(0.1x)/y$ $(1, 10.2)$ $h = 1.5$

Appendix: Answers to odd-numbered exercises

Section 1.1

1. $x_5 = 1.219$ 3. $x_6 = 2.172$
5. $x_4 = 5.488$ 7. $x_6 = 4.469$
9. $x_6 = 2.844$, best at $x_5 = 2.813$ 11. $x_4 = 2.438$
13. $x_5 = 1.344$, best at $x_3 = 1.375$

Section 1.2

15. $x_4 = 0.939$ 17. $x_3 = 7.511$ 19. $x_4 = -1.246$
21. $x_5 = 2.436$ 23. $x_5 = -0.544$ 25. $x_8 = 2.419$

Sections 1.1 and 1.2

27. bisection, $x_8 = 1.678$;   false position, $x_5 = 1.674$
29. bisection, $x_6 = 1.094$;   false position, $x_8 = 1.163$
31. bisection, $x_9 = -7.016$,
    best at $x_8 = -7.031$;   false position, $x_5 = -7.026$

Section 1.3

33. $x_5 = 1.149$ 35. $x_4 = 0.865$ 37. diverges; $x_5 < -5.757$ billion
39. $x_8 = 1.516$ 41. $x_6 = 2.242$ 43. diverges; $x_5 > 4.247$ billion
45. $x_6 = -0.651$

Section 1.4

47. $x_6 = 1.649$ 49. $x_5 = 5.705$ 51. $x_5 = 3.183$
53. diverges; the difference between consecutive x values approaches 1.255
55. $x_5 = 1.558$ 57. $x_3 = -0.584$
59. diverges; $x_5 < 0$ outside domain   61. $x_3 = 8.913$

Section 1.5

63. $x_6 = 2.209$ 65. $x_5 = -0.635$
67. $x_6 = 1.407$ 69. diverges; $x_2 < 0$, outside domain
71. $x_5 = -1.596$ 73. diverges; when $i = 4$, $f(x_4) = f(x_5)$; a zero denominator
    terminates the algorithm while $f(x_5) \neq 0$
75. $x_6 = 2.337$ 77. $x_4 = 17.304$

https://doi.org/10.1515/9783112221051-009

Section 1.6

79. $x_5 = 1.359$ 81. $x_6 = 2.657$
83. diverges; consecutive values differ in sign and increase in magnitude
85. $x_4 = 4.518$ 87. $x_4 = 3.063$
89. diverges; $x_2 < 0$ outside domain
91. $x_6 = 3.641$ 93. $x_4 = -2.684$

Sections 1.5 and 1.6

95. secant, $x_8 = 3.802$;   modified secant, $x_6 = 3.802$
97. secant, $x_9 = 2.791$;   modified secant, $x_6 = 2.791$
99. secant, $x_{12} = -0.418$;   modified secant, $x_5 = -0.418$

Sections 1.4 and 1.6

101. Newton–Raphson, $x_5 = 4.861$;   modified secant, $x_5 = 4.861$
103. Newton–Raphson, $x_5 = 1.105$;   modified secant, $x_4 = 1.105$
105. Newton–Raphson, $x_4 = -1.062$;   modified secant, $x_8 = -1.063$

Section 2.1

1. $x_5 = 1.718; f''(x_5) > 0$, min   3. $x_4 = -5.041; f''(x_4) < 0$, max
5. $x_5 = -2.545; f''(x_5) > 0$, min   7. $x_6 = 2.617; f''(x_6) < 0$, max
9. $x_5 = 4.512; f''(x_5) < 0$, max   11. $x_3 = 3.279; f''(x_3) > 0$, min
13. $x_6 = -1.871; f''(x_6) > 0$, min   15. $x_4 = -4.376; f''(x_4) < 0$, max

Section 2.2

17. $i = 6$ $x_1 = 1.459$ $f(1.459) = 4.299$
19. $i = 7$ $x_2 = -3.472$ $f(-3.472) = 2.268$
21. $i = 5$ $x_1 = 2.391$ $f(2.391) = -1.028$
23. $i = 4$ $x_2 = -1.087$ $f(-1.087) = -1.174$
25. $i = 8$ $x_2 = 1.372$ $f(1.372) = 0.963$
27. $i = 6$ $x_1 = -5.013$ $f(-5.013) = 2.531$
29. $i = 5$ $x_1 = 0.649$ $f(0.649) = -0.653$
31. $i = 6$ $x_2 = -4.369$ $f(-4.369) = -4.298$

Section 2.3

33.	$i = 6$	$x_4 = 4.037$	$f(4.037) = 1.762$
35.	$i = 5$	$x_4 = -1.518$	$f(-1.518) = 6.803$
37.	$i = 5$	$x_4 = -3.370$	$f(-3.370) = -5.538$
39.	$i = 7$	$x_4 = 0.913$	$f(0.913) = -4.197$
41.	$i = 7$	$x_4 = 3.535$	$f(3.535) = -5.835$
43.	$i = 4$	$x_4 = -6.460$	$f(-6.460) = -5.142$
45.	$i = 6$	$x_4 = -3.141$	$f(-3.141) = 1.823$
47.	$i = 4$	$x_4 = 2.592$	$f(2.592) = 0.949$

Section 3.1

1. $f_3(x) = 3 + 1.5(x - 2) + 0.3(x - 2)(x - 4) + 0.075(x - 2)(x - 4)(x - 7)$
3. $f_3(x) = 1 + 2x + 0.4x(x - 3) - 0.08x(x - 3)(x - 5)$
5. $f_3(x) = -2 + 1.7(x + 3) - 0.025(x + 3)(x + 1)(x - 7)$
7. $f_4(x) = 4.7 - 2.1(x+1) - 0.84x(x+1) + 1.84x(x+1)(x-2.5) + 0.18x(x+1)(x-2.5)(x-3.5)$
9. $f_4(x) = 5 + 3.4(x - 1) - 1.5(x - 1)(x - 1.5) + 0.54(x - 1)(x - 1.5)(x - 3)$
   $\qquad - 0.12(x - 1)(x - 1.5)(x - 3)(x - 4)$
11. $f_4(x) = 0.5 + 2.6x - 0.8x(x - 1.5) + 0.4x(x - 1.5)(x - 2)(x - 3)$
13. $f_5(x) = 4.8 - 2.5(x - 2) - 15.625(x - 2)(x - 2.4)(x - 2.6)(x - 3)$
   $\qquad + 78.125(x - 2)(x - 2.4)(x - 2.6)(x - 3)(x - 3.2)$
15. $f_5(x) = 6.2 + 1.9(x + 2)(x + 1.5) - (x + 2)(x - 1.5)(x - 0.5)$
   $\qquad - 0.6(x + 2)(x - 1.5)(x - 0.5)(x - 1)(x - 1.5)$
17. $f_3(x) = 2.3 - 3.2(x - 1.5) + 2.6(x - 1.5)(x - 2) - 0.76(x - 1.5)(x - 2)(x - 3)$
   $\qquad f(2.5) = 0.597, E_t = 0.007, E_{t,r} = 1.17\,\%, [0.590, 0.604]$
19. $f_3(x) = -15.9 + 2.74(x + 10) - 0.335(x + 10)(x + 4.5) + 0.07(x + 10)(x + 4.5)(x - 2)$
   $\qquad f(3.5) = -3.75, E_t = 1.296, E_{t,r} = 34.56\,\%, [-5.046, -2.454]$
21. $f_4(x) = -0.5 - 0.35(x - 1) + 0.375(x - 1)(x - 3) + 0.095(x - 1)(x - 3)(x - 7)$
   $\qquad - 0.045(x - 1)(x - 3)(x - 7)(x - 8)$
   $\qquad f(2) = 0.6, E_t = -0.540, E_{t,r} = 90.00\,\%, [0.060, 1.140]$
23. $f_5(x) = 7 + 1.2(x + 3) - 0.5x(x + 3) + 0.35x(x + 3)(x - 2)$
   $\qquad + 0.26x(x + 3)(x - 2)(x - 3) - 0.12x(x + 3)(x - 2)(x - 3)(x - 3.5)$
   $\qquad f(4) = 32.2, E_t = -0.392, E_{t,r} = 1.22\,\%, [31.808, 32.592]$
25. $f_2(x) = 12.2 - 3(x + 6) + 0.16(x + 6)(x + 2.5)$
27. $f_2(x) = 13 + 3x - 0.42x(x - 4.1)$
29. $f_3(x) = -3.7 - 2.9(x - 2.5) + 3.65(x - 2.5)(x - 3.8) + 1.28(x - 2.5)(x - 3.8)(x - 5)$
31. $f_4(x) = 1.8 + 2.05(x - 0.5) + 2.9(x - 0.5)(x - 0.9) - 1.6(x - 0.5)(x - 0.9)(x - 1.4)$
   $\qquad - 1.675(x - 0.5)(x - 0.9)(x - 1.4)(x - 2)$

Section 3.2

33. $L_2(x) = 0.3x^2 - 3.1x + 5.075$ 35. $L_2(x) = -4.5x^2 + 8.4x + 25.2$

37. $L_2(x) = 2.2x^2 - 17.02x + 16.5$ 39. $L_2(x) = -0.4x^2 + 13$

Sections 3.1 and 3.2

41. $L_2(x) = 1.1x^2 - 16.05x + 59.7 =$
$f_2(x) = 32 - 7.25(x - 2) + 1.1(x - 2)(x - 6)$

43. $L_2(x) = -0.32x^2 + 4.4x + 4.2 =$
$f_2(x) = 4.2 + 3.6x - 0.32x(x - 2.5)$

45. $L_2(x) = 0.25x^2 - 1.2x - 11.65 =$
$f_2(x) = -12.6 + 0.25(x - 1)(x - 3.8)$

47. $L_2(x) = -1.52x^2 - 10.76x - 18.72 =$
$f_2(x) = -31 + 6.2(x + 1) - 1.52(x + 1)(x - 4)$

Section 3.3.1

49. $p_1(x) = \begin{cases} 0.26x + 3, & 0 \le x < 5 \\ 0.8(x - 5) + 4.3, & 5 \le x < 8 \\ 1.8(x - 8) + 6.7, & 8 \le x < 11 \\ 1.06(x - 11) + 10.6, & 11 \le x \le 16 \end{cases}$

$p_1(4) = 4.04, \quad p_1(10) = 9.3$

51. $p_1(x) = \begin{cases} -0.52(x - 1.1) + 22.7, & 1.1 \le x < 3.6 \\ -4.5(x - 3.6) + 21.4, & 3.6 \le x < 5.4 \\ -2(x - 5.4) + 13.3, & 5.4 \le x < 7.3 \\ -1.75(x - 7.3) + 9.5, & 7.3 \le x \le 10.2 \end{cases}$

$p_1(5.1) = 14.65, \quad p_1(8.2) = 7.925$

53. $p_1(x) = \begin{cases} 2.3(x - 10.1) + 22.7, & 10.1 \le x < 10.2 \\ -0.72, & 10.2 \le x < 10.6 \\ 8.45(x - 10.6) - 0.72, & 10.6 \le x < 10.8 \\ 0.9(x - 10.8) + 0.97, & 10.8 \le x \le 11.1 \end{cases}$

$p_1(10.5) = -0.72, \quad p_1(10.9) = 1.06$

55. $p_1(x) = \begin{cases} -0.508(x - 5) + 1.64, & 5 \le x < 10 \\ -0.066(x - 10) - 0.9, & 10 \le x < 25 \\ -1.89, & 25 \le x < 35 \\ -0.014(x - 35) - 1.89, & 35 \le x \le 50 \end{cases}$

$p_1(8) = 0.116, \quad p_1(29) = -1.89$

57. $p_1(x) = \begin{cases} 1.5(x - 0.5) + 4.2, & 0.5 \le x < 1.9 \\ -6(x - 1.9) + 6.3, & 1.9 \le x < 2.6 \\ -4(x - 2.6) + 2.1, & 2.6 \le x < 3.7 \\ 0.5(x - 3.7) - 2.3, & 3.7 \le x < 5.1 \\ 2(x - 5.1) - 2.3, & 5.1 \le x \le 5.9 \end{cases}$

$p_1(2.4) = 3.3, \quad p_1(5.7) = -0.4$

59. $p_1(x) = \begin{cases} -14.4(x + 2) + 162, & -2 \le x < -1.5 \\ 3.8(x + 1.5) + 154.8, & -1.5 \le x < 0 \\ 10.08x + 160.5, & 0 \le x < 2.5 \\ 185.7, & 2.5 \le x < 3.5 \\ -22.4(x - 3.5) + 185.7, & 3.5 \le x < 5 \\ 1.4(x - 5) + 152.1, & 5 \le x \le 8.5 \end{cases}$

$p_1(-1.3) = 155.56, \quad p_1(4.2) = 170.02$

Section 3.3.2

61. $p_2(x) = \begin{cases} 3.2 + 4.8(x - 0.5), & 0.5 \le x < 1 \\ 5.6 + 4.8(x - 1) - 1.2(x - 1)^2, & 1 \le x < 4.5 \\ 7.7 - 3.6(x - 4.5) + 1.872(x - 4.5)^2, & 4.5 \le x \le 7 \end{cases}$

$p_2(2.5) = 10.1, \quad p_2(5) = 6.368$

63. $p_2(x) = \begin{cases} 29.7 - 5.325(x - 4), & 4 \le x < 8 \\ 8.4 - 5.325(x - 8) + 0.35(x - 8)^2, & 8 \le x < 24 \\ 12.8 + 5.875(x - 24) - 0.825(x - 24)^2, & 24 \le x \le 32 \end{cases}$

$p_2(12) = -7.3, \quad p_2(28.2) = 22.922$

65. $p_2(x) = \begin{cases} -3.9 - 10.5(x - 5), & 5 \le x < 5.2 \\ -6 - 10.5(x - 5.2) + 58.75(x - 5.2)^2, & 5.2 \le x < 5.6 \\ -0.8 + 36.5(x - 5.6) + 95(x - 5.6)^2, & 5.6 \le x \le 5.9 \end{cases}$

$p_2(5.1) = -4.95, \quad p_2(5.7) = 3.8$

67. $p_2(x) = \begin{cases} 0.9 + 1.925(x - 1), & 1 \le x < 9 \\ 16.3 + 1.925(x - 9) - 0.209(x - 9)^2, & 9 \le x < 14 \\ 20.7 - 0.165(x - 14) - 0.03(x - 14)^2, & 14 \le x \le 26 \end{cases}$

$p_2(13) = 20.656, \quad p_2(18.5) = 19.35$

69. $p_2(x) = \begin{cases} 13.5 - 4.05(x + 10), & -10 \le x < -6 \\ -2.7 - 4.05(x + 6) + 0.9(x + 6)^2, & -6 \le x < 0 \\ 5.4 + 6.75x - 4.275x^2, & 0 \le x < 2 \\ 1.8 - 10.35(x - 2) + 1.65(x - 2)^2, & 2 \le x \le 9 \end{cases}$

$p_2(1) = 7.875, \quad p_2(3.8) = -11.484$

71. $p_2(x) = \begin{cases} 3.2 + 1.55(x + 3), & -3 \le x < -1 \\ 6.3 + 1.55(x + 1) - 0.562(x + 1)^2, & -1 \le x < 4 \\ -4.07(x - 4) + 1.36(x - 4)^2, & 4 \le x < 6 \\ -2.7 + 1.37(x - 6) - 0.17(x - 6)^2, & 6 \le x \le 12 \end{cases}$

$p_2(-2.4) = 4.13, \quad p_2(10) = 0.06$

73. $p_2(x) = \begin{cases} 9.6 - 6(x - 2), & 2 \le x < 3.5 \\ 0.6 - 6(x - 3.5) + 8(x - 3.5)^2, & 3.5 \le x < 4.5 \\ 2.6 + 10(x - 4.5) - (x - 4.5)^2, & 4.5 \le x < 7 \\ 21.35 + 5(x - 7) - 2(x - 7)^2, & 7 \le x \le 11 \end{cases}$

min, $(3.875, -0.525)$ max, $(8.25, 24.475)$

75. $p_2(x) = \begin{cases} 17 + 13.5(x - 1.3), & 1.3 \le x < 3.3 \\ 44 + 13.5(x - 3.3) - 2.5(x - 3.3)^2, & 3.3 \le x < 10.8 \\ 4.625 - 24(x - 10.8) + 3(x - 10.8)^2, & 10.8 \le x < 20.8 \\ 64.625 + 36(x - 20.8) + 4.5(x - 20.8)^2, & 20.8 \le x \le 21.3 \end{cases}$

min, $(14.8, -43.375)$ max, $(6, 62.225)$

77. $p_2(x) = \begin{cases} -12 + (x - 10), & 10 \le x < 16 \\ -6 + (x - 16) + 0.75(x - 16)^2, & 16 \le x < 24 \\ 50 + 13(x - 24) + 0.5(x - 24)^2, & 24 \le x < 36 \\ 278 + 25(x - 36) - 2.5(x - 36)^2, & 36 \le x \le 50 \end{cases}$

max, $(41, 340.5)$

79. $p_2(x) = \begin{cases} 22 + 14(x - 1), & 1 \le x < 4 \\ 64 + 14(x - 4) - 5(x - 4)^2, & 4 \le x < 7 \\ 61 - 16(x - 7) - 1.75(x - 7)^2, & 7 \le x < 10 \\ -2.75 - 26.5(x - 10) + 12.5(x - 10)^2, & 10 \le x \le 13 \end{cases}$

min, $(11.06, -19.604)$ max, $(5.4, 73.8)$

Section 3.3.3

81. $p_3(4) = 3.382, p_3(9) = 24.234$

$p_3(x) = \begin{cases} 6 - 1.825(x - 2) + 0.129(x - 2)^3, & 2 \le x < 7 \\ 13 + 7.850(x - 7) + 1.935(x - 7)^2 - 1.784(x - 7)^3, & 7 \le x < 8 \\ 21 + 6.367(x - 8) - 3.418(x - 8)^2 + 0.285(x - 8)^3, & 8 \le x \le 12 \end{cases}$

83. $p_3(3.1) = 59.195, p_3(5) = 102.821$

$p_3(x) = \begin{cases} 14.2 + 22.011x - 0.940x^3, & 0 \le x < 2 \\ 50.7 + 10.727(x - 2) - 5.642(x - 2)^2 + 2.646(x - 2)^3, & 2 \le x < 4.5 \\ 83.6 + 32.130(x - 4.5) + 14.203(x - 4.5)^2 - 3.156(x - 4.5)^3, & 4.5 \le x \le 6 \end{cases}$

85. $p_3(1.6) = 10.810$, $p_3(4.1) = 9.378$

$$p_3(x) = \begin{cases} 12.8 - 6.698(x - 1.3) + 0.700(x - 1.3)^3, & 1.3 \leq x < 2.1 \\ 7.8 - 5.354(x - 2.1) + 1.680(x - 2.1)^2 + 0.813(x - 2.1)^3, & 2.1 \leq x < 3.3 \\ 5.2 + 2.192(x - 3.3) + 4.608(x - 3.3)^2 - 1.024(x - 3.3)^3, & 3.3 \leq x \leq 4.8 \end{cases}$$

87. $p_3(10.3) = 3.380$, $p_3(26.5) = 105.728$

$$p_3(x) = \begin{cases} 68.8 - 9.184(x - 1) + 0.025(x - 1)^3, & 1 \leq x < 9.4 \\ 6.4 - 3.917(x - 9.4) + 0.627(x - 9.4)^2 - 0.003(x - 9.4)^3, & 9.4 \leq x < 25.7 \\ 94.4 + 13.801(x - 25.7) + 0.460(x - 25.7)^2 - 0.014(x - 25.7)^3, & 25.7 \leq x \leq 36.5 \end{cases}$$

89. $p_3(4) = 26.479$, $p_3(6.3) = 14.892$

$$p_3(x) = \begin{cases} 38.1 - 12.086(x - 3) + 0.465(x - 3)^3, & 3 \leq x < 6 \\ 14.4 + 0.472(x - 6) + 4.186(x - 6)^2 - 0.985(x - 6)^3, & 6 \leq x < 9 \\ 26.9 - 0.998(x - 9) - 4.676(x - 9)^2 + 0.840(x - 9)^3, & 9 \leq x < 12 \\ 4.5 - 6.368(x - 12) + 2.886(x - 12)^2 - 0.321(x - 12)^3, & 12 \leq x \leq 15 \end{cases}$$

91. $p_3(5) = 36.185$, $p_3(7.3) = 12.178$

$$p_3(x) = \begin{cases} 13.5 - 2.05(x - 2.4) + 2.094(x - 2.4)^3, & 2.4 \leq x < 4.1 \\ 20.3 + 16.101(x - 4.1) + 16.101(x - 4.1)^2 - 9.951(x - 4.1)^3, & 4.1 \leq x < 5.3 \\ 37.8 - 1.261(x - 5.3) - 25.145(x - 5.3)^2 + 14.857(x - 5.3)^3, & 5.3 \leq x < 5.9 \\ 31.2 - 15.389(x - 5.9) + 1.598(x - 5.9)^2 - 0.222(x - 5.9)^3, & 5.9 \leq x \leq 8.3 \end{cases}$$

93. $p_3(12.4) = 23.415$, $p_3(30.2) = 39.950$

$$p_3(x) = \begin{cases} -3.9 + 14.842(x - 10.5) - 0.129(x - 10.5)^3, & 10.5 \leq x < 16.3 \\ 57 + 1.815(x - 16.3) - 2.246(x - 16.3)^2 + 0.197(x - 16.3)^3, & 16.3 \leq x < 22.7 \\ 28.2 - 2.761(x - 22.7) + 1.531(x - 22.7)^2 - 0.128(x - 22.7)^3, & 22.7 \leq x < 29.2 \\ 39.9 + 0.957(x - 29.2) - 0.959(x - 29.2)^2 + 0.052(x - 29.2)^3, & 29.2 \leq x \leq 35.4 \end{cases}$$

95. $p_3(11.8) = 33.042$, $p_3(19) = 66.502$

$$p_3(x) = \begin{cases} 21.2 + 7.827(x - 2) - 0.127(x - 2)^3, & 2 \leq x < 7 \\ 44.4 - 1.733(x - 7) - 1.912(x - 7)^2 + 0.374(x - 7)^3, & 7 \leq x < 11 \\ 30.8 + 0.915(x - 11) + 2.574(x - 11)^2 - 0.268(x - 11)^3, & 11 \leq x < 18 \\ 71.4 - 2.438(x - 18) - 3.053(x - 18)^2 + 0.593(x - 18)^3, & 18 \leq x < 21 \\ 52.6 - 4.748(x - 21) + 2.283(x - 21)^2 - 0.109(x - 21)^3, & 21 \leq x \leq 28 \end{cases}$$

97. $p_3(20.5) = 33.927$, $p_3(50.3) = 15.280$

$$p_3(x) = \begin{cases} 112.8 - 14.507(x - 10) + 0.063(x - 10)^3, & 10 \leq x < 20 \\ 31.2 + 4.533(x - 20) + 1.904(x - 20)^2 - 0.126(x - 20)^3, & 20 \leq x < 30 \\ 141.4 + 4.963(x - 30) - 1.861(x - 30)^2 + 0.069(x - 30)^3, & 30 \leq x < 40 \\ 74.1 - 11.487(x - 40) + 0.216(x - 40)^2 + 0.034(x - 40)^3, & 40 \leq x < 50 \\ 14.3 + 2.903(x - 50) + 1.223(x - 50)^2 - 0.041(x - 50)^3, & 50 \leq x \leq 60 \end{cases}$$

99. $p_3(10.8) = 52.743$, $p_3(20.2) = 49.974$

$$p_3(x) = \begin{cases} 31.8 + 10.108(x - 6.4) - 0.276(x - 6.4)^3, & 6.4 \leq x < 10.7 \\ 53.3 - 5.217(x - 10.7) - 3.564(x - 10.7)^2 + 0.621(x - 10.7)^3, & 10.7 \leq x < 14.6 \\ 15.6 - 4.675(x - 14.6) + 3.703(x - 14.6)^2 - 0.319(x - 14.6)^3, & 14.6 \leq x < 19.1 \\ 40.5 + 9.271(x - 19.1) - 0.604(x - 19.1)^2 + 0.005(x - 19.1)^3, & 19.1 \leq x < 23.9 \\ 71.7 + 3.852(x - 23.9) - 0.525(x - 23.9)^2 + 0.044(x - 23.9)^3, & 23.9 \leq x \leq 27.9 \end{cases}$$

101. min, $(4.172, 3.358)$ max, $(9.076, 24.249)$

103. no extrema

105. min, $(3.045, 4.927)$

107. min, $(12.597, 0.188)$

109. min, $(5.943, 14.384)$ max, $(8.888, 26.950)$ min, $(13.458, 0.356)$

111. min, $(2.971, 12.719)$ max, $(5.275, 37.817)$

113. max, $(16.728, 57.381)$ min, $(23.736, 26.840)$ max, $(29.721, 40.146)$

115. max, $(6.532, 44.850)$ min, $(10.813, 30.727)$

     max, $(17.576, 71.915)$ min, $(22.132, 49.993)$

117. min, $(18.761, 28.069)$ max, $(31.450, 144.894)$ min, $(48.704, 12.901)$

119. max, $(9.894, 55.345)$ min, $(15.293, 14.032)$

Sections 3.1, 3.3.1, and 3.3.3

121. (a) $f_5(x) = 63.8 - 10.6x + 1.725x(x - 2) + 0.65x(x - 2)(x - 4)$

         $- 0.35x(x - 2)(x - 4)(x - 6) + 0.075x(x - 2)(x - 4)(x - 6)(x - 8)$

$$\text{(b) } p_1(x) = \begin{cases} -10.6x + 63.8, & 0 \leq x < 2 \\ -3.7(x - 2) + 42.6, & 2 \leq x < 4 \\ 18.8(x - 4) + 35.2, & 4 \leq x < 6 \\ -10.3(x - 6) + 72.8, & 6 \leq x < 8 \\ -14.2(x - 8) + 52.2, & 8 \leq x \leq 10 \end{cases}$$

$$\text{(c) } p_3(x) = \begin{cases} 63.8 - 10.296x - 0.076x^3, & 0 \leq x < 2 \\ 42.6 - 11.208(x - 2) - 0.456(x - 2)^2 + 2.105(x - 2)^3, & 2 \leq x < 4 \\ 35.2 + 12.232(x - 4) + 12.176(x - 4)^2 - 4.446(x - 4)^3, & 4 \leq x < 6 \\ 72.8 + 7.590(x - 6) - 14.497(x - 6)^2 + 2.777(x - 6)^3, & 6 \leq x < 8 \\ 52.2 - 17.080(x - 8) + 2.162(x - 8)^2 - 0.360(x - 8)^3, & 8 \leq x \leq 10 \end{cases}$$

123. (a) $f_5(x) = 130 + 42.5(x - 1) + 6.5(x - 1)(x - 7)$

         $- 1.5(x - 1)(x - 7)(x - 12) + 0.1(x - 1)(x - 7)(x - 12)(x - 16)$

         $- 0.005(x - 1)(x - 7)(x - 12)(x - 16)(x - 25)$

$$(b)\ p_1(x) = \begin{cases} 42.5(x-1)+130, & 1 \le x < 7 \\ 114(x-7)+365, & 7 \le x < 12 \\ -30(x-12)+955, & 12 \le x < 16 \\ -27.4(x-16)+835, & 16 \le x < 25 \\ -58.62(x-25)+588.4, & 25 \le x \le 30 \end{cases}$$

$$(c)\ p_3(x) = \begin{cases} 130 + 9.302(x-1) + 0.922(x-1)^3, & 1 \le x < 7 \\ 385 + 108.896(x-7) + 16.599(x-7)^2 - 3.116(x-7)^3, & 7 \le x < 12 \\ 955 + 41.216(x-12) - 30.135(x-12)^2 - 3.083(x-12)^3, & 12 \le x < 16 \\ 72.835 - 51.896(x-16) + 6.857(x-16)^2 - 0.459(x-16)^3, & 16 \le x < 25 \\ 588.4 - 40.124(x-25) - 5.549(x-25)^2 + 0.370(x-25)^3, & 25 \le x \le 30 \end{cases}$$

125. $(a)\ f_5(x) = 520 - 1.8(x-10) - 0.04(x-10)(x-18)$
$$- 0.024(x-10)(x-18)(x-30) + 0.004(x-10)(x-18)(x-30)(x-35)$$
$$- 0.002(x-10)(x-18)(x-30)(x-35)(x-43)$$

$$(b)\ p_1(x) = \begin{cases} -1.8(x-10)+520, & 10 \le x < 18 \\ -2.6(x-18)+505.6, & 18 \le x < 30 \\ -13.48(x-30)+474.4, & 30 \le x < 35 \\ 13.3(x-35)+407, & 35 \le x < 43 \\ -217.2(x-43)+513.4, & 43 \le x \le 45 \end{cases}$$

$$(c)\ p_3(x) = \begin{cases} 520 - 4.648(x-10) + 0.045(x-10)^3, & 10 \le x < 18 \\ 505.6 + 3.896(x-18) + 1.068(x-18)^2 - 0.134(x-18)^3, & 18 \le x < 30 \\ 474.4 - 28.420(x-30) - 3.761(x-30)^2 + 1.349(x-30)^3, & 30 \le x < 35 \\ 407 + 35.175(x-35) + 16.480(x-35)^2 - 2.402(x-35)^3, & 35 \le x < 43 \\ 513.4 - 162.321(x-43) - 41.167(x-43)^2 + 6.861(x-43)^3, & 43 \le x \le 45 \end{cases}$$

127. $(a)\ f_5(x) = 110 + 25(x-1.2) - 43.75(x-1.2)(x-1.8)$
$$+ 37.5(x-1.2)(x-1.8)(x-2.6)$$
$$- 15.625(x-1.2)(x-1.8)(x-2.6)(x-3.6)$$
$$(+0(x-1.2)(x-1.8)(x-2.6)(x-3.6)(x-4))$$

$$(b)\ p_1(x) = \begin{cases} 25(x-1.2)+110, & 1.2 \le x < 1.8 \\ -36.25(x-1.8)+125, & 1.8 \le x < 2.6 \\ 47(x-2.6)+96, & 2.6 \le x < 3.6 \\ 92.5(x-3.6)+143, & 3.6 \le x < 4 \\ 25(x-4)+180, & 4 \le x \le 4.6 \end{cases}$$

$$(c)\ p_3(x) = \begin{cases} 110 + 42.637(x-1.2) - 48.992(x-1.2)^3, & 1.2 \le x < 1.8 \\ 125 - 10.274(x-1.8) - 88.185(x-1.8)^2 + 69.643(x-1.8)^3, & 1.8 \le x < 2.6 \\ 96 - 17.655(x-2.6) + 78.959(x-2.6)^2 - 14.305(x-2.6)^3, & 2.6 \le x < 3.6 \\ 143 + 97.348(x-3.6) + 36.044(x-3.6)^2 - 120.419(x-3.6)^3, & 3.6 \le x < 4 \\ 180 + 68.382(x-4) - 108.459(x-4)^2 + 60.255(x-4)^3, & 4 \le x \le 4.6 \end{cases}$$

Section 3.4

129. $s_1(x) = 6,$ $\quad -6x + 15.6 = 0,$ $\qquad\qquad\qquad\qquad x = 2.6$
$\quad\ \ s_3(x) = 6,$ $\quad -x^2 + 19x - 68.65 = 0,$ $\qquad\qquad x = 4.852$

131. $s_3(x) = 12,$ $\quad -1.75x^2 + 8.5x + 75.25 = 0,$ $\qquad x = 9.421$
$\quad\ \ s_4(x) = 12,$ $\quad 12.5x^2 - 276.5x + 1500.25 = 0,$ $\quad x = 12.578$

133. $s_3(x) = 48,$ $\quad -17.2 + 0.915(x - 11)$
$\qquad\qquad\qquad + 2.574(x - 11)^2 - 0.268(x - 11)^3 = 0, \quad x_4 = 13.838$

135. $s_2(x) = 132,$ $\quad -100.8 + 4.533(x - 20)$
$\qquad\qquad\qquad + 1.904(x - 20)^2 - 0.126(x - 20)^3 = 0, \quad x_5 = 28.724$
$\quad\ \ s_3(x) = 132,$ $\quad 9.4 + 4.963(x - 30)$
$\qquad\qquad\qquad - 1.861(x - 30)^2 + 0.069(x - 30)^3 = 0, \quad x_4 = 34.544$

Section 4.2

1. $r_1(x) = 3.42x + 11.8,$ $\qquad r_1(6) = 32.32,$ $\qquad r_1(18) = 73.36$
3. $r_1(x) = -2.616x + 28.579,$ $\quad r_1(3.4) = 19.685,$ $\quad r_1(8.2) = 7.128$
5. $r_1(x) = -0.119x + 2.657,$ $\quad r_1(4.3) = 2.145,$ $\qquad r_1(14.1) = 0.979$
7. $r_1(x) = 9.74x - 192.224,$ $\quad r_1(21.4) = 16.212,$ $\quad r_1(23.6) = 37.64$
9. $r_1(x) = 13.579x + 67.792,$ $\quad r_1(3.9) = 120.75,$ $\quad r_1(7.5) = 169.635$
11. $r_1(x) = -4.5x - 46.6,$ $\qquad r_1(9.9) = -91.15,$ $\quad r_1(12.7) = -103.75$

Section 4.3

13. $r_{\ln}(x) = 10.242 \ln x + 13.986$
$\quad\ \ r_{\ln}(1.6) = 18.8,$ $\qquad\qquad\qquad r_{\ln}(4.9) = 30.263$
15. $r_{\ln}(x) = -1.219 \ln x + 6.169$
$\quad\ \ r_{\ln}(20) = 2.517,$ $\qquad\qquad\qquad r_{\ln}(57) = 1.241$
17. $r_{\ln}(x) = 2.416 \ln x + 2.115$
$\quad\ \ r_{\ln}(11.9) = 8.098,$ $\qquad\qquad\quad r_{\ln}(14.75) = 8.617$
19. $r_{\ln}(x) = -3.993 \ln x + 25.882$
$\quad\ \ r_{\ln}(30.5) = 12.235,$ $\qquad\qquad\ r_{\ln}(84.9) = 8.147$
21. $r_{\ln}(x) = -0.96 \ln x + 36.938$
$\quad\ \ r_{\ln}(38.3) = 33.438,$ $\qquad\qquad\ r_{\ln}(66.1) = 32.914$
23. $r_{\ln}(x) = 6.632 \ln x + 7.685$
$\quad\ \ r_{\ln}(26.6) = 29.444,$ $\qquad\qquad\ r_{\ln}(70.8) = 35.936$

Section 4.4

25. $r_2(x) = -1.125x^2 + 19.282x + 7.461$
$r_2(3.5) = 61.167,$ $\qquad\qquad r_2(12.3) = 74.428$
27. $r_2(x) = 4.537x^2 - 44.116x + 122.585$
$r_2(5.1) = 15.601,$ $\qquad\qquad r_2(6.6) = 29.051$
29. $r_2(x) = -0.022x^2 - 0.602x + 88.346$
$r_2(13.8) = 75.849,$ $\qquad\qquad r_2(32) = 46.554$
31. $r_2(x) = 3.514x^2 - 99.871x + 741.515$
$r_2(14.9) = 33.58,$ $\qquad\qquad r_2(19.5) = 130.229$
33. $r_2(x) = 10.688x^2 - 60.461x + 104.864$
$r_2(2.3) = 22.343,$ $\qquad\qquad r_2(4.8) = 60.903$
35. $r_2(x) = -6.214x^2 + 46.327x - 45.22$
$r_2(3.4) = 40.458,$ $\qquad\qquad r_2(5.7) = 16.951$

Section 4.5

37. $r_1(x) = 6.475x - 2.164$ $\qquad E_a = 1.931$
$r_2(x) = -0.011x^2 + 6.604x - 2.507$ $\qquad E_a = 1.922$
39. $r_2(x) = -0.158x^2 + 8.831x - 27.232$ $\qquad E_a = 165.215$
41. $r_{\ln}(x) = 4.523 \ln x + 7.517$ $\qquad E_a = 1.985$
$r_2(x) = -0.004x^2 + 0.482x + 11.766$ $\qquad E_a = 9.214$
43. $r_{\ln}(x) = -8.988 \ln x + 45.997$ $\qquad E_a = 0.861$
$r_2(x) = 0.005x^2 - 0.622x + 29.804$ $\qquad E_a = 3.187$
45. $r_1(x) = -12.6x + 354$ $\qquad E_a = 77$
$r_2(x) = -0.084x^2 - 9.876x + 334.619$ $\qquad E_a = 30.937$
47. $r_2(x) = 0.007x^2 - 0.624x + 17.428$ $\qquad E_a = 1.503$
$r_{\ln}(x) = -5.759 \ln x + 24.993$ $\qquad E_a = 1.567$

Section 5.1.1

1. 2.254 $\qquad$ 3. 9.464 $\qquad$ 5. −1.929 $\qquad$ 7. 0.681

Section 5.1.2

no.	A_T	$E_{t,T}$	$A_{T,t}$	A_T over or under
9.	3.414	−0.567	2.847	over
11.	9.806	0.945	10.751	under
13.	−0.266	0.047	−0.219	under
15.	−0.362	−0.623	−0.985	over

Section 5.1.3

17. 5.963   19. 8.474   21. 0.404   23. −1.736

Section 5.1.4

no.	$A_T(n)$	$E_{t,T}(n)$	$A_{T,t}(n)$
25.	3.005	0.046	3.051
27.	1.802	−0.013	1.789
29.	12.906	0.078	12.984
31.	−4.485	−0.060	−4.545

Section 5.1.5

33. Analytic area: 125.754

Type	Error	Approx.	ε_T, %
A_T		171.440	36.33
$A_{T,t}$	−63.180	108.260	13.91
$A_T(4)$		129.758	3.18
$A_{T,t}(4)$	−3.949	125.809	0.04

35. Analytic area: 19.113

Type	Error	Approx.	ε_T, %
A_T		12.447	34.88
$A_{T,t}$	3.750	16.197	15.26
$A_T(5)$		18.962	0.79
$A_{T,t}(5)$	0.150	19.112	0.01

Section 5.2.1

37. 5.185   39. 6.374   41. −2.388   43. 15.572

Section 5.2.2

no.	A_S	$E_{t,S}$	$A_{S,t}$	A_S over or under
45.	7.893	−0.082	7.811	over
47.	0.971	−0.069	0.902	over
49.	4.005	−0.034	3.971	over
51.	34.260	0.052	34.312	under

53. 41.125   55. 16.92   57. −28   59. 54

Section 5.2.3

61. 13.153   63. 38.781   65. 21.224   67. 0.388

Section 5.2.4

no.	$A_S(n)$	$E_{t,S}(n)$	$A_{S,t}(n)$	$A_S(n)$ over or under
69.	23.122	−0.013	23.109	over
71.	41.517	0.028	41.545	under
73.	3.957	0.005	3.962	under
75.	2.455	−0.017	2.438	over

Section 5.2.5

77. Analytic area: 385.480

Type	Error	Approx.	ε_S, %
A_S		392.976	1.94
$A_{S,t}$	−5.882	387.094	0.42
$A_S(6)$		385.566	0.02
$A_{S,t}(6)$	−0.073	385.493	0.00

79. Analytic area: −153.121

Type	Error	Approx.	ε_S, %
A_S		−158.569	3.56
$A_{S,t}$	4.075	−154.494	0.90
$A_S(4)$		−153.393	0.18
$A_{S,t}(4)$	0.255	−153.138	0.01

Section 5.3

81.	21.287, 2.76 %	83.	−1.271, 7.08 %	85.	18.014, 1.42 %
87.	52.956, 0.53 %	89.	4.131, 9.44 %	91.	12.689, 4.26 %

Sections 5.2.4 and 5.3

no.	$A_S(n)$	$E_{t,S}(n)$	$A_{S,t}(n)$	Romberg
93.	60.229	0.010	60.239	60.238
95.	15.332	−0.043	15.289	15.320
97.	2.555	−0.029	2.526	2.544
99.	10.694	−0.006	10.688	10.689

Section 5.4

no.	M	Technique	Antiderivative	Integral
101.	3	u-substitution	$6\ln(x^2+4)$	4.159
103.	4.5	by parts	$-12(x+2)e^{-0.5x}$	14.940
105.	6	partial fractions	$3\ln|x| - 2\ln|x+4|$	6.248
107.	3.5	u-substitution	$-4e^{-x^2}$	3.813
	−3.5	$N = 450$ below x-axis		−1.721
			Total =	2.092
109.	4	by parts	$x^2(2\ln|x| - 1)$	1.365
	−1.5	$N = 375$ below x-axis		−0.764
			Total =	0.601
111.	6	partial fractions	$5\ln|x+2| - 15\ln|x+5|$	1.723
	−1	$N = 675$ below x-axis		−3.043
			Total =	−1.320

Section 6.1

1. 0.4   3. 2   5. −3.2   7. 1.2
9. 10.5   11. −8.75   13. −4   15. 1.5
17. $\mathrm{FDD}'_1(x_i)$, $s'(10) = 28.5$; $\mathrm{BDD}'_1(x_i)$, $s'(10.6) = 28.5$
19. $\mathrm{FDD}'_1(x_i)$, $s'(11.2) = 51$; $\mathrm{BDD}'_1(x_i)$, $s'(11.8) = 51$
21. $\mathrm{FDD}'_1(x_i)$, $s'(12.4) = 33$; $\mathrm{BDD}'_1(x_i)$, $s'(13) = 33$

Section 6.2.1

23. 0.6   25. 1.8   27. −2.6
29. 8.125   31. −9.875   33. −1.625

Section 6.2.2

35. 0.2   37. 4   39. −3.6
41. 3.375   43. −12.125   45. 3.125

Section 6.3

47. 63.75   49. 31.625

Section 6.4

51. −210   53. 45   55. 132.5
57. −7.5   59. 1.094   61. −0.625

Section 6.5

63. −32.5   65. 57.5   67. −20
69. 0.156   71. −0.625   73. 3.75

Sections 6.6 and 6.1

75. 0.6   77. 2.4   79. −2.533
81. 4.958   83. −11.375   85. 1.542

Sections 6.6 and 6.2.1

87. 1.467   89. −0.933   91. −11.667   93. −1.104

Sections 6.6 and 6.2.2

95. −0.567   97. 5.667   99. −9.875   101. 3.333

Sections 6.6 and 6.3

103. 29.873   105. 34.391

Sections 6.6 and 6.4

107. −275.833   109. 48.542   111. 1.966   113. −0.768

Sections 6.6 and 6.5

115. −247.083   117. 49.375   119. 1.315   121. −0.990

Section 6.7

123. 7.174   125. −1.923   127. 2.210   129. 6.497
131. −2.675   133. 0.283   135. −0.147

no.	x_0	x_1	x_2	$f'(x)$	no.	x_0	x_1	x_2	$f'(x)$
137.	10	11.7	13.1	2.037	139.	13.1	14.2	15.5	0.759
141.	15.5	16.3	17.4	−4.125	143.	1.12	1.45	1.78	−0.373
145.	2.81	3.22	3.54	1.094	147.	0	1.5	3.25	−8.207
149.	3.25	5.75	8	−5.669	151.	8	9.25	12	7.435
153.	28	32	35	0.254	155.	35	40	48	−1.239

Section 7.1

1.		3.		5.		7.	
x	y	x	y	x	y	x	y
3.25	3.000	0.9	0.529	9.05	4.041	5.8	15.255
3.50	4.159	1.2	0.636	9.10	2.030	5.9	28.467
3.75	5.477	1.5	0.818	9.15	3.705	6.0	43.069

9.		11.		13.		15.	
x	y	x	y	x	y	x	y
4.30	−1.684	2.5	7.214	6.8	1.172	0.2	8.800
4.45	−1.783	4.0	13.291	7.6	2.362	0.4	9.657
4.60	−1.927	5.5	22.542	8.4	3.589	0.6	10.575

no.	approximate $L(x_i)$	exact $y(x_i)$	DEQ analytic solution
17.	0.931	0.919	$y = 1.3\sqrt{x}$
19.	−1.454	−1.451	$y = -\frac{1}{3}\ln(0.964 - x^3)$
21.	1.053	1.046	$y = \sqrt[5]{2}\sin 2.5x$

Section 7.2

23.		25.		27.		29.	
x	y	x	y	x	y	x	y
2.6	6.889	0.17	0.837	−2.00	−0.994	4.5	2.894
2.8	10.203	0.19	1.047	−1.95	−0.795	5.0	4.075
3.0	13.512	0.21	1.399	−1.90	−0.499	5.5	5.290

31.		33.		35.		37.	
x	y	x	y	x	y	x	y
3.2	−1.124	−0.25	5.078	2.8	−20.298	3.60	7.055
3.5	−1.116	0.00	4.158	2.4	−18.779	3.95	7.747
3.8	−1.402	0.25	3.571	2.0	−15.821	4.30	8.474

no.	approximate $L(x_i)$	exact $y(x_i)$	DEQ analytic solution
39.	2.688	2.687	$y = 8e^{-3/x}$
41.	0.512	0.513	$y = \tan(0.1x^2 - 1.142)$
43.	5.046	5.048	$y = (2 + 0.5\tan x)^2$

Section 7.3

45.		47.		49.		51.	
x	y	x	y	x	y	x	y
4.35	3.574	−4.05	1.518	2.6	10.648	2.5	5.767
4.70	9.705	−3.95	1.296	2.8	13.543	3.0	7.187
5.05	24.857	−3.85	1.565	3.0	15.645	3.5	8.741

53.		55.		57.		59.	
x	y	x	y	x	y	x	y
5.15	4.792	−5.50	−7.324	1.6	0.008	3.90	5.949
5.30	2.932	−5.25	−8.142	2.2	−1.100	3.95	6.048
5.45	1.854	−5.00	−8.962	2.8	−2.590	4.00	6.148

no.	approximate $L(x_i)$	exact $y(x_i)$	DEQ analytic solution
61.	1.390	1.391	$y = e^{0.2e^{0.5x}}$
63.	0.943	0.946	$y = \ln(2 - \cos x)$
65.	4.728	4.725	$y = \sqrt{24\ln x - 35}$

Section 7.4

67.		69.		71.		73.	
x	y	x	y	x	y	x	y
2.25	1.079	−2.6	4.881	3.25	−2.554	10.3	16.550
2.75	−0.117	−2.2	6.910	3.50	−0.696	10.6	26.352
3.25	−3.438	−1.8	9.469	3.75	1.082	10.9	36.475

75.		77.		79.		81.	
x	y	x	y	x	y	x	y
1.6	−1.683	−2.10	0.989	6.75	2.790	7.15	0.979
1.8	−2.042	−2.45	1.461	7.50	−2.205	7.20	0.957
2.0	−1.872	−2.80	2.663	8.25	−7.207	7.25	0.935

no.	approximate $L(x_i)$	exact $y(x_i)$	DEQ analytic solution
83.	2.928	2.932	$y = (4x^{-2} - 0.1)^{-1/2}$
85.	6.518	6.521	$y = \sqrt{25 + 20\tan^{-1} x}$
87.	1.898	1.899	$y = 1.492x^x e^{-x}$

Section 7.5

89.		91.		93.		95.	
x	y	x	y	x	y	x	y
10.00	7.557	−1.5	−0.773	4.35	2.086	2.55	0.520
10.25	4.276	−1.0	−1.528	4.70	1.955	2.65	1.087
10.50	1.099	−0.5	−1.583	5.05	1.922	2.75	2.135

97.		99.		101.		103.	
x	y	x	y	x	y	x	y
4.25	3.888	6.2	5.515	3.9	8.619	1.75	−1.808
4.55	4.196	6.6	8.843	3.7	8.800	2.50	−2.012
4.85	5.123	7.0	12.378	3.5	8.347	3.25	−2.308

no.	approximate $L(x_i)$	exact $y(x_i)$	DEQ analytic solution
105.	3.650	3.649	$y = \sqrt{2x^4 + 2x + 5}$
107.	2.426	2.426	$y = (20\ln^2 x + 1.6)^{0.4}$
109.	1.059	1.060	$y = \tan(0.5x^2 - 7)$

Sections 7.2, 7.3, 7.4, and 7.5

111.

x	Midpoint $L(x_8)$	Heun's $L(x_8)$	Ralston's $L(x_8)$	RK $L(x_8)$
5.6	0.546	0.522	0.543	0.506
error	0.040	0.016	0.037	0.000

$y = (1.2 + \cos x)^{-1}$, $(5.6, 0.506)$

113.

x	Midpoint $L(x_5)$	Heun's $L(x_5)$	Ralston's $L(x_5)$	RK $L(x_5)$
1.75	0.997	0.990	0.993	0.997
error	0.000	0.007	0.004	0.000

$y = \tan^{-1}(0.5x^2 + 0.016)$, $(1.75, 0.997)$

115.

x	Midpoint $L(x_6)$	Heun's $L(x_6)$	Ralston's $L(x_6)$	RK $L(x_6)$
2.0	0.963	0.954	0.958	0.959
error	0.004	0.005	0.001	0.000

$y = \sqrt{\ln^2 x + 0.44}$, $(2.0, 0.959)$

117.

x	Midpoint $L(x_8)$	Heun's $L(x_8)$	Ralston's $L(x_8)$	RK $L(x_8)$
0.20	8.435	8.208	8.311	8.456
error	0.032	0.259	0.156	0.011

$y = (x \sin x + \cos x + 1.89)^2$, $(0.20, 8.467)$

Section 8.1, 8.2, 8.3, and 8.4

1. $y = 1/\sqrt{2e^x - 1.84}$

3. $y = \ln(2.5x^2 + 0.13)$

5. $y = \sqrt{0.35x^4 + 11.9}$

7. $y = (2.39 \tan^{-1}(x - 10))^2$

9. $y = 1/\sqrt{1.895 - 0.5\ln(4x + 1)}$

11. $y = \sqrt[5]{x^3 + 17.3x - 5.985}$

Index

https://doi.org/10.1515/9783112221051-010

Bibliography

[1] A. Ralston and P. Rabinowitz, *A First Course in Numerical Analysis*. New York, NY: McGraw-Hill, 1978.

[2] A. Yew. [Handout]. https://www.dam.brown.edu/people/alcyew/handouts/numdiff.pdf (accessed May, 2024).

Further Reading

[3] O. Aberth, *Precise Numerical Analysis*. Dubuque, IA: Wm. C. Brown Publishers, 1988.

[4] S. Chapra and R. Canale, *Numerical Methods for Engineers*. New York, NY: McGraw-Hill Education, 2021.

[5] J. B. Fraleigh, *Calculus with Analytic Geometry*. Reading, MA: Addison-Wesley Publishing Company, 1985.

[6] Y. Hodge, *The Concise Encyclopedia of Statistics*. Springer, 2009.

[7] R. Larson, *Elementary Linear Algebra*. Boston, MA: Cengage Learning, 2017.

[8] W. H. Press, S. A. Teukolsky, W. T. Vetterling, and B. P. Flannery, *Numerical Recipes 3rd Edition: The Art of Scientific Computing*. New York, NY: Cambridge University Press, 2007.

https://doi.org/10.1515/9783112221051-011

Acknowledgments

What began as a routine goals meeting on February 16th, 2024 morphed into a challenge to write a Numerical Analysis textbook tailored to high school students fluent in a year of college calculus. I am indebted to Dr. Rachel Wright, the director of Moravian Academy in Bethlehem, Pennsylvania (my school of 27 years and counting), for the initial idea and confidence in placing this project before me.

My contacts at De Gruyter Brill, Inc. Publishing paved the smoothest path to publication for a first-time author. Acquisition editor Dr. Ranis N. Ibragimov immediately and continually provided encouragement and assurance of approval during the early, self-doubting stages. Ms. Marie Hammerschmidt was the initial project manager who cleverly incorporated my childhood classroom on the back cover. Mr. Felix Schaefer, content editor, quickly refined my initial manuscript into the ideal final version. Ms. Vilma Vaičeliūnienė of VTeX provided clear and timely instructions for proofing the final copy.

Dr. John McKay, a former colleague and teacher for whom I have the highest respect, graciously and thoroughly critiqued the initial draft. His knowledge, attention to details, and explanations were invaluable in shaping the final product. Specifically, his straightforward observations of when the truncation error adjustment of Simpson's rule is counterproductive led to the inclusion of Section 5.2.6. He helped craft the necessary, precise language regarding the concepts in the realm of Statistics. And during my single day of utter despair, he quickly verified I had indeed uncovered an analysis error in one of my source texts, which is why all my attempts to prove the untrue general claim naturally failed.

In 2023, eleven brave souls registered for a new course without any prior knowledge other than its instructor. I couldn't have handpicked a more qualified group, and the enthusiasm of the inaugural class certainly prompted the completion of this work. Thank you Francesca (Chessie) Bartolacci, Devon Berkove, Noah Breithaupt, Emma Derby, Aman Desai, Kaveh Fayyazi, Kanchan Gupta, Krish Gupta, Rayna Malhotra, Ryan Paul, and Vasiliki (Valia) Tsirukis.

A small 2024 class endured the embryonic stages of this textbook. In particular, Grayson Domzalski unearthed many typos and offered valuable formatting and layout feedback until it was "just right". By 2025, a class of thirteen seriously test-drove the complete initial version of the text. Thanks to Shaan Ailawadi, Aveer Chadha, Misha Desai, Mara Dubacher, Jake Herron, Laura Li, Charlene Liu, Hailey Migut, Carsten Rothbauer, Nathaniel Schmidt, Nikhil Skandan, and Maxwell (Max) Williams. A noteworthy thank you to Elena Koprowski, whose commitment to completing every exercise preceding submission made her a de factor editor on whom I relied.

From secondary education through graduate school I have been blessed with instructors and professors who were rigorous, demanding, encouraging, and excited by what they taught. I have endeavored to emulate the best qualities of my Mount Rushmore of mathematicians, Mr. Francis McGouldrick of Allentown (Pennsylvania) Central

https://doi.org/10.1515/9783112221051-012

Catholic High School, Mr. Robert Harwick and Dr. Jonathan Sevy of DeSales University (formerly Allentown College of St. Francis DeSales), and the late Dr. J. David Walker of Lehigh University. Dr. Sevy also served as an enthusiastic higher-level editor on the initial draft. Lastly, my notes scribed from Mr. William Finley's thorough undergraduate Numerical Analysis lectures at Allentown College were referenced several times to refresh my own memories of the subject.

My "lil sis" Christie is forever a willing and patient sounding board for anything, and a life role model. My late 19.5-year-old Shih Tzu Pookie was an emotional support by either sitting near, against, or on me for the majority of my writing.

Lastly, my parents, Jeanette and Leon Sr., have supported me unconditionally since the day I was born. This work is a culmination of all the lessons they taught and every sacrifice they made on my behalf. Any success I enjoy I credit to them. Thanks, Mom and Dad!

About the author

Leon Galitsky has taught advanced level mathematics courses, including Multivariable Calculus, Linear Algebra, and Numerical Analysis, for the past 27 years at Moravian Academy (secondary education). He received a B.S., summa cum laude, from DeSales University with a dual major in mathematics and computer science, and an M.S. from Lehigh University in applied mathematics in the Department of Mechanical Engineering. Publications include a contributing author to a chapter in Survey of Text Mining: Clustering, Classification, and Retrieval and an editor of "A Survey of Emerging Trend Detection in Textual Data Mining". Both were early research forays into what is now considered AI. Other textual data mining research was conducted as an intern at the IBM research facility in Almaden, CA. Additional graduate research at Lehigh University involved clustering algorithms in Bioinformatics.

https://doi.org/10.1515/9783112221051-013